Springer Series in Materials Science

Volume 256

The Springer Series in Materials Science covers the complete spectrum of materials physics, including fundamental principles, physical properties, materials theory and design. Recognizing the increasing importance of materials science in future device technologies, the book titles in this series reflect the state-of-the-art in understanding and controlling the structure and properties of all important classes of materials.

More information about this series at http://www.springer.com/series/856

Stanislav I. Sadovnikov · Andrey A. Rempel
Aleksandr I. Gusev

Nanostructured Lead, Cadmium, and Silver Sulfides

Structure, Nonstoichiometry and Properties

 Springer

Stanislav I. Sadovnikov
Institute of Solid State Chemistry
Ural Branch of the Russian Academy of
  Sciences
Ekaterinburg
Russia

Aleksandr I. Gusev
Institute of Solid State Chemistry
Ural Branch of the Russian Academy of
  Sciences
Ekaterinburg
Russia

Andrey A. Rempel
Institute of Solid State Chemistry
Ural Branch of the Russian Academy of
  Sciences
Ekaterinburg
Russia

ISSN 0933-033X   ISSN 2196-2812 (electronic)
Springer Series in Materials Science
ISBN 978-3-319-56386-2   ISBN 978-3-319-56387-9 (eBook)
DOI 10.1007/978-3-319-56387-9

Library of Congress Control Number: 2017947831

Printed on acid-free paper

This Springer imprint is published by Springer Nature
The registered company is Springer International Publishing AG
The registered company address is: Gewerbestrasse 11, 6330 Cham, Switzerland

Preface

In 1998, a monograph "Nanocrystalline Materials: Methods of Production and Properties" by Gusev [1] was published in Ekaterinburg. The monograph became the first Russian and one of the first world's generalizations of experimental results and theoretical concepts about the structure and properties not only of dispersed, but also of compact solid state with nanometer dimensions of particles, grains, crystallites or other microstructure elements.

The monograph "Nanocrystalline Materials" (Cambridge International Science Publishing) by Gusev and Rempel [2] was published in 2004 in Cambridge. This monograph was devoted to one of the most topical problems lying at the interface of material science, physics and solid state chemistry – nanocrystalline state of matter. The objects described in the monograph are metals, alloys, intermetallic compounds, oxides, carbides, nitrides, borides, carbon nanostructures, nanoporous materials and nanocomposites.

Extensive research over the two latest decades revealed that the greatest effect of particle (grain, crystallite) size reduction down to tens of nanometers and less is observed for semiconducting compounds. This is due to the commensurability of the size of semiconducting particles with such physical parameter having a dimensionality of length as the exciton diameter.

The new monograph offered to the readers is focused on simultaneous analysis of structure, nonstoichiometry and properties of semiconducting lead, cadmium and silver sulfides in the form of nanocrystalline powders, colloidal solutions, quantum dots, isolated nanoparticles and heteronanostructures. These nanostructured sulfides, along with other chalcogenides, attract the greatest attention owing to their potential application in electronics, biology and medicine.

The synthesis and properties of nanostructured chalcogenides were described in a number of books [3–6]. However, too broad range of objects inevitably made the discussion too brief. Nanostructured lead, cadmium, and silver sulfides are not mentioned almost in this book. Nonstoichiometry of discussed nanostructured semiconductors not considered at all.

Nonstoichiometry, i.e. deviation from stoichiometric composition, is a fundamental characteristic of inorganic substances, which affects the structure and properties of compounds, on the one hand, and depends on the size of structural elements (particles, crystallites, domains) of substances, on the other hand. Until recently, the relation and interdependence between nonstoichiometry and particle size in the nanometer scale has been scarcely examined or discussed only on the example of strongly nonstoichiometric compounds (carbides, oxides, nitrides of transition metals). And certainly the nonstoichiometry in nanostructured sulfides, which in the conventional state are traditionally considered as stoichiometric compounds, has never been discussed. Indeed, at present the number of studies devoted to nonstoichiometry of sulfide nanopartricles is extremely limited. The investigation of the relationship between the size of nanoparticles and their non-stoichiometry is a topical research area since nonstoichiometry affects the properties of nanoparticles and can be used for additional control over the functional properties of nanoparticles and nanomaterials.

Modern solid state physics, physical material science and electronics are inconceivable without semiconducting heterostructures. Semiconducting heterostructures, especially quantum wells, quantum wires and quantum dots, allow to control such fundamental parameters of semiconducting crystals as forbidden band width, effective mass and mobility of charge carriers and electronic energy spectrum.

Heteronanostructures combining the properties of semiconductors in nanocrystalline state, on the one hand, and nonstoichiometry, on the other hand, are the next step in the development of quantum electronics.

The production (preparation, synthesis) of nanostructured sulfides is immediately connected with the development and application of nanotechnologies. The essence of nanotechnology consists in the possibility to work at the atomic and molecular level, in the length scale between 1 and 100 nm in order to produce and use materials and devices having new properties and functions. Therefore much attention in the monograph is devoted to different methods of synthesis of nanostructured sulfides with allowance for their structure and composition.

The authors of this monograph tried to take into account both the purely scientific, fundamental interest in the problem of nanosized sulfides and some applied aspects of this problem that are of considerable importance for practical application of these substances.

The monograph includes much essential information about nanostructured lead, cadmium and silver sulfides in the form of nanocrystalline powders and isolated nanoparticles with different morphology, colloidal solutions, quantum dots and heteronanostructures. Writing it, the authors used a large number of original studies beginning with 1828 and up to 2017 inclusive. More than 80% of all references are given to the works performed after 2000. Thus, the monograph reflects the state of the art in the research of nanostructured lead, cadmium and silver sulfides. It will be useful and interesting for a wide range of specialists dealing with condensed matter physics, solid stare chemistry, physical chemistry and material science, as well as for engineers involved in the production and application of nanocrystalline semiconducting materials.

The study of nanostructured silver sulfide was supported by the Russian Science Foundation (project No. 14-23-00025) via the Institute of Solid State Chemistry, Ural Branch of the Russian Academy of Sciences.

Ekaterinburg, Russia

Stanislav I. Sadovnikov
Andrey A. Rempel
Aleksandr I. Gusev

References

[1] Gusev, A.I.: Nanocrystalline materials: Methods of production and properties, 200 pp. Ural Branch of the Russian Academy of Sciences, Ekaterinburg (1998) (in Russian)
[2] Gusev A.I., Rempel A.A.: Nanocrysnalline materials, 351 pp. Cambridge Intern. Science Publication, Cambridge (2004)
[3] Bimberg, D. (ed.): Semiconductor nanostructures, 357 pp. Springer, Berlin, Heidelberg (2008)
[4] Reithmaier, J.P., Petkov, P., Kulisch, W., Popov, C. (eds.): Nanostructured Materials for Advanced Technological Applications, 547 pp. Springer, Netherlands (2009)
[5] Granitzer, P. Rumpf, K. (eds.): Nanostructured Semiconductors: From Basic Research to Applications, 700 pp. CRC Press, New York (2014)
[6] Qurashi, A. (ed.): Metal chalcogenide nanostructures for renewable energy applications, 320 pp. Wiley, New York (2015)

Contents

Main Notations

a_B	Bohr radius
a_{B1}	Lattice constant of cubic unit cell with $B1$ structure
$\boldsymbol{a}, \boldsymbol{b}, \boldsymbol{c}$	The fundamental translation vectors of crystal lattice
a, b, c, β	Unit cell parameters
B	Bulk modulus
c	Concentration
c_ℓ, c_t	Velocities of propagation of longitudinal and transverse elastic vibrations
C_p, C_v	Heat capacity at constant pressure and at constant volume
D	Size or average size (diameter) of particle (grain, crystallite, domain)
E_g	Band gap
f_j	Atomic scattering factor
FWHM	Full line width at half maximum
F_{hkl}^2	Structural factor
G	Free energy
$g(\theta)$	The Gauss function
$g(\omega)$	Frequency distribution function
h, k, l	Miller indices
$h, \hbar = h/2\pi$	Plank constant
I	Relative intensity
k_B	Boltzmann constant
K_{hkl}	Scherrer constant
$l(\theta), \ell(\theta)$	The Lorentz (Cauchy) function
m_0	Free electron mass
m_e	Effective electron mass
m_h	Effective hole mass
N_A	Avogadro's number
$N(E)$	Density of electronic states
p	Pressure

P	Probability
P_{hkl}	Multiplicity
R	Gas constant
R_{ex}	Bohr radius of the exciton
R_j	Radius of the coordination sphere
$R(\theta)$	Angle resolution function
S_{dis}	Configurational entropy
S_{sp}	Specific surface area
t	Time
T	Temperature
T_{melt}	Melting temperature
T_{trans}	Phase transformation (transition) temperature
v_m	Molar volume
x_I , y_I , z_I	Coordinates of the j-th atom
y	Relative content of interstitial atoms X in nonstoichiometric compounds MX_y
α	Thermal expansion coefficient
α_j	Short-range order parameter for j-th coordination sphere
β	Broadening of diffraction reflection
β_d	Deformation broadening (broadening caused by the lattice microstrains)
β_h	Inhomogeneity broadening (broadening due to the inhomogeneity of nonstoichiometric compounds or solid solutions)
β_s	Size broadening (broadening due to the small size of particles)
Δy	Inhomogeneity (deviation from the homogeneity)
γ	Grüneisen constant
ε_j	Correlation parameter for the j-th coordination sphere
$\varepsilon(\omega)$	Dielectric constant
$\zeta(m)$	Riemann zeta function
λ	Radiation wavelength
λ_i	Multiplicity of i-configuration
μ_{ex}	Reduced exciton mass
η	Viscosity
θ	Diffraction angle
ρ	Density
$\omega_{max}, \omega_{min}$	Upper and lower boundaries of the phonon spectrum
$\square$	Vacant site (structural vacancy)

Chapter 1
Size Characterization of Nanostructured Materials

Present monograph is concerned with one of the most important current scientific problems, which is common for materials science, solid state physics and solid state chemistry, namely the nanocrystalline state of matter.

Up to now, the main scientific data on the nanocrystalline state of different compounds including chalcogenides are published in conference proceedings, scientific journals and collected papers rather than monographs. The authors of this monograph present to the reader information on numerous original studies of the nanocrystalline semiconducting sulfides, including the special features of these investigations, and pay the attention on the most interesting and practically important effects of the nanocrystalline state of sulfides.

The term *nano* derived from the Greek word νάνος which means dwarf, designates a milliardth (10^{-9}) fraction of a unit. Thus, the nanostructures and nanomaterials science or nanoscience deals with very small objects of condensed substance; these objects have a size from 1 to 100 nm.

The special physical properties of small particles have been utilised for a very long time, although this has been carried out unknowingly. Suitable examples are dye pigments used in different historical periods, ancient Egypt glasses and Chinese porcelains which were colored with colloidal particles of metals. The chaotic movement of flower pollen particles, suspended in a liquid, which was discovered in 1827 by the Scottish botanist Robert Brown is the first scientific mention of the small particles. This phenomenon is referred to as Brownian motion. Its article [1] on the microscopic observation of this movement laid foundations to many studies. The theory of Brownian motion, developed independently by Einstein [2, 3] and Smoluchowski [4] at the beginning of the twentyth century, is the basis of one of experimental methods of determining the size of small particles. The scattering of light by colloidal solutions and glasses was studied by Faraday [5] between 1850 and 1860.

The starting points of examination of the nanostructured state of different substances were the studies in the field of colloidal chemistry, which were already quite extensive since the middle of the 19th century. At the beginning of the 20th century,

© Springer International Publishing AG 2018
S.I. Sadovnikov et al., *Nanostructured Lead, Cadmium, and Silver Sulfides*,
Springer Series in Materials Science 256, DOI 10.1007/978-3-319-56387-9_1

a significant contribution to the experimental verification of the theory of Brownian motion, to the development of colloidal chemistry and examination of dispersed substances, and to the determination of the size of colloidal particles was provided by the Swedish scientist T. Svedberg. In 1919, Svedberg developed a new method of separating colloidal particles from a solution using an ultracentrifuge [6, 7]. In 1926, he was awarded the Nobel Prize in chemistry for his work on dispersed systems.

The twentyth century was characterized by extensive investigations and application of heterogeneous catalysis, ultrafine powders and thin films. These investigations have raised a question about the influence of the small particles (grains) size on the properties of studied materials. At present, the nanostructured materials involve nanopowders of metals, alloys, and intermetallics, nanopowders of such compounds as oxides, carbides, nitrides, borides, and chalcogenides, and the same consolidated substances with the grains (crystallites) of the nanometer size, together with nanocomposites, nanopolymers, carbon nanostructures, nanoporous materials, and biological nanomaterials. The development of nanomaterials is directly associated with the improvement and application of nanotechnologies. The study of nanocrystalline materials has revealed a large number of grey areas in the fundamental knowledge of the nature and stability of the nanocrystalline state under different conditions. On the whole, the field of nanoscience, nanomaterials and nanotechnologies is very wide and at present has no distinct contours.

The unique structure and properties of small aggregations of atoms are very interesting because they represent an intermediate state between the structure and properties of isolated atoms, on the one hand, and bulk solids, on the other hand. However, the main problem at what stage of association of atoms in a single substance the properties of bulk crystals is formed completely has not yet been solved. It is not clear how the contributions of surface (associated with the interfaces) and bulk (associated with the particles size) effects to the nanocrystalline materials properties can be separated. The studies in this field were carried out for a long time on isolated clusters, consisting from two atoms up to some hundreds of atoms, small particles with a size of more than 1–2 nm, and also ultrafine powders with grain size about 100 nm. The transition from the properties of isolated nanoparticles to the properties of bulk crystalline substances remained completely unexplored since an intermediate state, namely bulk solid with the nanometer grains, was absent and there were no any methods for its artifical creation. The methods for preparation of bulk nanocrystalline substances were developed only after 1985–1990, and from this time works were started to fill this gap in the knowledge of solids. However, this problem is not solved yet in full.

The spectrum of properties of solid substances can be enormously enhanced if nanosized grains, particles or crystallites are agglomerated to a bulk material so that in addition to the nanosized crystallites this material consists of a large portion of interfaces with a disordered structure. In very small particles with a size of a few nanometers, new properties appear due to quantum size effects or scaling laws.

The scientific interest to the nanocrystalline state of the solids is associated mainly with expectation of various size effects on the properties of the nanoparticles and nanocrystallites whose size is comparable or smaller than the characteristic correlation scale of a specific physical phenomenon or the characteristic length, which are present in the theoretical description of some property or process. Such characteristic lengths are the size of the exciton in semiconductors, the free path of the electrons, the length of coherence in superconductors, the wavelength of elastic oscillations, the size of the magnetic domain in ferromagnetics, etc.

The industrial interest to the nanostructured materials is caused by the possibility of large modification and even principal changes in the properties of the conventional materials at transition to the nanocrystalline state, and by new possibilities opened and offered by nanotechnologies in the creation of materials, devices and wares from structural elements with the nanometer size. The essence and main content of nanotechnology is the possibility to work at the atomic and molecular level, in the length scale between 1 and 100 nm, in order to produce and use materials and devices with such properties and functions that caused the small scale of their structure. Thus, the term "nanotechnology" relates to the small sizes of structural elements in particular. Nanocrystalline products are playing an important role in almost all branches of industry. The fields of application of these nanoproducts are immense and include more efficient catalysts, thin films for microelectronics, new magnetic materials, protective coatings on metals, plastics and glasses. In the next decades, nanostructured materials and nanosized devices will be used and operate in biology and medicine. The successes of nanotechnologies are manifested most efficiently in modern electronics as a result of continuous miniaturisation of semiconducting electronic devices and arrays.

The monograph in the generalized and concentrated form includes the most important data on the preparation, structure and properties of the most applied nanostructured sulfides. A very large number of original studies, starting from 1828 up to the year 2017, is used in writing the monograph.

Semiconducting chalcogenides are widely used in various fields of modern technique.

Energy conversion and storage devices (high-capacity cells and energy storage systems, fuel cells, photo-electrochemical cells for production of hydrogen from water, conversion units for direct solar to electrical energy) play an important role in the sustainable technology development and in the solution of energetic and environmental problems. The successful use of optoelectronic and energy conversion devices, and their design is dependent on the synthesis of new nanostructured materials with enhanced performance characteristics. The investigations of the last 20–25 years shown that nanostructured chalcogenides are the most promising materials for the design of high-performance energy conversion and storage devices, detection of impurities in the gas and aqueous media, detection of radiation over a broad wavelength range, and for biomedical applications. The synthesis and properties of nanostructured chalcogenides were described and generalized in a number of reviews and monographs [8–13]. However, too broad range of objects (sulfides, arsenides, phosphides, antimonides) inevitably made the discussion too brief.

Sulfur S, arsenic As, selenium Se and tellurium Te are the four key elements which form various chalcogenide compounds. Chalcogenide nanocrystals find wide use in electronic technology. Wide use of nanostructured chalcogenides is due to simple synthesis and unique structural, physical, optical and chemical properties (semiconductor or ionic conductivity, thermal stability, variety of crystal structures) that can be controlled by changing the size of nanoparticles. Indeed, the potential advantages of chalcogenide nanocrystals in such applications as light-emitting diodes, photoelectric devices, field effect transistors, etc., are caused by the possibilities to control the optical and electrical properties by changing the nanoparticles size and shape. Therefore the studies of the last decade concerned with the synthesis conditions of nanostructured semiconductor chalcogenides were stimulated by the above-listed potential technological applications of these compounds.

Most often, nanostructured chalcogenides are synthesized by "bottom-up" methods; among them, the most efficient process resulting in the formation of nanoparticles with well controllable size and low size dispersion is the solvo- or hydrochemical synthesis of colloidal solutions.

Sulfides of lead Pb, silver Ag, copper Cu, cadmium Cd, and zinc Zn, solid solutions of these sulfides, also arsenides of gallium Ga and indium In and solid solutions involving them are of most interest among chalcogenide compounds.

1.1   Concept of Nanostructured Material

Before passing to the discussion of the structure and properties of nanostructured sulfides and sulfide nanomaterials, we will briefly consider the concept of nanomaterials.

In this monograph, we use the conventional definition of nanomaterials as materials having a characteristic length scale less than 100 nm. This length scale could be a radius of particle, size of grain, thickness of layer, or width of conducting lines in electronic chip.

Materials, the average particle (grain, crystallite) size of which does not exceed 100 nm, are usually referred to as nanocrystalline or nanostructured [14–17]. In recent decades, it is proved that the decreasing particle size of a solid less than some threshold magnitude can induce considerable changing the properties, which is called size effect [15, 17–22]. The threshold size when the properties of nanoparticles change abruptly ranges from 1 to 50–60 nm for most of the materials known to date. The size effects become most pronounced when the nanoparticle size is smaller than 10 nm. Since these values are in the nanometer range, the materials that exhibiting size effects are called nanomaterials. The particle size is a highly important characteristic of nanomaterials determining the details of their properties and the scope of their applications.

Formally, any materials in which the size of structure elements is of the order of some nanometer can be called nanomaterial. In this case, the essence of the term "nanomaterials" narrows down to only determining the size of structure elements.

Meanwhile, the nanomaterial concept should include not only the size of structure elements but also the stepwise change in the properties resulting from the small size of these structure elements [19–21].

The nanocrystalline state is intermediate between a molecule and the solid state. The distances at which the main interaction forces in a substance act vary in the range from 1 to about 100 nm; therefore, the stepwise change in the properties occurs especially in the nanometer range [21]. Various types of interactions (electron-phonon, phonon-phonon, electron-electron, magnon-magnon, etc.) cover different distances in the same substance. Therefore, the size effects for various properties may observe at different nanoparticle sizes in the same substance.

Nanomaterials are often referred to as nanocrystalline materials. This is not always justified. A material is considered to be nanocrystalline material if its structure elements (building blocks, i.e., particles or grains) are crystals. As a rule, nanomaterials are thermodynamically non-equilibrium systems; therefore, their structure elements do not necessarily possess a good crystal structure [21]. Moreover, the building blocks in many nanomaterials have a highly defective structure, often approaching an amorphous state. In other words, the most nano-materials have the structure elements where the long-range order is markedly violated or absent, and the many-particle correlations in the arrangement of atoms are provided mainly by the short-range order. Therefore, the term "nanomaterials" should be understood more specifically as "nanostructured materials".

There are different approaches to nanomaterial classification. The most appropriate classification is based on the dimensionality of structure elements that constitute the nanomaterials. The main types of nanostructured materials are zero– (0D), one– (1D), two– (2D) and three-dimensional (3D) [19, 20]. By zero-dimensional nanomaterials are meant nanocluster material and nanodispersions, i.e., materials in which the nanoparticles are isolated from one another, and also heteronanostructures, in par-ticular, quantum dots (QDs). One-dimensional materials are nanofibrous and nan-otubular materials, including quantum wires. The length of fibres, rods or tubes 5–40 nm in diameter can range from 100 nm to some tens of micrometers. Nanometer-thick films and semiconductor heteronanostructures such as quantum wells are classified as two-dimensional nanomaterials. The nanoparticles of 0D, 1D and 2D nanomaterials can occur either in the liquid or solid matrix or on a substrates. Three-dimensional nanomaterials are powders, fibrous, multilayer and polycrystalline materials in which zero-, one-and two-dimensional particles closely adjoin one another. As a result, interfaces are formed between them. A typical example of three-dimensional nanomaterial is a polycrystal with nanosized grains. In this nanomaterial, the whole bulk is filled by nanocrystalline grains, which are virtually devoid of free surface and have only interfaces between the grains.

The physical and chemical properties of semiconducting nanomaterials differ considerably from the same properties of coarse-grained materials [23, 24], which is caused by three main reasons [15, 20]. First, the nanoparticle size is comparable with the diameter of excitons in semiconductors, which affects optical, luminescent, electronic and other properties of nanomaterials. Second, smaller size of grains leads to increasing area of surface; when the grain size is equal from 100 to 1 nm, the surface contains from 1 to 50% of atoms of the solid, and surface atoms make a considerable contribution to all characteristics of the nanomaterials. Third, the proper nanoparticle size approaches the size of a molecule.

Such semiconductor heteronanostructures as quantum wires, quantum wells and quantum dots provide the possibility of controlling fundamental characteristics of semiconductors such as the width of the forbidden band (energy band gap) E_g, the effective masses and mobilities of charge carriers, the electron energy spectrum [20–26].

The density of states (DOS) $N(E)$ for three-dimensional (3D) semiconductor is a continuous function of energy E (Fig. 1.1).

A decrease in the dimensionality of the electron gas results in a change of the continuous energy spectrum to discrete energy spectrum as a result of its splitting. The quantum well is a two-dimensional (2D) nanostructure in which charge carriers move freely in the plane of the layer and are restricted in the direction normal to layers. In quantum wires, the charge carriers are restricted in two directions and can move freely only along the wire axis. The quantum dots are a zero-dimensional (0D) objects where the charge carriers are already restricted in all three directions and are characterized by a completely discrete energy spectrum. It is clear that quantum confinement creates new energy states and results in the modification of DOS and all the properties (especially, optoelectronic) of semiconductors.

The smaller quantum dots (1–20 nm), whose surface is protected by organic molecules preventing aggregation of the particles, are produced by colloidal chemistry methods [25].

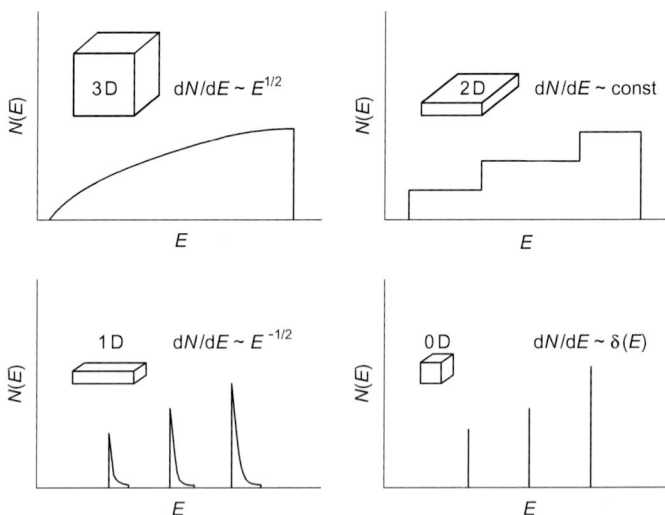

Fig. 1.1 Density of electronic states, $N(E)$ in one band as a function of dimension of a semiconductor: (3D) three-dimensional (bulk) semiconductor, (2D) quantum well, (1D) quantum wire, (0D) quantum dot

Quantum dots, i.e. heteronanostructures with spatially limited charge carriers in all three directions realize the limiting case of size quantization in semiconductors, when modifications of DOS and the electronic properties of the material are most strongly expressed [26]. The electron spectrum, i.e. density of states $N(E)$ of a perfect quantum dots (see Fig. 1.1) represents a set of discrete levels separated by the forbidden bands. This spectrum corresponds to the electron energy spectrum of an isolated single atom. However, the real quantum dot (or "superatom") can consist of some hundreds of thousands of atoms and more. Nanometer-scale quantum dots with a discrete atom-like electron spectrum have major advantages for use in modern optoelectronic devices. The properties of quantum dots have already resulted in many breakthroughs and advances especially in the semiconductor lasers.

Thus, the "nanostructured material" concept covers a broad range of objects – from colloidal solutions and isolated nanoparticles to heteronanostructures of different dimensionality, nanofilms and nanopowders.

In accordance with this, we will discuss in our monograph the synthesis methods, the special features of structures and properties of colloidal solutions, single nanoparticles, quantum dots, nanofilms, nanopowders, and heteronanostructures of such semiconductors as PbS, CdS and Ag_2S sulfides. Until recently, no survey publications devoted to nanostructured semiconducting sulfides are available from scientific literature.

1.2 Methods of Evaluation of the Size of Small Particles

Studies of substances with minute particles show that the decreasing particle size down to some threshold magnitude leads to a large changing the properties. Size effect appears if the particle size does not exceed of 100 nm, and manifests themselves most considerably when the particle size is less than 10 nm.

The particle size is the main parameter that determines the peculiarities of the properties and the application field of the nanomaterials. Particles, grains or crystallites of what sizes can be called nanoparticles? Where is the limit behind which we achieve the nanocrystalline state?

As a first step, one can accept the conventional division of any substances on the particle size. On this level of understanding, materials with an average particle (grain) size of more than 1 μm are referred to as coarse-grained materials. Polycrystalline materials having an average particle size from 300 to 40 nm are usually referred to as submicrocrystalline materials. Polycrystalline materials with an average particle size below 40 nm are called as nanocrystalline or nanostructured materials. The conventional classification of materials on the size D of particles (grains) is shown in Fig. 1.2. However, this conventional division is not scientifically justified.

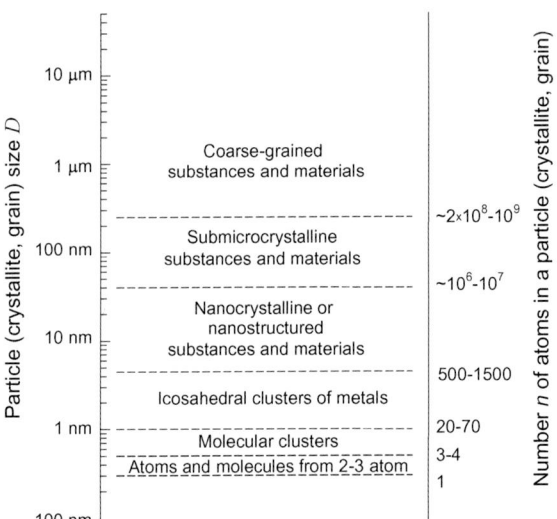

Fig. 1.2 Size classification of substances and materials

The justified determination of the nanocrystalline state requires deep physical understanding of the problem. As was noted, the transition to the nanocrystalline state is associated with the appearance of size effects on the properties. While there are no size effects, there is no nanocrystalline state. Actually, if the length of the structure elements of a substance in one, two or three dimensions are comparable or smaller than the characteristic correlation length of a specific physical phenomenon or physical parameter, used in the theoretical description of some property or process, size effects will be detected on the appropriate properties. Such physical parameters as the magnetic domain size in ferromagnetic, the electron free path, the de Broglie wavelength, the coherence length in superconductors, the wavelength of elastic oscillations, the size of the exciton in semiconductors can be used for evaluating the transition to the nanocrystalline state. Thus, the size effects refer to the set of phenomena associated with the change in the properties of a substance as a result of (1) a decrease of the particle size, (2) a simultaneous increase in the fraction of the surface contribution to the general properties of a substance, and (3) a commensurability of a particle size with physical parameters which have dimension of length. The substance is nanocrystalline substance when the size of its particles is equal or smaller than some characteristic physical parameters which have the length dimension. Only in this case it is possible to detect the changes in the properties, which are determined by the size effects.

Now we understand what the nanocrystalline state is from the physical viewpoint. Let us consider the main methods for determining size of small particles. It should be mentioned that the size and shape of particles to a large extent are determined by the method of preparing nanomaterial.

1.2.1   *Electron Microscopy*

The most widely used method of determining the size of small particles and objects is microscopy. In fact, the invention of the optical microscope already resulted in a large number of inventions in crystallography, medicine, biology and made microscopy very popular and well known. The development of electron microscopy made it possible to examine objects, which are considerably smaller than 1 μm and resulted in its extensive application in solid state physics, crystallography, solid state chemistry, materials science and mineralogy. The first electronic microscopes have appeared in England and Germany in 30th years of the XX-th century, soon after publication of work [27]. One of its authors, E. Ruska, has been awarded the Nobel Prize on the physicist 1986. Second half of this award has been received by G. Binnig and H. Rorer which have invented a scanning tunnel microscope [28]. Already in the late forties of the XX-th century the electronic microscopy was used in material science for studying of small particles. Ability to distinguish separate columns of atoms in a crystal has led to occurrence of the term "high-resolution transmission electron microscopy". At present, high-resolution transmission electron microscopy [29–32] may be used to observe the distribution of atomic columns, to study the distribution of atoms in crystalline and colloidal particles, various defects in the crystal lattice, the mutual location of the molecules in biological objects. Electron microscopy is used for the direct measurement of the size of nanoparticles and nanocrystallites. It is used widely in investigations of nanomaterials and in nanotechnologies for advanced electronics.

Electron microscopy is the only technique for determining the composition of the nanoparticles produced from colloidal solutions. Figure 1.3 shows HRTEM image of PbS nanowire with a diameter about 60 nm [33]. Parallel rows of atoms for cubic lead sulfide structure with interplanar distance about 0.3 nm are seen clearly. Observed interplanar distance 0.3 nm coincides with the distance between (200) atomic planes of cubic (space group $Fm\bar{3}m$) lead sulfide.

Fig. 1.3 HRTEM image of part of separate PbS nanowire [33]

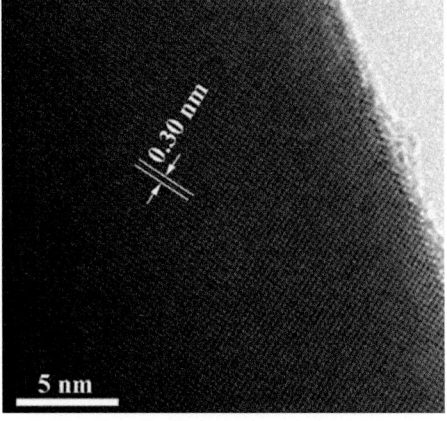

The ability to characterize nanostructured materials has been increased greatly by the invention of the scanning tunneling microscope (STM), the atomic force microscope (AFM), the magnetic force microscope, and the optical near-field microscope.

Scanning probe microscopy is widely used for studying atomic structure of a surface. It has the high resolution, allows estimating the sizes of observable objects, to visualize separate atoms and molecules, and to build three-dimensional images. Its versions are scanning tunnelling microscopy [28] and atomic force microscopy [34]. The description of principles of action and the basic designs of scanning probe microscopes can be found in [35–37]. The possibility to determine the position of every atom on the surface makes STM the optimal tool for surface growth studies. Near-field scanning optical microscopes overcome the limitation of standard conventional optics by making use of the evanescent field that exists within the first several nanometers of the surface of light pipes, and can have a resolution down to 12 nm using 500 nm wavelength light.

The image of InAs quantum dots with a size of 10–20 nm on a surface (001) of GaAs [38] is shown in Fig. 1.4 as an example of using scanning probe microscopy.

With all advantages of electron microscopy as a method for determination of the size of particles, it must be taken into account that it is a local method and provides information on the size of the object only in the field of observation. However, the observed area may not be representative, i.e. not characteristic for the entire volume of the substance. Therefore, electron microscopy examination must be carried out on several areas of substance under investigation in order to obtain statistically averaged out information of the entire substance.

Electron microscopy is the only direct method for determining the size of very small particles. All other methods are indirect, because the information on the average nanoparticle size is extracted from the data on the variation of some property of the substance or process parameter. Indirect methods include

Fig. 1.4 Layer of spontaneously ordered InAs quantum dots with a size of 10–20 nm on a (001) surface of GaAs quantum well with a thickness of 12 nm [38]. The image is obtained by means of scanning probe microscopy

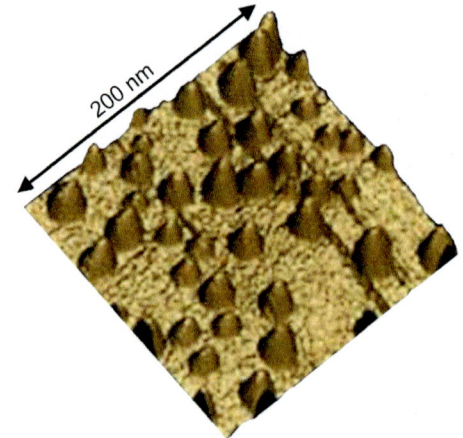

diffraction, magnetic, sedimentation, photon correlation spectroscopy, and gas-adsorption methods.

The combination of scanning and high-resolution transmission electronic microscopy with X-ray diffraction is optimum for studying of a structure of nanoparticles and determination of the nanoparticle sizes. Transmission electron microscopy (TEM) and X-ray diffraction (XRD) provide mutually complementary structural data on nanocrystalline particles. TEM characterizes single structures with atomic resolution, whereas XRD yields information on statistically averaged large atomic ensembles.

1.2.2   X-Ray Diffraction Method

The method of X-ray diffraction (XRD) occupies the main position among indirect methods for determining the nanoparticle size. At the same time, this method is simplest and accessible because the X-ray examination of the crystal structure is used very widely and efficient equipment is available.

The size of determined by X-ray diffraction method can be a little underestimated in comparison with results of electronic microscopy. However XRD gives a particle size which is averaged on the investigated volume of studied substance whereas the electronic microscopy is a local method and determines the objects size only in the limited field of observation.

The XRD method to measure diameter of a nanoparticle was invented by P. Scherrer. 100 years ago he has proposed the equation that combines the full width at half maximum (FWHM) of an XRD line with the diameter of small particle [39]. A little bit later similar formula connecting the FWHM of X-ray diffraction reflection and the diameter of a nanoparticle was independently derived by Russian expert in the field of X-ray analysis Seljkow [40]. Therefore this formula is sometimes named after Seljkow and Scherrer. Nevertheless the priority of this formula belongs to P. Scherrer without any doubts.

In diffraction experiments on powder or polycrystalline specimens, all the crystal lattice defects are studied on the base of quantitative analysis of changes in the diffraction reflections.

Any defects or imperfections of the crystal lattice are connected with the atomic displacements from the sites of perfect crystal lattice.

The intensity and the shape of diffraction reflections are conventionally considered as a function of different defects of a crystal lattice. M.A. Krivoglaz separated all the crystal defects into the 1st and 2nd groups, which are responsible only for the decrease in the intensity of the diffraction reflections and only for the broadening of the diffraction reflections, respectively [41]. The small size D of coherent scattering regions (CSRs) and microstrains ε are the defects of the 2nd group. The size broadening β_s caused by the small particles size D and stacking faults, is proportional to sec θ, where θ is the diffraction angle. The microstrains ε

and edge and screw dislocations with random distribution cause the strain broadening β_d which is proportional to tan θ.

An inhomogeneity is the 2nd group's defect in crystal solids also. It is caused by the non-uniform composition of a substance, i.e. a variation of the substance composition over its volume. This broadening type is characteristic of the X-ray and neutron diffraction patterns of compounds with atom-atom or atom-vacancy substitution, i.e., in the nonstoichiometric compounds $MX_y\square_{1-y}$ with vacancies $\square$ and substitutional solid solutions (alloys) A_yB_{1-y}. The composition of such inhomogeneous substances varies in the range from $(y - \Delta y)$ to $(y + \Delta y)$, where Δy is a deviation from the homogeneity (in other words, degree of inhomogeneity). The inhomogeneous broadening β_h is proportional to $(\sin^2\theta)/\cos \theta$ [42–46].

Separating all the types of broadening and determining the nanoparticle size D, microstrains ε, and the degree of inhomogeneity Δy is possible due to different dependences of the size, strain, and inhomogeneous broadenings on the diffraction angle 2θ.

Sometimes the diffraction patterns of nanocrystalline and the same coarse-crystalline substance is compared for determination of diffraction reflection broadening. Such broadening determination and estimation of the particle size, microstrains, and inhomogeneity leads to large errors from few tens to hundreds of percent.

For correct determination of broadening, it is necessary to compare the measured diffraction reflection width with the diffraction reflection width from a crystal without any defects of the 2nd type. This means that the width of every reflection of studied substance should be compared with the instrumental width of the same reflection, i.e., with the diffractometer resolution function.

The separation of the considered contributions into the broadening is based on a quantitative analysis of the X-ray or neutron diffraction patterns.

According to [19, 20, 47, 48], the diffraction determination of the average particle size, microstrains, and inhomogeneity from the broadening of diffraction reflections includes 5 steps:

(1) the experimental recording the XRD pattern of a reference substance with following quantitative analysis of the diffraction reflections profile (shape) and calculation of a diffractometer resolution function;
(2) the recording the XRD pattern of the studied substance, quantitative analysis of the diffraction reflection profiles, and determining reflections width;
(3) the determination of the angular dependences of the width and broadening of diffraction reflections for the substance examined;
(4) the partition of broadening on separate contributions which are caused by the small particle size, microstrains, and inhomogeneity;
(5) the calculation of an average size of nanoparticles, values of microstrains and inhomogeneity.

XRD method for determining the small particle size, microstrains and degree of inhomogeneity was considered earlier [19, 20, 47–50]. This method has been tested repeatedly on nonstoichiometric carbides and oxides, carbide solid solutions and chalcogenides including sulfides.

The determination of the size of small particles and microstrains by the diffraction method is most widely used and accessible. Therefore special features of using this method will be discussed in detail here.

The profile of diffraction reflections has a bell-shaped form therefore for the analytical description of diffraction reflections it is necessary to use the functions which have bell-shaped image. Such functions are the Gauss and the Lorentz functions, and therefore any of them or their combination can be used for the quantitative analysis of diffraction reflections.

The profile of diffraction reflection can be approximated by the Gauss function

$$g(\theta) = A \, \exp[-(\theta - \theta_0)^2/(2\theta_G^2)] \tag{1.1}$$

or by the Lorentz function (in mathematics, this function is known as the Cauchy distribution)

$$\ell(\theta) = A \left[1 + \frac{(\theta - \theta_0)^2}{\theta_L^2} \right]^{-1}. \tag{1.2}$$

The most accurate approximation of the profile of diffraction reflections is the pseudo-Voigt function [51, 52] which is a linear superposition of the Gauss and the Lorentz distributions (1.1) and (1.2):

$$V(\theta) = cA \left[1 + \frac{(\theta - \theta_0)^2}{\theta_L^2} \right]^{-1} + (1 - c)A \, \exp\left[-\frac{(\theta - \theta_0)^2}{2\theta_G^2} \right]. \tag{1.3}$$

In the Gauss, Lorentz, and pseudo-Voigt functions (1.1)–(1.3), A is the normalizing factor, θ_0 is the angle position of the maximum of the function, and c is the relative contribution of the Lorentz function to the pseudo-Voigt function $V(\theta)$. All parameters of the pseudo-Voigt function $V(\theta)$ (1.3) which describes the selected diffraction reflection are determined using the least-squares method.

If the XRD pattern is collected using radiation with two wavelengths λ_1 and λ_2, then each reflection is a doublet. Therefore, each XRD reflection (hkl) is considered as a sum of two pseudo-Voigt functions. It follows from the Wulff-Bragg law $2d_{hkl} \sin \theta_{hkl} = n\lambda$ that maxima θ_1 and θ_2 of the pseudo-Voigt functions are connected through the interplanar distance d_{hkl}, corresponding to the given reflection (hkl), by the relationship

$$\theta_2 = 2 \arcsin\left[\frac{\lambda_2}{\lambda_1} \sin(\theta_1/2) \right], \tag{1.4}$$

where the angles θ_1 and θ_2 correspond to the maxima of the doublet lines of examined (hkl) diffraction reflection. The numerical refinement should be calculated taking into account the ratio of intensities of $K\alpha_2$- and $K\alpha_1$-lines in the doublet (for CuK$\alpha_{1,2}$-radiation, the intensity of the $K\alpha_2$-line is equal to 0.497 of the $K\alpha_1$-line intensity).

The quantitative description of diffraction lines by the model pseudo-Voigt function (1.3) allows accurate determining angle position and integral intensity for each line, and the full widths at half maximum. By knowing FWHMs, it is possible to calculate the broadening for all diffraction reflections. The found angular dependence of broadening allows to separate different contributions to the experimental broadening and to determine the CSRs average size, microstrains, and the degree of inhomogeneity.

Let us consider special features of the Gauss and Lorentz distributions (Fig. 1.5), which are required for further analysis of diffraction reflections.

It follows from (1.1), that the maximum value of the Gauss function $g(\theta)$ is A. Parameter θ_G is the second momentum, expressed in angles θ, of the Gauss function. In other words, parameter θ_G is the value of the argument that correspons to inflection point of the function where $\partial^2 g(\theta)/\partial \theta^2 = 0$. If the value of $g(\theta)$ function (1.1) is equal to half of its height, i.e. $h_G \exp[-(\theta - \theta_0)^2/(2\theta_G^2)] = h_G/2$, then $\exp[-(\theta - \theta_0)^2/(2\theta_G^2)] = 1/2$. From this it follows that $(\theta - \theta_0) = \theta_G \sqrt{2 \ln 2}$. It is seen from Fig. 1.5a, the full width of Gauss function at half of its height is $\mathrm{FWHM}_G = 2(\theta - \theta_0) = 2\theta_G \sqrt{2 \ln 2} \approx 2.355\theta_G$.

For the Lorentz function $\ell(\theta)$, the full width at half maximum FWHM_L is found in the same manner as for the Gauss function. The Lorentz function $\ell(\theta)$ at the maximum is equal A. Parameter θ_L of the Lorentz distribution coincides with the half width of this function at half height. Let it be that the Lorentz function (1.2) has

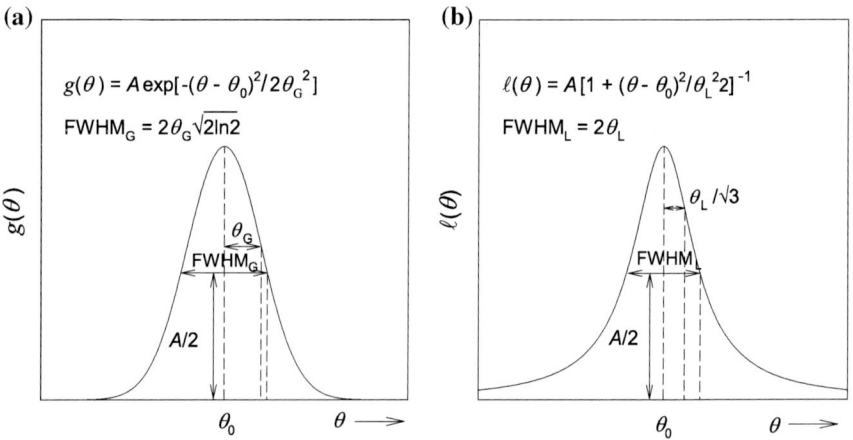

(a)

$g(\theta) = A \exp[-(\theta - \theta_0)^2/2\theta_G^2]$

$\mathrm{FWHM}_G = 2\theta_G \sqrt{2\ln 2}$

θ_G

FWHM_G

$A/2$

θ_0 $\theta \longrightarrow$

(b)

$\ell(\theta) = A[1 + (\theta - \theta_0)^2/\theta_L^2]^{-1}$

$\mathrm{FWHM}_L = 2\theta_L$

$\theta_L /\sqrt{3}$

FWHM_L

$A/2$

θ_0 $\theta \longrightarrow$

Fig. 1.5 The Gauss $g(\theta)$ **(a)** and Lorentz $\ell(\theta)$ **(b)** distributions, used for model description of the profile of the diffraction reflections

the value, which is equal to half of the height, i.e. $\ell(\theta) = h_L/2$ (Fig. 1.5b). The value of the argument, which corresponds to this value of the Lorentz function, is determined from the equation $[1 + (\theta - \theta_0)^2/\theta_L^2]^{-1} = 1/2$. From this it follows that $(\theta - \theta_0)^2/\theta_L^2 = 1$, $\theta = \theta_0 + \theta_L$, and the FWHM$_L$ for the Lorentz function (1.2) is equal to $2\theta_L$ (Fig. 1.5b). The 2nd momentum of the $\ell(\theta)$ function, i.e. the value of the argument θ, which corresponds to inflection point of the function, can be determined from the condition $\partial^2\ell(\theta)/\partial\theta^2 = 0$. Calculation shows that the 2nd momentum for the Lorentz distribution (1.2) is $\theta_L/\sqrt{3}$.

The full width at half maximum FWHM$_V$ for the pseudo-Voight function (1.3) is the superposition of FWHMs of the Lorentz and Gauss distributions and has the form

$$\text{FWHM}_V = c\text{FWHM}_L + (1 - c)\text{FWHM}_G = 2c\theta_L + 2\sqrt{2\ln 2}(1 - c)\theta_G$$
$$\equiv 2c\theta_L + 2.355(1 - c)\theta_G. \tag{1.5}$$

If the parameters θ_L and θ_G are equal to b, then exact value of the full width at half maximum FWHM$_V$ for the pseudo-Voigt $V(\theta)$ function can be determined by the solution to (1.3) at $V(\theta_L) = A/2$. In this case, (1.3) has the form

$$c\left[1 + \left(\frac{\theta - \theta_0}{b}\right)^2\right]^{-1} + (1 - c)\exp\left[-\frac{1}{2} \cdot \left(\frac{\theta - \theta_0}{b}\right)^2\right] = \frac{1}{2}. \tag{1.6}$$

Equation (1.6) is nonlinear function of $(\theta - \theta_0)$ argument and therefore has no exact analytical solution. From (1.6) it follows that the FWHM$_V$ of the pseudo-Voigt function $V(\theta)$ is equal to $2(\theta - \theta_0)$. As the parameter b is constant, the (1.6) can be solved in numerical form at different values of parameter c, which changes from 0 to 1. According to [47, 48], on the one hand, the found graphical dependence FWHM$_V$ = $f(c)$ was approximated with enough high accuracy by the following quadratic function

$$\text{FWHM}_V = b(2.355 - 0.276c - 0.079c^2). \tag{1.7}$$

On the other hand, taking into account (1.5), the FWHM$_V$ for the pseudo-Voigt function with good approximation is described as additive sum of FWHM$_G$ and FWHM$_L$ for the Gauss and Lorentz functions, and is non-linear function of c parameter:

$$\text{FWHM}_V = c \cdot \text{FWHM}_L + (1 - c) \cdot \text{FWHM}_G = 2b[c + (1 - c)\sqrt{2\ln 2}]. \tag{1.8}$$

Maximum difference between the magnitudes of FWHM$_V$ calculated from relationships (1.7) and (1.8) is 0.9% and is observed for $c = 0.5$.

In order to find XRD broadening, the experimental width of reflection should be compared with the instrumental angular resolution function of the diffractometer.

The instrumental angular resolution function is determined in a special diffraction experiment by measuring the FWHM of the diffraction reflections for a substance containing no defects of the second group that can lead to an additional broadening. A well annealed (without microstrains) and completely homogeneous substance with average particle size more than 1 μm should be used as a reference substance. In this case, an additional broadening caused by defects of the second group is absent, and only instrumental diffraction reflections broadening is observed. Cubic lanthanum hexaboride LaB_6 (NIST Standard Reference Powder 660a) with the lattice constant $a = 415.69162 \pm 0.00097$ pm, hexagonal aluminum oxide Al_2O_3 (NIST Standard Reference Material SRM 676) with the lattice constants $a = 475.90914 \pm 0.00101$ pm and $c = 1299.1779 \pm 0.00327$ pm, and hexagonal silicon carbide 6H-SiC are used for determining the instrumental resolution function of the diffractometer.

The instrumental resolution function has the following conventional analytical form

$$\text{FWHM}_R(2\theta) = \sqrt{(u\tan^2\theta + v\tan\theta + w)^{1/2}}. \tag{1.9}$$

Cagliotti et al. [53] proposed this analytical form of angular resolution function, and the beginning of wide scientific application of relationship (1.9) was initiated by by Rietveld [54]. The description of the experimental dependence $\text{FWHM}_R(2\theta)$ by (1.9) allows calculating the parameters u, v, and w of the instrumental resolution function.

In order to find the diffraction reflections broadening, the X-ray or neutron diffraction pattern of the studied substance is recorded, and every reflection is described by the pseudo-Voigt function (1.3). Further, the experimental FWHM_{exp} is calculated using formula (1.5), (1.7) or (1.8).

According to [19], the largest particles (grains, crystallites) size which cause a measurable reflection broadening is equal about 200 nm. Thus, the diffraction method makes it possible to determine the particle size only for substances which consist of particles less than 200 nm.

The broadening of the diffraction reflections caused by nanoparticles with a size less than 10 nm is very large, and the number of such observed reflections is small and not enough for the construction of angular dependence of broadening. The full width at half maximum of these reflections can only be measured with a very large error. Therefore, the size of 8–10 nm is the minimum nanoparticles size which can be determined by XRD method.

Thus, the XRD method makes it possible to determine the size of nanoparticles in the range from 10 to 200 nm.

For the arbitrary shape particles, their average size $\langle D \rangle \approx V^{1/3}$ (V is volume of the particle) should be determined using the Debye-Scherrer equation [19, 20, 39]:

$$\langle D \rangle = \frac{K_{hkl} \cdot \lambda}{\cos\theta \cdot \text{FWHM}_{\text{exp}}(2\theta)} \equiv \frac{K_{hkl} \cdot \lambda}{2\cos\theta \cdot \text{FWHM}_{\text{exp}}(\theta)}, \tag{1.10}$$

where λ is the radiation wavelength. The coefficient $K_{hkl} \approx 1$ is the Scherrer's constant which value depends on the particle shape and on the indices (hkl) of diffraction reflection.

In the substances with cubic lattice, the crystallites have the comparable size in three perpendicular directions [55], and Scherrer's constant K_{hkl} for reflections (hkl) of a cubic crystal lattice is calculated as

$$K_{hkl} = 6|h|^3 / \left[\left(6h^2 - 2|hk| + |kl| - 2|hl| \right) \left(h^2 + k^2 + l^2 \right)^{1/2} \right]. \tag{1.11}$$

It is clear from (1.11) that the same value of K_{hkl} corresponds to any of diffraction lines of one family. Values of K_{hkl} for the diffraction lines (hkl) of some families of face-centered cubic (fcc) lattice are presented in Table 1.1.

Experimental diffraction reflection is always broadened due to a finite resolution of a diffractometer and cannot be less than the instrumental width of reflection. It means that the reflection broadening β with respect to the instrumental width FWHM_R should be used for the calculation of average particle size. If the experimental and the instrumental widths FWHM_{exp} and FWHM_R are described by the Gauss $g(\theta)$ or the pseudo-Voight $V(\theta)$ functions with large content of the $g(\theta)$ function, then physical broadening β is

$$\beta(2\theta) = \left[\left(\text{FWHM}_{\text{exp}} \right)^2 - \left(\text{FWHM}_R \right)^2 \right]^{1/2}. \tag{1.12}$$

Just such description of the experimental broadening and the instrumental angular resolution function is the most widespread.

If the FWHM_{exp} and FWHM_R are described by the Lorentz functions only, then the physical broadening of reflection is equal to a difference of the experimental and the instrumental widths of reflection, i.e. $\beta(2\theta) = \text{FWHM}_{\text{exp}} - \text{FWHM}_R$.

According to [41], the size broadening of XRD lines caused by the small particles size is described by the function

$$\beta_s(2\theta) = \frac{K_{hkl}\lambda}{\langle D \rangle \cos\theta} \approx \frac{\lambda}{\langle D \rangle \cos\theta} \quad [\text{rad}]. \tag{1.13}$$

Table 1.1 Scherrer's constant K_{hkl} for the (hkl) lines from cubic crystals with fcc lattice

(hkl)	(111), (222), (333)	(200), (400)	(220), (440)	(311)	(331)	(420)	(422)	(511)
K_{hkl}	1.155	1.000	1.061	1.136	1.126	1.073	1.153	1.102

For (333) and (511) lines, average K_{hkl} is 1.211

Deformation distortions and induced non-uniform atom displacements may appear in the case of random distribution of the dislocations in the volume of the specimen. In this case, the atom displacements are determined by the superposition of displacements from every dislocation. This may be regarded as a local variation of interplanar distances. In other words, the distance between the planes continuously changes from $(d_0 - \Delta d)$ to $(d_0 + \Delta d)$ (d_0 is interplanar distance for the ideal crystal, and Δd is average variation of the distance between the (hkl) planes in all volume V of the crystal). In this case, the quantity $\varepsilon = \Delta d/d_0$ is the microstrain of the lattice which characterises the uniform strain averaged out with respect to the crystal. The diffraction maximum from the regions of the crystal with the changed interplanar distance appears at angle θ, which slightly differs from angle θ_0 for the ideal crystal. Therefore, the reflection broadens. The equation for the reflection broadening that caused by the lattice microdeformations, can be easily derived by differentiation of the Wullf-Bragg equation $d = n\lambda/(2\sin\,\theta)$; the result is $\Delta d/\Delta\theta = -(n\lambda/2) \times (\cos\,\theta/\sin^2\theta) = -d/\tan\,\theta$. The reflection broadening to one side of the maximum of the reflection, corresponding to the interplanar distance d, with the variation of the interplanar distance by $+\Delta d$ is $\Delta\theta = -(\Delta d/d)\times\tan\,\theta$, and in the case of variation by $-\Delta d$, it is $\Delta\theta = (\Delta d/d) \times \tan\,\theta$. The total broadening of reflection due to the strain-induced crystal lattice distortions (strain broadening) is equal to the sum of these broadenings and its value with respect to angle θ is $\beta_d(\theta) = (2\Delta d/d) \times \tan\,\theta = 2\varepsilon\,\tan\,\theta$. Taking into account $\beta_d(2\theta) \equiv 2\beta_d(\theta)$, the strain broadening with respect to angle 2θ can be written as

$$\beta_d(2\theta) = 4\varepsilon\,\tan\,\theta \;\; [\text{rad}]. \tag{1.14}$$

where ε is the microstrain value which is averaged over a crystal volume.

The anisotropy of microstrains should be takes into consideration when the diffraction reflections broadening of strongly deformed crystals is studied. According to [56, 57], strain broadening $\beta_d(2\theta)$ is equal to

$$\beta_d(2\theta) = 4\varepsilon_{hkl}\,\tan\,\theta, \quad [\text{rad}] \tag{1.15}$$

where $\varepsilon_{hkl} = \sigma/E_{hkl} = k_p C_{hkl}^{1/2}$ is the effective microstrain that allowing for crystal deformation anisotropy; σ is the FWHM of the strain distribution function; E_{hkl} is the Young's modulus which depends on the $[hkl]$ direction in a crystal; and k_p is the constant which depends on the density of edge and screw dislocations and the Burgers vector for a specimen studied, i.e. on the changing the interplanar distances and atom displacements. For cubic crystals, the anisotropic Young's modulus E_{hkl} is determined through the elastic constants c_{11}, c_{12} and c_{44} or through elastic deformation tensor components s_{11}, s_{12} and s_{44} as

$$E_{hkl} = \frac{1}{s_{11} - (2s_{11} - 2s_{12} - s_{44})H}, \tag{1.16}$$

where $H = (h^2k^2 + k^2l^2 + h^2l^2)/(h^2 + k^2 + l^2)^2$ is the anisotropy dislocation factor. The coefficient C_{hkl} takes into account the relative contents f_E and $f_S = (1 - f_E)$ of edge and screw dislocations in the deformed crystal. For cubic crystals, the coefficient C_{hkl} has the form [58]:

$$C_{hkl} = f_E C_{hkl,E} + (1 - f_E)C_{hkl,S} = [f_E A_E + (1 - f_E)A_S] + [f_E B_E + (1 - f_E)B_S] \cdot H$$
$$= A + BH,$$

$$(1.17)$$

where A and B are the constants which depend on the density and content of edge and screw dislocations in a studied specimen.

With allowance for $\varepsilon_{hkl} = \sigma/E_{hkl} = k_\rho C_{hkl}^{1/2}$ and (1.17), expression (1.15) can be written as

$$\beta_{\mathrm{d}}(2\theta) = 4k_\rho C_{hkl}^{1/2} \tan\theta \equiv 4k_\rho(A + BH)^{1/2}\tan\theta.   (1.18)$$

From (1.17) and (1.18) it follows that microstrains anisotropy is caused a greater extent by dislocations or dislocation-like defects than to atom displacements.

The value of microstrains $\varepsilon_{\mathrm{aver}}$ averaged over the crystal volume is

$$\varepsilon_{\mathrm{aver}} = \left(\sum \varepsilon_{hkl} P_{hkl}\right)\bigg/\sum P_{hkl},   (1.19)$$

where P_{hkl} is the multiplicity factor.

With allowance for (1.13), the average particle size D is determined by the Warren method [59]

$$\langle D \rangle = \frac{K_{hkl}\lambda}{\cos\theta \cdot \beta_{\mathrm{s}}(2\theta)} \equiv \frac{K_{hkl}\lambda}{2\cos\theta \cdot \beta_{\mathrm{s}}(\theta)}.   (1.20)$$

Note that $\beta(2\theta) \equiv 2\beta(\theta)$.

As a rule, the diffraction reflections broadening in nanocrystalline materials is caused by both the small particles size D and the microstrains ε. In order to separate contributions of size and strain broadening to the experimental broadening of diffraction reflections, the Williamson-Hall extrapolation [60] is used. In this case, the dependence of the reduced broadening $\beta^*(2\theta) = [\beta(2\theta)\cos\theta]/\lambda$ of the (hkl) reflection on the scattering vector $s = (2\sin\theta)/\lambda$ is constructed. The base of the Williamson-Hall extrapolation is the assumption that size β_{s} and strain β_{d} broadenings are described by the Lorentz functions $\ell(\theta)$ only or by the pseudo-Voight functions $V(\theta)$ with large contribution $c \geq 0.7$ of the Lorentz function. According to [60], in this case total broadening of reflection is a sum of size and strain broadenings, i.e.

$$\beta(2\theta) = \beta_s(2\theta) + \beta_d(2\theta) = \lambda/[\langle D \rangle \cos\theta] + 4\varepsilon\tan\theta. \tag{1.21}$$

Without taking into account of anisotropy of microstrains, the reduced total reflection broadening is

$$\beta^*(2\theta) = [\beta(2\theta)\cos\theta]/\lambda = 1/\langle D \rangle + 2\varepsilon s = 1/\langle D \rangle + s \times \tan\varphi \tag{1.22}$$

with $2\varepsilon \equiv \tan\varphi$. It is seen that the reduced broadening $\beta^*(2\theta)$ is a linear function of the scattering vector s, and φ is a slope angle of the straight line approximating the function (1.22)

The average size $\langle D \rangle$ of particles is determined by the extrapolation of the linear dependence $\beta^*(2\theta) = 1/\langle D \rangle + 2\varepsilon s$ to the value $s = 0$; that is,

$$\langle D \rangle = \frac{1}{\beta^*(2\theta)|_{s=0}} \tag{1.23}$$

The average value of the microstrains ε can be estimated from the slope angle φ as

$$\varepsilon = [(\tan\phi)/2] \cdot 100\% \tag{1.24}$$

If the physical broadening is determined as a sum of size and strain broadenings, i.e. $\beta(2\theta) = \beta_s(2\theta) + \beta_d(2\theta)$ (see (1.21)) then after the substitution of (1.13) and (1.18) in expression (1.21) and replacements of broadening $\beta(2\theta)$ to reduced broadening $\beta^*(2\theta)$, we obtain

$$\beta^*(2\theta) = \beta_s^*(2\theta) + \beta_d^*(2\theta) = 1/\langle D \rangle + 2k_\rho(A + BH)^{1/2}s, \tag{1.25}$$

where $k_\rho(A + BH)^{1/2} = \varepsilon_{hkl}$. The reduced size broadening $\beta_s^*(2\theta)$ is inversely proportional to the average size $\langle D \rangle$ of the small particles (crystallites) and gives a constant contribution into total broadening that does not depend on the vector s. The strain $\beta_d^*(2\theta)$ broadening is function of the scattering vector s.

Consideration the anisotropy of microstrains ε_{hkl} is important for describing diffraction reflection broadening in strongly deformed substances.

At low deformation distortions of the lattice, the anisotropy of the microstrains is negligibly small ($B \rightarrow 0$), and (1.25) takes the conventional form (1.22) where $k_\rho A^{1/2}$ is average value of the microstrains ε.

For illustration, the quantitative analysis of the XRD reflection profile is performed and the reflection width is found for a nanocrystalline powders of lead sulfide PbS with cubic (space group $Fm\overline{3}m$) structure. These nanopowders were synthesized by the hydrochemical bath deposition (see Sect. 2.1.3). Figure 1.6 shows the XRD patterns of PbS powders with average particle size 70 and 100 nm. For comparison, the XRD pattern of the reference specimen LaB_6 is shown in Fig. 1.6. It is seen that all XRD reflections of PbS nanopowders are broadened.

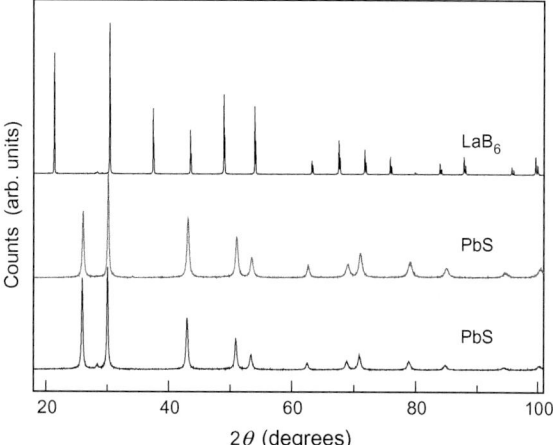

Fig. 1.6 XRD patterns of the PbS nanopowders with cubic (space group $Fm\overline{3}m$) structure. XRD pattern of the reference specimen LaB_6 is shown for comparison. Diffraction reflections of the PbS nanopowders are broadened. XRD patterns were collected using $CuK\alpha_{1,2}$ radiation

As an example, the doublet XRD reflections $(220)_{B1}$, $(311)_{B1}$ and $(222)_{B1}$ of the nanocrystalline lead sulfide PbS and their quantitative description by two pseudo-Voigt functions (1.3) are shown in Fig. 1.7. The least-squares method was used for calculation of parameters of the approximate pseudo-Voigt functions. It is seen that the approximation accuracy is high.

The microstrains are small in these PbS nanopowders, synthesized by the chemical deposition, therefore the anisotropy of microstrains ε can be neglected. The dependences of $\beta^*(2\theta)$ of the XRD reflections on the scattering vector s for the PbS powders with different average particle size from 9 to 100 nm are presented in Fig. 1.8. If the reflection broadening is due to only small particle size D, the experimental points are scattered near horizontal lines. It is seen that the XRD broadening increases with length of vector s. This means the presence of microstrains ε in the particles.

Fig. 1.7 Experimental doublet XRD reflections $(220)_{B1}$, $(311)_{B1}$ and $(222)_{B1}$ (crosses) of the nanocrystalline PbS and their numerical approximation by two pseudo-Voigt functions. XRD patterns are collected in $CuK\alpha_{1,2}$ radiation

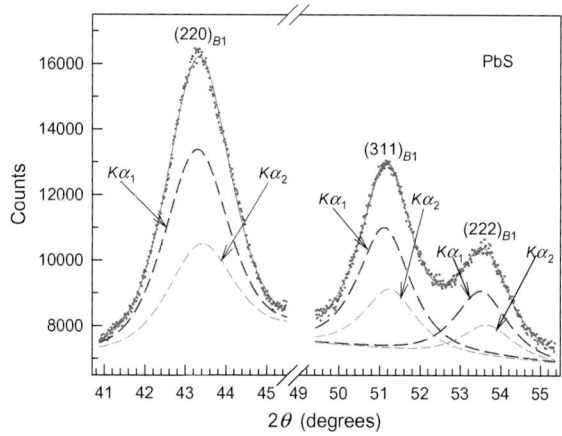

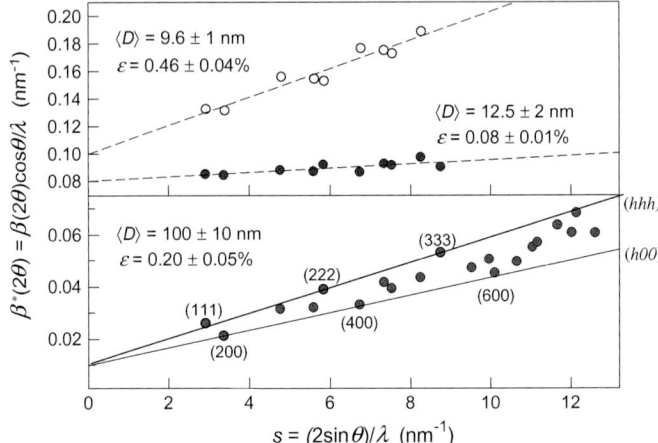

Fig. 1.8 Estimation of the average particle size D and microstrains ε by the Williamson-Hall extrapolation from the dependence of $\beta^*(2\theta) = [\beta(2\theta)\cos\theta]/\lambda$ on the vector $s = (2\sin)/\lambda$ for the PbS powders with particle size of 9, 12 and 100 nm

Figure 1.8 shows that experimental points of dependence $\beta^*(s)$ for PbS powder with $D = 100$ nm are scattered in the sector between two straight lines. The upper and bottom lines pass through family of (hhh) reflections, and family of $(h00)$ reflections, respectively. In this case the reduced reflection broadening should be calculated by the Debye-Scherrer formula $\beta^*(2\theta) = [\beta(2\theta)\cos\theta]/K_{hkl}\lambda$. Coefficients K_{hkl} for cubic PbS powder were calculated by the formula (1.11) (see also Table 1.1). The broadening of the reflections for PbS nanopowders with a particle size ≤ 20 nm is very large, therefore the number of observed reflections does not exceed 10. At a small number of observed reflections and in the absence of the anisotropy of microstrains, the reduced broadening β^* is approximated by a single linear function $\beta^*(2\theta) = 1/\langle D\rangle + 2\varepsilon s$.

According to the calculations, the size of separate nanoparticles (more exactly, size of CSR) in studied PbS powders is 100 ± 10, 12 ± 2, and 9 ± 1 nm. The value of microstrains ε for these PbS powders amount to 0.20, 0.08, and 0.46%, respectively.

Such reason of broadening as inhomogeneity, i.e., a non-uniform composition of a substance over its volume, is considered in works [19, 20, 42–46]. In our book, the reflection broadening β_h which is caused by inhomogeneity is not considered because an inhomogeneity of nanostructured sulfides of lead, cadmium and silver is negligibly small and does not influence on the broadening β_h practically.

1.2.3   Sedimentation, Photon Correlation Spectroscopy, Gas Adsorption, and Gas Filtration

Determination of the size of small particle by the sedimentation is based on measuring the time during which a particle travels the fixed distance S in a liquid medium with known viscosity η. Other version of the sedimentation is determination of the particles distribution on a height in a liquid or gas medium.

The condition for uniform movement of a particle is the equality of gravity force and friction force. Friction force is the force of resistance of liquid or gas to the particle movement. Viscosity determines the friction force. At a low movement rate $V = S/t$ and a small size of the particle, the force of resistance to movement of the particle with an average size R is described by the well-known Stokes equation

$$F = (4\pi/\alpha)\eta RV = (4\pi/\alpha)\eta RS/t \qquad (1.26)$$

or (by a means of the volume $v \sim R^3$)

$$F = (4\pi/\alpha)hv^{1/3}S/t \qquad (1.27)$$

In a general case, coefficient α depends on the shape of falling particle (body) and has different values. For spherical particles $\alpha = 2/3$. For a steady rate of movement, the falling particle is affected, in addition to the resistance force, by the force of gravity $P = mg = v\rho g$, where g is the gravitational acceleration, and the mass m of the particle is equal to the product of the volume v by density ρ. To improve the accuracy of calculations, a correction for the injecting Archimedes force is introduced. Thus, the force causing the particle to fall in the given liquid or gas medium is

$$F_d = P - v\rho_m = vg(\rho - \rho_m), \qquad (1.28)$$

where ρ_m is the density of the medium. From the equality of the force F_d and the force of resistance to fall, F, it is easy to find the volume and averaged out size of the falling particle:

$$(4\pi/\alpha)\eta v^{1/3}S/t = vg(\rho - \rho_m). \qquad (1.29)$$

It follows that

$$v = \left[\frac{(4\pi/\alpha)\eta S}{g(\rho - \rho_m)t} \right]^{3/2} \qquad (1.30)$$

and

$$R \sim \sqrt{\frac{(4\pi/\alpha)\eta S}{g(\rho - \rho_m)\,t}} \sim \sqrt{k/t}. \tag{1.31}$$

For spherical particle $F = 6\pi\eta RS/t$ and $F_d = vg(\rho - \rho_m) = (4/3)\pi R^3 g(\rho - \rho_m)$. Therefore it follows from equality $F = F_d$ that

$$R_{sph} = \sqrt{\frac{9\eta S}{2g(\rho - \rho_m)\,t}}. \tag{1.32}$$

Sedimentation rate V for spherical particles is

$$V = \frac{mg}{6\pi\eta R}\frac{\rho - \rho_m}{\rho} \tag{1.33}$$

where m is the mass of a particle.

Equations (1.27), (1.31) and (1.32) are valid only for the solid particles moving with uniform low rate in a medium, which is an infinite large in comparison with the size of falling particle. The distance between the particles should be large enough to avoid interparticle interaction.

In the sedimentation method, the size of the falling particle is inversely proportional to the time which the particle travels a fixed path. In currently available sedimentographs, the travel of the particles and their number is recorded by a means of laser radiation. This technique is used to determine not only the size of separate particles but also the size distribution of the particles, i.e. the dispersion of the particle size. The volume concentration of particles in a liquid medium is typically less than 1%.

A disadvantage of the sedimentation is that it cannot be used to measure the size of very small (less than 50 nm) particles. In addition to this, in order to obtain high precision of measurements, the liquid must efficiently wet the surface of the particles. In poor wetting of particles, the surface tension force will maintain the small particles on the surface, and a gas shell will surround the large particles. During sedimentation measurements it is also necessary to avoid coagulation and agglomeration of the separate particles which may greatly distort the results. Detailed description of sedimentation methods is given in monograph [61].

An efficient approach for determining the small particles size is based on using of Brownian movement [1], on the one hand, and analysis of the spectral composition of the light scattered by the suspension or a colloid solution, on the other hand. A. Einstein invented this approach for determination of the small particles size. In fact, his first article [62] from a serious of studies into the theory of Brownian motion [2, 3, 62, 63] is referred to as "Eine neue Bestimmung der Moleküldimensionen" ("New determination of the size of molecules"). It is interesting to note that this article was in fact an Einstein's dissertation for the title of doctor of philosophy. The study was presented in 1905 at the Natural Mathematics section of the Higher Faculty of Philosophy of the Zurich University. Almost independently and at the same time as Einstein, the theory of Brownian motion was developed by Polish scientist M. Smoluchowski [4].

The size of small particles can be measured with the use of the well-known Einstein equation, describing the Brownian movement of spherical particles in a liquid:

$$R = k_B T / (6\pi\eta D_{diff}), \tag{1.34}$$

where η is the liquid shear viscosity, D_{diff} is the Brownian particles diffusion coefficient. The diffusion coefficient of Brownian particles is determined measuring the full width of a non-displaced (central) spectral line in the spectrum of scattered light using an optical displacement spectrometer [64]:

$$\text{FWHM} = 2D_{diff}K^2, \tag{1.35}$$

where K is the variation of wave vector during light scattering. This method for determination of the small particle size is known as photon correlation spectroscopy (PCS) [64–68].

Photon correlation spectroscopy (another name is the dynamic light scattering (DLS)) is based on analysis of the spectral composition of the light, scattered by the examined specimen. Photon correlation spectroscopy is used widely for measuring the sizes of submicrocrystalline and nanosized particles in transparent suspensions and colloidal systems. Measurements are carried out using the radiation of a He–Ne laser with a wavelength of $\lambda = 633$ nm, laser power up to 10 mW. Light radiation is focused in the centre of a cuvette containing a liquid with a specimen. The cuvette is placed in a thermostat for temperature stabilisation. The light is scattered under the given angle θ and is received by a photoelectric multiplier. Then the autocorrelation function $G(\tau)$ of scattered light is measured by a multichannel correlator. Approximation of autocorrelation function by the exponential dependence makes it possible to determine the width of the spectral line and then find the diffusion coefficient and the particle size. An increasing radiation power results in an improvement of the accuracy and accelerates the measurements.

Industrial analysers of the particle size, using the PCS (DLS) method, take measurements under several angles and to determine the particles size in the range from 3 nm to 3 μm and also the particle size distribution. For example, the size (hydrodynamic diameter D_{dls}) of colloidal sulfide nanoparticles is determined by dynamic light scattering on a Zetasizer Nano ZS facility (Malvern Instruments Ltd) with the He–Ne laser and the back-scattering light detection angle equal to 173°.

In order to determine the small particles size in colloidal solutions and suspensions, it is also efficient to use the mass distribution of particles using an ultracentrifuge. This method of separation of small particles by a means of a centrifugal force in an ultracentrifuge was developed the Swedish scientist T. Svedberg [7]. If the particle density is known, measuring the mass of the particle it is easy to find its volume and linear size.

The gas adsorption is a special method for determining the specific surface area S_{sp} of the powder. Measured value of the specific surface area may be used to estimate the mean particle size. To measure specific surface, helium or argon-helium mixture is passed through the powder in a special chamber at a temperature of liquid nitrogen (77 K). Helium atoms are adsorbed by the particle surface. Helium-saturated powder is heated to remove the entire amount of adsorbed helium and the amount of adsorbed helium is determined from the variation of the volume or mass. Assuming that helium atoms form a monolayer on the surface of the particles, from the volume of the adsorbed gas it is possible to determine the total surface area of the particles and the specific surface area of the powder (in the units of area per units of mass, for example, $m^2\ g^{-1}$).

At present, a more reliable equation obtained in theory polymolecular adsorption by Brunauer, Emmett and Teller (BET) is used for treatment of experimental isotherms of adsorption to determine the S values. This theory is based on the model of the adsorption process, proposed by Langmuir, but it takes into account the possibility of adsorption is not only the first but also the second, third and subsequent layers of already adsorbed molecules, i.e. the possibility of multilayer adsorption.

Adsorption isotherms measure at a temperature of liquid nitrogen equals 77 K. For the fine and ultrafine powders with a specific surface from 10 to 500 $m^2\ g^{-1}$, the sample mass is equal from 0.2 to 0.02 g.

If all the particles have the same size and identical shape, then a simple relationship can be derived between the specific surface S_{sp} and the linear particle size R. For example, the mass of a spherical particle is $m = (4\pi/3)R^3\rho$ and the surface area is $s = 4\pi R^2$, therefore its specific surface is

$$S_{sp} = s/m = 3/\rho R. \qquad (1.36)$$

An average particle size $D = 2R$ is

$$D = 6/\rho S_{sp}. \qquad (1.37)$$

Thus, as the specific surface increases, the particle size decreases. If a function describing the size distribution of the particles is available, then from the measured specific surface it is possible to determine the particle size, taking size distribution function into account.

The described methods of determination of the size of nanoparticles are used most widely.

References

1. Brown, R.A.: Brief account of microscopical observations made in the months on June, July, and August, 1827, on the particles contained in the pollen of plants; and on the general existence of active molecules in organic and inorganic bodies. Phil. Mag. **4**, 161–173 (1828)
2. Einstein, A.: Über die von der molekularkinetischen theorie der wärme geforderte bewegung von in ruhenden flüssigkeiten suspendierten teilchen. Ann. Phys. **322**(8), 549–560 (1905)
3. Einstein, A.: Zur theorie der brownschen bewegung. Ann. Phys. **324**(2), 371–381 (1906)
4. Smoluchowski, M.: Zur kinetischen theorie der brownschen molekularbewegung und der suspensionen. Ann. Phys. **326**(14), 756–780 (1906)
5. Faraday, M.: Experimental relations of gold (and other metals) to light. Philosoph. Trans. Roy. Soc. (London) **147**, 145–181 (1857)
6. Svedberg, T., Nichols, J.B.: Determination of size and distribution of size of particles by centrifugal methods. J. Am. Chem. Soc. **45**(12), 2910–2917 (1923)
7. Svedberg, T.: Colloid Chemistry. Chemical Catalog Co., New York (1924)
8. Zakery, A., Elliott, S.R.: Optical properties and applications of chalcogenide glasses: a review. J. Non-Cryst. **330**, 1–12 (2003)
9. Fu, H., Tsang, S.-W.: Infrared colloidal lead chalcogenide nanocrystals: synthesis, properties, and photovoltaic applications. Nanoscale. **4**(7), 2187–2201 (2012)
10. Gao, M.-R., Xu, Y.-F., Jianga, J., Yu, S.-H.: Nanostructured metal chalcogenides: synthesis, modification, and applications in energy conversion and storage devices. Chem. Soc. Rev. **42**(7), 2986–3017 (2013)
11. Kozhevnikova, N.S., Rempel A.A.: Physical Chemistry of Aqueous Solutions. Theoretical Basic and Synthesis of Perspective Semiconducting Optical Materials, p. 157. Ural State Techn. University–Urals Polytechn. Institute, Ekaterinburg (2006) (in Russian)
12. Markov, V.F., Maskaeva, L.N., Ivanov, P.N.: Hydrochemical Synthesis of Metal Sulfide Films: Modeling and Experiment, p. 217. Publ. of the Ural Division of the RAS, Ekaterinburg (2006) (in Russian)
13. Zimin, S.P., Gorlachev, E.S.: Nanostructured Lead Chalcogenides, p. 230. Yaroslavl's State University, Yaroslavl (2011) (in Russian)
14. Edelstein, A.S., Cammarata, R.C. (eds.): Nanomaterials: Synthesis, Properties and Applications, p. 596. IOP, Bristol (1996)
15. Gusev, A.I.: Effects of the nanocrystalline state in solids. Uspekhi Fiz. Nauk. **168**(1), 55–83 (1998) (in Russian) (Engl. Transl.: Physics–Uspekhi. **41**(1), 49–76 (1998))
16. Gleiter, H.: Nanostructered materials: basic concepts and microstructute. Acta Mater. **48**(1), 1–29 (2000)
17. Gusev, A.I.: Nanocrystalline Materials: Production and Properties, p. 200. Ural Division of the RAS, Ekaterinburg (1998) (in Russian)
18. Gusev, A.I., Rempel, A.A.: Nanocrystalline Materials. p. 224. Nauka–Fizmatlit, Moscow (2000) (in Russian)
19. Gusev, A.I., Rempel, A.A.: Nanocrysnalline Materials, p. 351. Cambridge International Science Publishing, Cambridge (2004)
20. Gusev, A.I.: Nanomaterials, Nanostructures, and Nanotechnologies, 3rd edn., p. 416. Nauka–Fizmatlit, Moscow (2009) (in Russian)
21. Rempel, A.A.: Nanotechnologies. Properties and applications of nanostructured materials. Uspekhi Khimii. **76**(5), 474–500 (2007) (in Russian) (Engl. Transl.: Rus. Chem. Rev. **76**(5), 435–461 (2007))
22. Schaefer, H.-E.: Nanoscience. The Science of the Small in Physics, Engineering, Chemistry, Biology and Medicine, p. 772. Springer, Heidelberg (2010)
23. Khairutdinov, R.F.: Chemistry of semiconductor nanoparticles. Uspekhi Khimii. **67**(2), 125–139 (1998) (in Russian) (Engl. Transl.: Russ. Chem. Rev. **67**(2), 109–122 (1998))
24. Alferov, ZhI: Nobel Lecture: The double heterostructure concept and its applications in physics, electronics, and technology. Rev. Modern Phys. **73**(3), 767–782 (2001)

25. Alivisatos, A.P.: Semiconductor clusters, nanocrystals and quantum dots. Science **271**(5251), 933–937 (1996)
26. Ledentsov, N.N., Ustinov, V.M., Shchukin, V.A., Kop'ev, P.S., Alferov, ZhI, Bimberg, D.: Quantum dot heterostructures: fabrication, properties, lasers. Semiconductors **32**(4), 343–365 (1998)
27. Knoll, M., Ruska, E.: Das Elektronenmikroscop. Z. Phys. **78**(5–6), 318–339 (1932)
28. Binnig, G., Rohrer, H., Gerber, Ch., Weibel, E.: Surface studies by scanning tunneling microscopy. Phys. Rev. Lett. **49**(1), 57–61 (1982)
29. Hirch, P.B., Howie, A., Nicholson, R.B., Pashley, D.W., Whelan, M.J.: Electron Microscopy of Thin Crystals, p. 453. Krieger Publ. Co., Huntington, New York (1977)
30. Tomas, G., Goringe, M.J.: Transmission Electron Microscopy of Materials, p. 404. Wiley, New York (1990)
31. Spence, J.C.H.: High-Resolution Electron Microscopy, 3rd edn, p. 400. Clarendon Press, Oxford (1981)
32. Shindo, D., Hiraga, K.: High-Resolution Electron Microscopy for Materials Science, p. 190. Springer, Tokyo (1998)
33. Ding, B., Shi, M., Chen, F., Zhou, R., Deng, M., Wang, M., Chen, H.Z.: Shape-controlled syntheses of PbS submicro-/nano-crystals via hydrothermal method. J. Cryst. Growth. **311**, 1533–1538 (2009)
34. Binnig, G., Quate, C.F., Gerber, H.: Atomic force microscope. Phys. Rev. Lett. **56**(9), 930–933 (1986)
35. Wiesendanger, R.: Scanning Probe Microscopy and Spectroscopy: Methods and Applications, p. 659. Cambridge University Press, Cambridge (1994)
36. Cohen, S.H., Lightbody, M.L. (eds.): Atomic Force Microscopy/Scanning Tunneling Microscopy, vol. 3, p. 218. Kluwer Academic/Plenum Publishers, Dordrecht (1999)
37. Goldstein, J., Newbury, D.E., Joy, D.C., Lyman, C.E., Echlin, P., Lifshin, E., Sawyer, L.C., Michael, J.R.: Scanning Electron Microscopy and X-ray Microanalysis, 3rd edn, p. 586. Kluwer Academic/Plenum Publishers, Dordrecht (2003)
38. Patanè, A., Polimeni, A., Eaves, L., Main, P.C., Henini, M., Belyaev, A.E., Dubrovskii, Y.V., Brounkov, P.N., Vdovin, E.E., Khanin, Y.N., Hill, G.: Modulation of the luminescence spectra of InAs self-assembled quantum dots by resonant tunneling through a quantum well. Phys. Rev. B. **62**(20), 13595–13598 (2000)
39. Scherrer, P.: Bestimmung der grösse und der inneren struktur von kolloidteilchen mittels röntgenstrahlen. Nachr. Ges. Wiss. Göttingen, Math. Phys. **2**, 98–100 (1918)
40. Seljkow, N.: Eine röntgenographische Methode zur messung der absoluten dimensionen einzelner kristalle in körpern von fein-kristallinischem bau. Z. Phys. **31**(5–6), 439–444 (1925)
41. Krivoglaz, M.A.: Theory of X-ray and Thermal-Neutron Scattering by Real Crystals, p. 405. Plenum Press, New York (1969)
42. Rempel, A.A., Rempel, S.V., Gusev, A.I.: Quantitative assesment of homogeneity of non-stoichiometric compounds. Dokl Phys. Chem. **369**(4–6), 321–325 (1999)
43. Rempel, A.A., Gusev, A.I.: Preparation of disordered and ordered highly nonstoichiometric carbides and evaluation of their homogeneity. Fiz. Tverd. Tela. **42**(7), 1243–1249 (2000) (in Russian) (Engl. Transl.: Phys. Solid State. **42**(7), 1280–1286 (2000))
44. Gusev, A.I., Rempel, A.A.: Nonstoichiometry, Disorder and Order in Solids, p. 580. Ural Division of the Rus. Acad. Sci., Ekaterinburg. (2001) (in Russian)
45. Gusev, A.I., Rempel, A.A., Magerl, A.J.: Disorder and Order in Strongly Nonstoichiometric Compounds: Transition Metal Carbides, Nitrides and Oxides, p. 608. Springer, Berlin (2001)
46. Gusev, A.I.: Nonstoichiometry, Disorder, Short-Range and Long-Order in Solids, p. 856. Nauka–Fizmatlit, Moscow (2007) (in Russian)
47. Kurlov, A.S., Gusev, A.I.: Determination of the particle sizes, microstrains, and degree of inhomogeneity in nanostructured materials from X-ray diffraction data. Fiz. Khim. Stekla. **33**(3), 383–392 (2007) (in Russian) (Engl. Transl.: Glass Phys. Chem. **33**(3), 276–282 (2007))
48. Gusev, A.I., Kurlov, A.S.: Characterization of nanocrystalline materials by the size of particles (grains). Metallofizika i Nov. Tekhnologii. **30**(5), 679–694 (2008) (in Russian)

49. Kozhevnikova, N.S., Kurlov, A.S., Uritzkaya, A.A., Rempel, A.A.: Diffraction analysis of nanocrystalline particle size of lead and cadmium sulfides prepared by chemical deposition from aqueous solutions. Zh. Strukt. Khimii. **45**, 147–153 (2004) (in Russian) (Engl. Transl.: J. Struct. Chem. **45**, S154–S159 (2004))

50. Sadovnikov, S.I., Gusev, A.I.: Chemical deposition of nanocrystalline lead sulfide powders with controllable particle size. J. Alloys Comp. **586**, 105–112 (2014)

51. Dasgupta, P.: On use of pseudo-Voigt profiles in diffraction line broadening analysis. Fizika A (Croatia). **9**(2), 61–66 (2000)

52. Puerta, J., Martin, P.: Three and four generalized Lorentzian approximations for the Voigt line shape. Appl. Optics. **20**(22), 3923–3928 (1981)

53. Cagliotti, G., Paoletti, A., Ricci, F.P.: Choice of collimators for a crystal spectrometer for neutron diffraction. Nuclear Instrum. Methods. **3**(3), 223–228 (1958)

54. Rietveld, H.M.: A profile refinement method for nuclear and magnetic structures. J. Appl. Crystallogr. **2**(2), 65–71 (1969)

55. James, R.W.: The Optical Principles of the Diffraction of X-Rays, p. 624. G. Bell & Sons Ltd., London (1954)

56. Hall, W.H.: X-ray line broadening in metals. Proc. Phys. Soc. London A. **62**(11), No 359A, 741–743 (1949)

57. Stokes, A.R., Wilson, A.J.C.: The diffraction of X rays by distorted crystal aggregates–I. Proc. Phys. Soc. **56**(3), 174–181 (1944)

58. Scardi, P., Ortolani, M., Leoni, M.: WPPM: microstructural analysis beyond the Rietveld method. Mater. Sci. Forum. **651**, 155–171 (2010)

59. Warren, B.E., Averbach, B.L., Roberts, B.W.: Atomic size effect in the X-ray scattering by alloys. J. Appl. Phys. **22**(12), 1493–1496 (1951)

60. Williamson, G.K., Hall, W.H.: X-ray line broadening from filed aluminium and wolfram. Act. Metal. **1**(1), 22–31 (1953)

61. Bernhardt, C.: Particle Size Analysis Classification and Sedimentation Methods, p. 448. Kluwer Academic Publishers, Dordrecht (1994)

62. Einstein, A.: Eine neue bestimmung der moleküldimensionen. inaugural dissertation. zürich universität. bern: buchdruck. k.j. wyss, 1905. Ann. Phys. **324**(2), 289–306 (1906)

63. Einstein, A.: Elementare theorie der brownschen bewegung. Z. Elektrochem. **14**(17), 235–239 (1908)

64. Cummins, H.Z., Pike, E.R. (eds.): Photon Correlation and Light Beating Spectroscopy, p. 584. Plenum Press, New York (1974)

65. Anisimov, M.A., Kiyachenko Yu, F., Nikolaenko, G.L., Yudin, I.K.: Measurement of liquid viscosity and sizes of suspended particles by means of correlation spectroscopy of light beating. Inzhenerno-Fiz. Zh. **38**(4), 651–655 (1980) (in Russian)

66. Dahneke, B.E. (ed.): Measurement of Suspended Particles by Quasi-Elastic Light Scattering. p. 584. Wiley, New York (1983)

67. Pike, E.R., Abbiss, J.B. (eds.): Light Scattering and Photon Correlation Spectroscopy, p. 470. Kluwer Academic Publishers, Dordrecht (1997)

68. Yudin, I.K., Nikolaenko, G.L., Kosov, V.I., Agayan, V.A., Anisimov, M.A., Sengers, J.V.: A compact photon-correlation spectrometer for research and education. Int. J. Thermophys. **18** (10), 1237–1248 (1997)

Chapter 2
Nanostructured Lead Sulfide PbS

Lead sulfide is a representative of the abundant family of semiconducting chalcogenides. Lead sulfide PbS is a single compound existing in the Pb–S system (Fig. 2.1). Metallic lead Pb has the face-centered cubic (space group $Fm\bar{3}m$) structure of $A1$ type with a lattice constant 0.4948 nm. Lead sulfide contains ~ 50 at.% S, has the melting temperature 1391 K, and possesses narrow region of homogeneity [1]. Indeed, the detailed analysis of phase diagram of the Pb–S system near ~ 50 at.% S (Fig. 2.2) shows that the maximum melting temperature of lead sulfide corresponds to nonstoichiometric composition $PbS_{0.9998}$ [2]. It follows from data [2] that lead sulfide has very narrow homogeneity region from 49.9905 to 50.003 at.% S at a temperature of 1300 K.

In normal conditions, bulk and coarse-grained PbS is a direct narrow-gap semiconductor with a $B1$ cubic crystal structure. At the temperature 300 K, the band gap of PbS single crystals and polycrystalline coarse-grained PbS films is equal to 0.41–0.42 eV [3–5]. Lead sulfide PbS is one of the most demanded narrow-gap semiconductors. It is used in a number of devices such as photodetectors operating in a broad [from infrared (IR) to ultraviolet (UV)] wavelength range, high efficiency solar cells, thermoelectric transducers and optical switches. Transition from coarse-grained to nanostructured PbS has considerably extended the scope of its application, especially in those cases where the particle size is comparable with (or less than) the lead sulfide exciton radius, which is 18 nm.

The latest achievements in the synthesis of colloidal nanocrystals of lead chalcogenides (PbS and PbSe) and nanocrystals for solar cells can be regarded as a claim for the design of photoelectric and infrared devices based on nanostructured lead chalcogenides. The most efficient process resulting in the formation of nanocrystals with well controllable size and low size dispersion is the chemical synthesis of colloid chalcogenide solutions.

Large Bohr exciton radius (18 nm) and small masses for electrons and holes promote effective charge delocalization in PbS thin films and therefore high charge

© Springer International Publishing AG 2018
S.I. Sadovnikov et al., *Nanostructured Lead, Cadmium, and Silver Sulfides*,
Springer Series in Materials Science 256, DOI 10.1007/978-3-319-56387-9_2

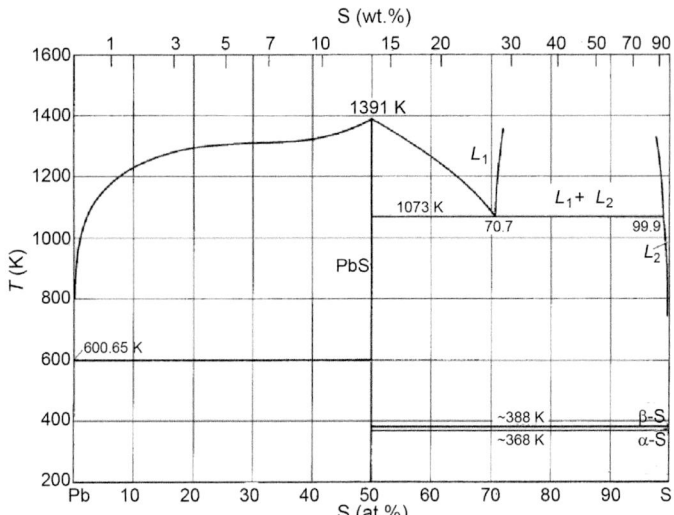

Fig. 2.1 Phase diagram of the Pb–S system [1]

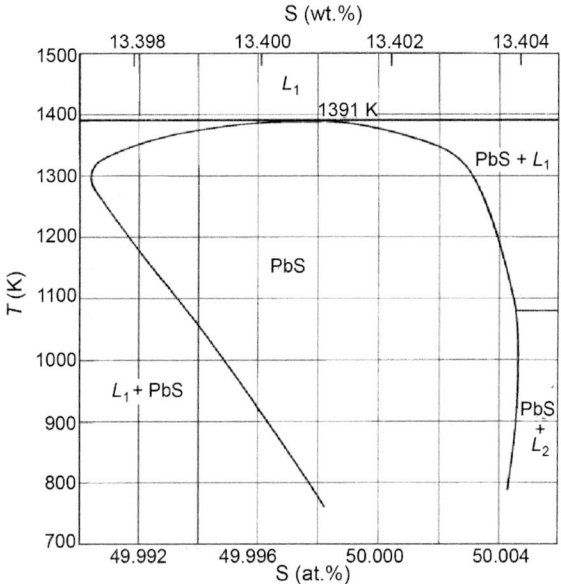

Fig. 2.2 Phase diagram of the Pb–S system near 50 at.% S [2]

carrier mobilities can be anticipated. The formation of multiple electron–hole pairs per absorbed photon has the potential to significantly improve the power conversion efficiencies in photovoltaic devices based on nanostructured lead sulfide.

Lead, silver, cadmium and zinc sulfides are of most interest among chalcogenides. Out of these nanostructured sulfides, lead sulfide is especially demanded owing to the extensive application.

It is shown in Sect. 1.1 that such concept as the "nanostructured material" covers a broad range of objects from colloidal solutions and isolated nanoparticles to heteronanostructures of different dimensionality, nanofilms and nanopowders. In line with this, we discuss in this chapter the preparation methods, structures and properties of lead sulfide colloidal solutions, single nanoparticles, nanofilms and nanopowders. Until recently, there were no generalizing works on nanostructured lead sulfide in the scientific literature. Mention only may be made of a small chapter in the monograph [6] devoted to chemical deposition of PbS films. The first fundamental review of structure and properties of nanostructured lead sulfide [7] was published in 2016.

The principal methods of synthesis described in this chapter are general and applicable to the synthesis of any chalcogenide, and the key properties of nanostructured lead sulfide are also typical of other chalcogenides.

2.1 Methods for the Preparation of Nanostructured Lead Sulfide: From Colloidal Solutions to Thin Films

The most important factors that have to be taken into account in the synthesis of nanoparticles are the following:

(1) non-equilibrium nature of systems; virtually all nanosystems are thermodynamically unstable; they are prepared under conditions far from equilibrium; this provides active nucleation and, owing to the enormous number of nuclei which accumulate the major part of material, the possible growth of the formed nanoparticles can be appreciably restricted;

(2) high chemical homogeneity; the nanomaterial homogeneity is achieved if no component separation, either within a single nanoparticle or between the particles, takes place during the synthesis;

(3) monodispersity; the properties of nanoparticles strongly depend on the particle size; therefore, for the manufacture of materials with good performance characteristics, it is necessary to synthesize particles with rather narrow size distribution.

The preparation processes of nanostructured materials can be classified into two groups according to the way the nanostructure is formed [8], namely, the bottom-up approach implies the assembly of nanoparticles from atoms and molecules, i.e, the initial species are enlarged to give nanoparticles; the top-down processes are based on disintegration of larger particles or powders to the nanosize.

2.1.1 Colloidal Solutions of Lead Sulfide Nanoparticles

The usual method for the preparation of colloidal solutions of sulfides consists in the chemical reaction between solution components (water-soluble metal salt, sulfidizer and complexing agent) and termination of the reaction at a definite time point. For certain conditions of synthesis, the dispersion system thus formed can be preserved in the liquid state, i.e, as a colloidal solution. By transferring the liquid colloidal dispersion system to the solid dispersed state, it is possible to precipitate nanosized (or larger) particles from the solution. In the general case, the preparation of a chalcogenide colloidal solution can be conventionally divided into four stages [9]: (1) formation of an equilibrium homogeneous complexing agent – water system; (2) formation and dissociation of the metal complex $[ML_i]^{n+ik}$ (L is ligand, n is the oxidation state of metal M, i is the number of ligands, k the ligand charge); (3) hydrolysis of the chalcogenizing agent; and (4) formation of metal chalcogenide. At the final stage, the dispersed chalcogenide solid phase precipitates or is deposited on the substrate as a film.

A colloidal solution should be considered as stable if during a long period of storage (more than 100 h), no nanopowder precipitation occurs and the size of nanoparticles present in the solution remains invariable or varies within the error of measurements.

In the vast majority of publications devoted to the synthesis of chalcogenide nanoparticles by chemical bath deposition, the issue of stability of colloidal solutions is not addressed, although only stable colloidal solutions can serve as sources of the smallest-size nanoparticles. An exception is a paper of Deng et al. [10] who consider solubilization, that is, transformation of oil-soluble PbS quantum dots into water- soluble quantum dots, and mention the stability of aqueous colloidal solutions.

Currently, most of colloidal solutions of lead sulfide are synthesized using organometallic compounds [11–14]. Soluble metal complexes are sulfided with hydrogen sulfide gas or hydrosulfuric acid H_2S, sodium thiosulfate $Na_2S_2O_3$, sodium sulfide Na_2S, amide (thioacetamide) $CH_3C(S)NH_2$ or diamide N_2H_4CS of thiocarbonic acid and their derivatives – thiocarbazide $(NH_2)_2(NH)_2CS$, acetylthiocarbamide $C_3H_6N_2OS$ and so on. In some cases (usually in the case of solvothermal synthesis), elemental sulfur is used for sulfiding [15]. The particle size is controlled by changing the number of sulfide phase nuclei as they are formed. The use of toxic organometallic lead compounds and organic ligands (thioglycerol, heptane, trioctylphosphine) to prevent agglomeration and growth of PbS particles results in toxicity of the obtained colloidal solutions and restricts the scope of their applicability by ruling out the biological and medical applications.

From the standpoint of a potential biomedical application, hydrochemical synthesis using aqueous solutions of reactants is more acceptable [16–19]. However, in most cases, stability of aqueous colloidal solutions, including lead sulfide solutions, is moderate, ranging from several minutes to 1–2 h.

The conditions of synthesis of stable non-toxic aqueous PbS colloidal solutions were elucidated by Sadovnikov et al. [19]. Lead acetate $Pb(CH_3COO)_2 \equiv Pb(AcO)_2$ and sodium sulfide Na_2S served as the sources of Pb^{2+} and S^{2-} ions, while disodium ethylenediaminetetraacetate Na_2H_2-EDTA$\equiv C_{10}H_{18}N_2Na_2O_{10}$ (Trilon B) or sodium cit-rate $Na_3C_6H_5O_7 \equiv Na_3Cit$ functioned as the complexing agents. The initial solutions of $Pb(AcO)_2$, Na_2S, Na_3Cit, and Trilon B had the same concentration 50 mmol l^{-1}. The initial concentrations of $Pb(OAc)_2$, Na_2S, sodium citrate and Trilon B in the reaction mixtures varied from 0.25 to 25 mmol l^{-1}. The molar concentration ratio for the reactants in the mixtures was stoichiometric: $[Pb^{2+}]:[S^{2-}]:[TrB, Cit^{3-}] = 1:1:1$, where $[Pb^{2+}]$, $[S^{2-}]$, $[TrB]$, and $[Cit^{3-}]$ are the concentrations of lead ions, sulfide ions, and the complexing agents, respectively. The colloidal solutions were prepared at room temperature and pH from 4.75 to 6.25. Synthesis comprised the following sequence of steps: to lead acetate was added with constant stirring a complexing agent. Next, the volume of the mixture was brought to 100 mL, and the resultant solution was mixed with 100 mL of a Na_2S solution. According to Sadovnikov and Gusev [20], the order of addition of the complexing agent to lead acetate or sodium sulfide solution does not affect the solution stability or nanoparticle size but can give rise to impurity phases. Indeed, in an aqueous solution containing hydrated Pb^{2+}, AcO^- and S^{2-} ions, two dozens of products can appear upon hydrolysis, apart from the colloidal PbS particles [21].

Synthesis conditions of all colloidal solutions and particle size are given in Table 2.1.

The colloidal solution of lead sulfide nanoparticles is formed upon the following reaction:

$$Pb(OAc)_2 + Na_2S \xrightarrow{Na_3Cit \text{ or } Na_2H_2 - EDTA} PbS + 2NaOAc. \qquad (2.1)$$

Table 2.1 Concentration of reactants in the reaction mixtures, stability time t of colloidal solutions, and the size D_{DLS} of PbS nanoparticles [19]

Concentration of reactants in the reaction mixtures (mmol l^{-1})				t (day)	D_{DLS} (nm)
$Pb(CH_3COO)_2$	Na_2S	$Na_3C_6H_5O_7$	Trilon B		
25	25	0	0	a	7–20[b]
0.8	0.8	0	0	a	5[b]
1.25	1.25	0	1.25	4	120
0.8	0.8	0	0.8	4	15
0.5	0.5	0	0.5	4	18
0.25	0.25	0	0.25	>6	15
1.25	1.25	1.25	0	30	13
0.8	0.8	0.8	0	30	13
0.5	0.5	0.5	0	5	11
0.25	0.25	0.25	0	5	10

[a]Stability of these colloidal solutions less than 1 h
[b]Particle size in deposited powders was determined by XRD method

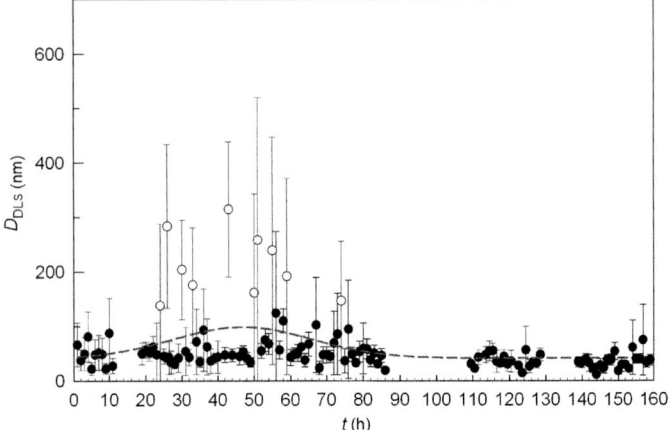

Fig. 2.3 Ranges of $D_{DLS\ (nm)}$ values for PbS agglomerates (1) and nanoparticles (2) in a colloid solution as functions of solution storage time at room temperature [19]. The enveloping curve shows the particle size distribution depending on the storage time

In order to estimate the effect of complexing agents on the stability of colloidal solutions, similar solutions were prepared without addition of sodium citrate or Trilon B. In particular, PbS nanoparticles precipitated from reaction mixtures with equal concentrations $[Pb(AcO)_2]=[Na_2S]$, either 0.8 or 25 mmol l^{-1}, within 10 and 60 min after the synthesis, respectively. No stable colloidal solution formed.

The stability of colloidal solutions was evaluated in study [19] by considering the change in the colloidal nanoparticle size (hydrodynamic diameter D_{DLS}) depending on the time (t) of solution storage. The D_{DLS} value of PbS nanoparticles directly in the colloidal solution was determined by dynamic light scattering.

When Trilon B was added to the reaction mixture, stability of the colloidal solutions increased to 4 days, and by using sodium citrate, it was possible to prepare colloidal solutions that were stable for more than 30 days (they did not change their properties during this period) (Fig. 2.3). The PbS particle size in colloidal solutions depends on the initial concentrations of lead acetate and sodium sulfide and on the type of complexing agent. By varying these parameters, it is possible to obtain stable colloidal solutions with a specified particle size ranging from 10 to 20 nm [19].

This process for preparation of colloidal solutions of PbS is protected by the patent [22].

2.1.2 Analysis of the Formation Conditions of Lead Sulfide in Aqueous Solutions

It was noted above that, along with colloidal PbS particles, numerous products of hydrolysis are formed in an aqueous solution containing hydrated Pb^{2+}, AcO^- and

S^{2-} ions. The important role of metal hydroxide during chemical deposition was discussed in numerous publications (see, for example [15, 23, 24]. For example, upon chemical deposition from aqueous solutions of thiourea N_2H_4CS, metal sulfides are formed as thin films [15, 23, 25, 26] on non-metallic surfaces only in the thermodynamic stability region of metal hydroxide in the solution (hydroxide region). The microstructure of chemically deposited sulfide films depends on the pH value and on the presence of metal hydroxide in the reaction mixture [27]. Metal hydroxide is significant for the assembly of core-shell type sulfide heteronanostructures [28]. Therefore, calculation of the concentration region of formation of metal hydroxide, which makes it possible to predict the possibility for its appearance in the solution before the sulfide synthesis, is an important task faced by a researcher who performs chemical deposition from aqueous solutions. This calculation is based on comparison of the ionic product with the solubility product of metal hydroxide, $K_{sp} = [M^{m+}] \cdot [OH^-]^m$, where $[M^{m+}]$ and $[OH^-]$ are the equilibrium concentrations of free metal ions and hydroxide ions, respectively.

The data on the conditions of formation of lead hydroxide in the $Pb^{2+}-H_2O$ system taking account of the existence of multinuclear hydroxo complexes are scarce. Therefore, a calculation method was proposed in studies [29, 30] and a distribution diagram for complexes in the $Pb^{2+}-H_2O$ system taking account of all polynuclear hydroxo complexes for which the formation constants are known was presented [31].

In aqueous solution the metal ions together with an insoluble hydroxide form soluble hydroxo complexes, the compounds in which hydroxy ions are the substituting ligands. Many metals (e.g., lead, iron, nickel, bismuth, tin, beryllium, etc.) form with the OH^- ions not only mononuclear hydroxocomplexes $M(OH)_n$, but also polynuclear neutral or charged hydroxo complexes $M_p(OH)_n$ ($n \geq 1, p \geq 2$, both are integers). The studies [24, 32–34] on the determination of conditions for the formation of the hydroxide in the process of chemical deposition of sulfides and selenides take into account the mononuclear hydroxo complexes of metals, or only some of the polynuclear complexes. In the polynuclear complexes the number of ligands associated with the metal atom depends both on the concentrations of ligand and metal. For this reason, the equations describing the equilibrium conditions are not linear with respect to the analytical concentration of the metal ions M^{m+} that greatly complicates or even makes it impossible to solve the equations analytically. Apparently, this is a reason for taking into account only mononuclear hydroxo complexes in the calculations of the concentration regions of formation [24, 32, 34], although in recent years the software have appeared for the calculations and simulation of ionic equilibrium in aqueous solutions. Another reason for considering only the mononuclear hydroxo complexes lies in the fact that in the systems with ligands that form more stable complexes than hydroxo complexes, the impact of polynuclear hydroxo complexes on the equilibrium is rather weak. However, the exclusion from the calculation of polynuclear hydroxo complexes leads, as shown below, to a significant shift of the boundaries of the concentration regions.

There is only limited published information on the formation of lead hydroxide in the system of $Pb^{2+}-H_2O$ which accounts for the existence of polynuclear hydroxo complexes. The data are controversial due to discrepancy in the numerical values of the constants of formation of polynuclear complexes.

The equilibrium state in a saturated solution of lead hydroxide is described as

$$K_{sp} = \left[Pb^{2+}\right] \cdot [OH^-]^2, \qquad (2.2)$$

where $K_{sp} = (1.0-1.2) \times 10^{-15}$ [35, 36] is the solubility product of lead hydroxide at the normal conditions. In accordance with (2.2) the equilibrium concentration $[Pb^{2+}]$ of free ions of lead is

$$\left[Pb^{2+}\right] = K_{sp}/[OH^-]^2 \equiv K_{sp}[H^+]^2/K_w^2, \qquad (2.3)$$

where $K_w = [H^+] \cdot [OH^-] = 10^{-14}$ is the ion product of water at 298 K.

Many studies of hydrolysis suggest that the distribution of lead(II) between its individual hydroxo complexes depends on three most important variables: the concentration of the lead salts, the medium pH, and the temperature. For example, according to [37, 38], in very dilute solutions with a total concentration of Pb(II) which equals 1×10^{-9} M, the hydrolysis is detected at pH > 4, and increase in pH to 9 leads to the formation of PbOH$^+$ only. With further increase in pH the PbOH$^+$ fraction in solution decreases, while the accumulation occurs of di- and trihydroxo complexes Pb(OH)$_2$ and Pb(OH)$_3^-$. At pH up to ~ 10.5 both complexes are formed in comparable amounts, and in more alkaline environment the fraction of trihydroxo complex rapidly increases, so at pH > 12 it becomes the only form of the Pb(II) existence. With the increase in the total concentration of lead, the polynuclear hydroxo complexes form in solution [31, 37, 38]. At the Pb(II) concentration equal to 5×10^{-5} M the solution contains complexes Pb$_3$(OH)$_4^{2+}$, and at a higher concentration, of 1×10^{-3} M, the solution contains the complexes Pb$_4$(OH)$_4^{4+}$ and Pb$_6$(OH)$_8^{4+}$. When the concentration of Pb(II) is above 0.5 M the hydroxo complex Pb$_2$(OH)$^{3+}$ appears [37]. The complexes Pb$_3$(OH)$_3^{3+}$ [37] and Pb$_3$(OH)$_5^+$ [38, 39] were detected also, but, according to [40], the experimental evidence of their existence is not reliable enough.

Let us calculate the ionic composition of the "Pb^{2+}–Cit^{3-}–OH$^-$–H$_2$O" system taking into account not only mononuclear, but also the polynuclear Pb(II) hydroxo complexes in aqueous solution. In describing the equilibria in solutions of complex lead ions we use the equilibrium constant β_{ij} recommended in [31, 41, 42] (Table 2.2). Note that in alkaline solution the ion OH$^-$ competes with such a stable in an alkaline environment ligand as citrate ion C$_6$H$_5$O$_7^{3-}$ (Cit^{3-}) forming enough stable complexes with the lead ions [42].

For the citrate system "Pb^{2+}–Cit^{3-}–OH$^-$–H$_2$O" the condition of material equilibrium in all eleven soluble forms of lead Pb^{2+}, Pb(OH)$^+$, Pb(OH)$_2$, Pb(OH)$_3^-$, Pb$_3$(OH)$_4^{2+}$, Pb$_4$(OH)$_4^{4+}$, Pb$_6$(OH)$_8^{4+}$, Pb$_2$OH^{3+}, PbCit$^-$, Pb(Cit)$_2^{4-}$, Pb(OH)Cit^{2-} can be written as

Table 2.2 The conditions of equilibrium and equilibrium constants in the systems "Pb^{2+}–OH$^-$–H$_2$O" [31, 41] and "Pb^{2+}–OH$^-$–Cit^{3-}–H$_2$O" [41, 42]

Equilibrium conditions	Equilibrium constant at 298.15 K, 10^5 Pa, and ionic strength $I_m = 0$ mol·kg^{-1}
Pb^{2+} + H$_2$O ↔ Pb(OH)$^+$ + H$^+$	$\beta_{11} = [\text{Pb(OH)}^+][\text{H}^+]/[\text{Pb}^{2+}] = 3.47 \times 10^{-8}$
Pb^{2+} + 2H$_2$O ↔ Pb(OH)$_2$ + 2H$^+$	$\beta_{12} = [\text{Pb(OH)}_2][\text{H}^+]^2/[\text{Pb}^{2+}] = 1.15 \times 10^{-17}$
Pb^{2+} + 3H$_2$O ↔ Pb(OH)$_3^-$ + 3H$^+$	$\beta_{13} = [\text{Pb(OH)}_3^-][\text{H}^+]^3/[\text{Pb}^{2+}] = 9.33 \times 10^{-29}$
3Pb^{2+} + 4H$_2$O ↔ Pb$_3$(OH)$_4^{2+}$ + 4H$^+$	$\beta_{34} = [\text{Pb}_3\text{(OH)}_4^{2+}][\text{H}^+]^4/[\text{Pb}^{2+}]^3 = 9.77 \times 10^{-24}$
4Pb^{2+} + 4H$_2$O ↔ Pb$_4$(OH)$_4^{4+}$ + 4H$^+$	$\beta_{44} = [\text{Pb}_4\text{(OH)}_4^{4+}][\text{H}^+]^4/[\text{Pb}^{2+}]^4 = 2.69 \times 10^{-21}$
6Pb^{2+} + 8H$_2$O ↔ Pb$_6$(OH)$_8^{4+}$ + 8H$^+$	$\beta_{68} = [\text{Pb}_6\text{(OH)}_8^{4+}][\text{H}^+]^8/[\text{Pb}^{2+}]^6 = 1.29 \times 10^{-43}$
2Pb^{2+} + H$_2$O ↔ Pb$_2$OH^{3+} + H$^+$	$\beta_{21} = [\text{Pb}_2\text{OH}^{3+}][\text{H}^+]/[\text{Pb}^{2+}]^2 = 5.25 \times 10^{-8}$
Pb^{2+} + Cit^{3-} ↔ PbCit$^-$	$\beta_1 = [\text{PbCit}^-]/[\text{Pb}^{2+}][\text{Cit}^{3-}] = 3.07 \times 10^6$
Pb^{2+} + 2Cit^{3-} ↔ Pb(Cit)$_2^{4-}$	$\beta_{12\text{cit}} = [\text{Pb(Cit)}_2^{4-}]/[\text{Pb}^{2+}][\text{Cit}^{3-}]^2 = 3.03 \times 10^8$
Pb^{2+} + OH$^-$ + Cit^{3-} ↔ Pb(OH)Cit^{2-}	$\beta_1' = [\text{Pb(OH)Cit}^{2-}]/[\text{Pb}^{2+}][\text{OH}^-][\text{Cit}^{3-}] = 5.25 \times 10^{13}$

$$C_{Pb,\Sigma}^{cit} = [Pb^{2+}]\left\{1 + \frac{[Pb(OH)^+]}{[Pb^{2+}]} + \frac{[Pb(OH)_2]}{[Pb^{2+}]} + \frac{[Pb(OH)_3^-]}{[Pb^{2+}]} + \frac{3[Pb_3(OH)_4^{2+}]}{[Pb^{2+}]}\right.$$

$$+ \frac{4[Pb_4(OH)_4^{4+}]}{[Pb^{2+}]} + \frac{6[Pb_6(OH)_8^{4+}]}{[Pb^{2+}]} + \frac{2[Pb_2OH^{3+}]}{[Pb^{2+}]} + \frac{[PbCit^-]}{[Pb^{2+}]} \quad (2.4)$$

$$\left. + \frac{[Pb(Cit)_2^{4-}]}{[Pb^{2+}]} + \frac{[Pb(OH)Cit^{2-}]}{[Pb^{2+}]}\right\}$$

Let us represent the ratio of the concentrations of complexes to the concentration $[Pb^{2+}]$ in (2.4) through the equilibrium constant (see Table 2.2) and transform it into a form more convenient for further analysis:

$$C_{Pb,\Sigma}^{cit} = [Pb^{2+}]\left\{1 + \frac{\beta_{11}}{[H^+]} + \frac{\beta_{12}}{[H^+]^2} + \frac{\beta_{13}}{[H^+]^3} + \frac{3\beta_{34}[Pb^{2+}]^2}{[H^+]^4} + \frac{4\beta_{44}[Pb^{2+}]^3}{[H^+]^4}\right.$$

$$\left. + \frac{6\beta_{68}[Pb^{2+}]^5}{[H^+]^8} + \frac{2\beta_{21}[Pb^{2+}]}{[H^+]} + \beta_1[Cit^{3-}] + \beta_{1,2}[Cit^{3-}]^2 + \beta_1^*\frac{K_w[Cit^{3-}]}{[H^+]}\right\}$$

$$(2.5)$$

If the citrate ion concentration $[Cit^{3-}] = 0$, then (2.5) is converted to a particular form, which describes the condition of the material balance of the lead soluble forms in a citrate-free "Pb^{2+}–OH^-–H_2O" system:

$$C_{Pb,\Sigma} = [Pb^{2+}]\left\{1 + \frac{\beta_{11}}{[H^+]} + \frac{\beta_{12}}{[H^+]^2} + \frac{\beta_{13}}{[H^+]^3} + \frac{3\beta_{34}[Pb^{2+}]^2}{[H^+]^4} + \frac{4\beta_{44}[Pb^{2+}]^3}{[H^+]^4}\right.$$

$$\left. + \frac{6\beta_{68}[Pb^{2+}]^5}{[H^+]^8} + \frac{2\beta_{21}[Pb^{2+}]}{[H^+]}\right\} \quad (2.6)$$

In accordance with (2.6) the fractional concentration of free Pb^{2+} ions in the "Pb^{2+}–OH^-–H_2O" system is

$$\alpha_{Pb^{2+}} = \frac{[Pb^{2+}]}{C_{Pb,\Sigma}} = \left\{1 + \frac{\beta_{11}}{[H^+]} + \frac{\beta_{12}}{[H^+]^2} + \frac{\beta_{13}}{[H^+]^3} + \frac{3\beta_{34}[Pb^{2+}]^2}{[H^+]^4} + \frac{4\beta_{44}[Pb^{2+}]^3}{[H^+]^4}\right.$$

$$\left. + \frac{6\beta_{68}[Pb^{2+}]^5}{[H^+]^8} + \frac{2\beta_{21}[Pb^{2+}]}{[H^+]}\right\}^{-1} \quad (2.7)$$

Using (2.7), the fractional concentrations $\alpha_i = C_i / C_{Pb,\Sigma}$ of all Pb(II) hydroxo complexes in the "Pb^{2+}–OH^-–H_2O" system can be written as the functions of $\alpha_{Pb^{2+}}$ and the concentrations of H^+ and Pb^{2+} ions:

$$\alpha_{Pb(OH)^+} = \beta_{11}\alpha_{Pb^{2+}}/[H^+], \tag{2.8a}$$

$$\alpha_{Pb(OH)_2} = \beta_{12}\alpha_{Pb^{2+}}/[H^+]^2, \tag{2.8b}$$

$$\alpha_{Pb(OH)_3^-} = \beta_{13}\alpha_{Pb^{2+}}/[H^+]^3, \tag{2.8c}$$

$$\alpha_{Pb_3(OH)_4^{2+}} = 3\beta_{34}\alpha_{Pb^{2+}}[Pb^{2+}]^2/[H^+]^4, \tag{2.8d}$$

$$\alpha_{Pb_4(OH)_4^{4+}} = 4\beta_{44}\alpha_{Pb^{2+}}[Pb^{2+}]^3/[H^+]^4, \tag{2.8e}$$

$$\alpha_{Pb_6(OH)_8^{4+}} = 6\beta_{68}\alpha_{Pb^{2+}}[Pb^{2+}]^5/[H^+]^8, \tag{2.8f}$$

$$\alpha_{Pb_2(OH)^{3+}} = 2\beta_{21}\alpha_{Pb^{2+}}[Pb^{2+}]/[H^+] \tag{2.8g}$$

Equations (2.6) and (2.7) are of 6th degree with respect to the concentration of Pb^{2+} ions and do not have analytical solutions of a general form. Still, the problem of hydrolysis of the Pb^{2+} ion described by a set of (2.7), (2.8a, b, c, d, e, f, g) can be solved numerically.

Indeed, at a known total concentration $C_{Pb,\Sigma}$ of the soluble forms of lead the dependence of $[Pb^{2+}]$ on the concentration of protons H^+ (that is, the solution pH) can be found by means of self-consistent numerical solution of (2.7) by the successive approximation procedure. In other words, at a given concentrations $C_{Pb,\Sigma}$ we can calculate the dependence of the fraction of uncomplexed lead ions on the pH. For this purpose, using a software package SigmaPlot 2001 for Windows [43] a special program was written. With the numerical relationship $[Pb^{2+}] = f([H^+])|_{C_{Pb,\Sigma}=const}$ found from (2.7) it is easy to obtain fractional concentrations α_i of free ions and all the Pb(II) hydroxo complexes at different pH values.

Figure 2.4 shows the fractional concentrations of free ions and the Pb(II) hydroxo complexes in solution for the system "Pb^{2+}–OH^-–H_2O" depending on the pH value, calculated with the (2.7), (2.8a, b, c, d, e, f, g) for three values of $C_{Pb,\Sigma}$ 0.001, 0.01, and 0.1 M. As seen, the hydrolysis of lead ions begins at pH $\geq$ 6. The increase in the analytical concentration of lead ion shifts the Pb^{2+} hydrolysis in the region of lower pH values (Fig. 2.4a). When $C_{Pb,\Sigma}$ = 0.001 M, at pH = 8–10 lead exists mainly in the form of complex ions, $Pb_3(OH)_4^{2+}$ and $Pb_6(OH)_8^{4+}$, at pH $\geq$ 11 sharply increased the fraction of $Pb(OH)_2$ and $Pb(OH)_3^-$ (Fig. 2.4c, d). When $C_{Pb,\Sigma}$ = 0.1 M, at pH from 7 to 12 lead forms in solution the ion $Pb_6(OH)_8^{4+}$ (Fig. 2.4h). Thus, an increase in $C_{Pb,\Sigma}$ expands the pH range of existence of complexes $Pb_3(OH)_4^{2+}$ and $Pb_6(OH)_8^{4+}$, and the fraction of the complex $Pb_6(OH)_8^{4+}$ in solution increases. The only polynuclear complex, the relative content of which is negligible, is Pb_2OH^{3+} (this is consistent with the results [33]). It exists in a narrow range 3 < pH < 8 (Fig. 2.4g). At pH $\geq$ 12 for all concentrations used in the calculations of $C_{Pb,\Sigma}$ the main form of existence of lead in

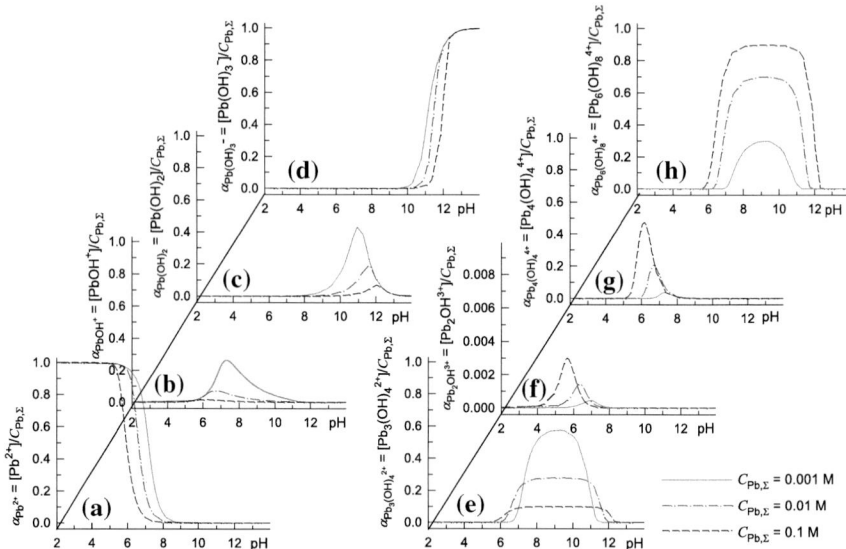

Fig. 2.4 Dependence of partial concentration α_i of **a** the free ions Pb^{2+}, mononuclear **b** $Pb(OH)^+$, **c** $Pb(OH)_2$, **d** $Pb(OH)_3^-$, and polynuclear **e** $Pb_3(OH)_4^{2+}$, **f** $Pb_2(OH)^{3+}$, **g** $Pb_4(OH)_4^{4+}$, and **h** $Pb_6(OH)_8^{4+}$ hydroxo complexes of lead on the pH in the solution of the "$Pb^{2+}-OH^--H_2O$" system at $T = 298.15$ K for three values of the total analytical concentration of lead in solution $C_{Pb,\Sigma}$: 0.001, 0.01 and 0.1 M

solution is a mononuclear complex $Pb(OH)_3^-$ (Fig. 2.4d). Thus, for the precise analysis of ionic equilibria in aqueous solutions of lead salts the polynuclear hydroxo complexes should be taken into account, which are the main form of existence of lead in solution in the pH range from 7 to 12.

Note that all dependences (2.7) and (2.8a, b, c, d, e, f, g) of the fractional concentrations α_i are power functions of $[H^+]$ (or pH), that is, smooth functions with no break point or discontinuity. However, in [31] (see Fig. 2.4a) the break points were observed in the calculated dependences of the fractional concentrations of $Pb_3(OH)_4^{2+}$ and $Pb_6(OH)_8^{4+}$ on the pH in the "$Pb^{2+}-OH^-$" system with $C_{Pb,\Sigma} = 0.00048$ M at pH $\sim$ 9.25. This is due to the fact that in work [31] the fractional concentrations were calculated taking into account the abrupt change in the total concentration of lead in solution due to precipitation of lead oxide at pH > 9.3.

Let us now determine the stability region of lead hydroxide $Pb(OH)_2$ in the citrate-free system "$Pb^{2+}-OH^--H_2O$". The conditions of formation of solid phase $Pb(OH)_2$ in water solution can be found from the following relation

$$K_{sp} < [Pb^{2+}][OH^-]^2. \qquad (2.9)$$

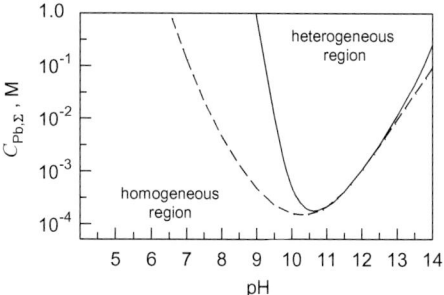

Fig. 2.5 Effect of accounting for the polynuclear hydroxo complexes of lead on the position of the boundary of the lead hydroxide Pb(OH)$_2$ formation in the system "Pb^{2+}–OH$^-$–H$_2$O" at the different pH values at 298.15 K: *solid line* corresponds to considering all mono- and polynuclear hydroxo complexes, dashed line corresponds to considering only the mononuclear hydroxo complexes. The total concentration of lead, $C_{Pb,\Sigma}$, in the solution is shown in a logarithmic scale

Substituting the expression (2.3) in (2.6) for the concentration [Pb^{2+}], we obtain the relation, which is the equation of the boundary of the Pb(OH)$_2$ solid phase deposition

$$
C_{Pb,\Sigma} = \frac{K_{sp}[H^+]^2}{K_w^2}\left\{1 + \frac{\beta_{11}}{[H^+]} + \frac{\beta_{12}}{[H^+]^2} + \frac{\beta_{13}}{[H^+]^3} + \frac{3\beta_{34}K_{sp}^2}{K_w^4} + \frac{4\beta_{44}K_{sp}^3[H^+]^2}{K_w^6} \right.
$$
$$
\left. + \frac{6\beta_{68}K_{sp}^5[H^+]^2}{K_w^{10}} + \frac{2\beta_{21}K_{sp}[H^+]}{K_w^2}\right\}
$$

(2.10)

The positions of the Pb(OH)$_2$ formation boundary in the "Pb^{2+}–HO$^-$–H$_2$O" system determined considering only mononuclear complexes or considering both mono- and polynuclear lead hydroxo complexes are shown in Fig. 2.5. The calculation is made according to (2.10).

As can be seen, the effect of polynuclear complexes on the equilibrium position of the heterogenic equilibrium "Pb(OH)$_2$ hydroxide–Pb(II) ions" is especially pronounced in the region of pH < 10.5. In the alkaline region due to amphoteric nature of lead hydroxide Pb(OH)$_2$ the deposition begins and occurs at pH $\leq$ 13, which agrees well with the reference and experimental data. Sulfide formation in the form of thin films on nonmetallic substrates by chemical deposition from aqueous solutions of thiocarbamide occurs only in the region of stability of the metal hydroxide [16, 23]. Figure 2.5 clearly shows that at accounting for all mononuclear and polynuclear hydroxo complexes the PbS films can be obtained in alkaline solutions with 9 < pH < 13, which fully agrees with the experimental data, for example, [16, 23, 25, 26]. If the polynuclear complexes are neglected, the left boundary of the region of formation of lead hydroxide is displaced to the acidic region, where the hydroxide Pb(OH)$_2$ is not really formed. Thus, the consideration of polynuclear hydroxo complexes is important not only theoretically, but also practically, for the synthesis of the films.

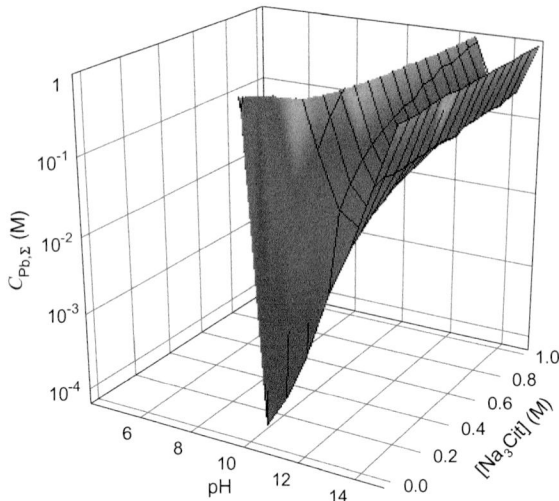

Fig. 2.6 Solubility of Pb(OH)$_2$ as a function of pH of the medium and of citrate ion concentration in the "Pb^{2+}–Cit^{3-}–OH$^-$–H$_2$O" system [30] [solubility was calculated on the (2.11)]. The three-dimensional surface is the boundary of the heterogeneous "Pb(OH)$_2$–Pb^{2+}" equilibrium. Below this surface, there is the homogeneous region in which lead exists only in the dissolved form, while above the surface lies the heterogeneous region in which Pb(OH)$_2$ is also formed and, upon the introduction of sulfide ions, PbS deposition as a film is possible. The lead concentration $C_{\text{Pb},\Sigma}$ in the solution is presented on the logarithmic scale

We now define the stability domain of lead hydroxide Pb(OH)$_2$ in the solution, where, in addition to hydroxo complexes the citrate complexes are present. In this case, the condition of the material balance is described by (2.5), and the equation of the deposition boundary, taking into account (2.3) and (2.5), has the form

$$C_{\text{Pb},\Sigma}^{\text{cit}} = \frac{K_{\text{sp}}[\text{H}^+]^2}{K_{\text{w}}^2}\left\{1 + \frac{\beta_{11}}{[\text{H}^+]} + \frac{\beta_{12}}{[\text{H}^+]^2} + \frac{\beta_{13}}{[\text{H}^+]^3} + \frac{3\beta_{34}K_{\text{sp}}^2}{K_{\text{w}}^4} + \frac{4\beta_{44}K_{\text{sp}}^3[\text{H}^+]^2}{K_{\text{w}}^6}\right.$$
$$\left. + \frac{6\beta_{68}K_{\text{sp}}^5[\text{H}^+]^2}{K_{\text{w}}^{10}} + \frac{2\beta_{21}K_{\text{sp}}[\text{H}^+]}{K_{\text{w}}^2} + \beta_1[\text{Cit}^{3-}] + \beta_{12\text{cit}}[\text{Cit}^{3-}]^2 + \beta_1'\frac{K_{\text{w}}[\text{Cit}^{3-}]}{[\text{H}^+]}\right\}$$

$$(2.11)$$

The dependence of the solubility of lead hydroxide on the medium pH in the presence of sodium citrate, as Fig. 2.6 shows, defines the boundary conditions of existence of homogeneous and heterogeneous systems. Homogeneous region is the region of the lead existence only in the dissolved form, that is, free form and the complex ions. This region is located below the boundary surface of the heterogeneous equilibrium "Pb(OH)$_2$ hydroxide–Pb(II) ions". Above this surface is a heterogeneous region where lead hydroxide Pb(OH)$_2$ is also formed. As can be seen, the solubility of lead hydroxide decreases with increasing pH, and at [Na$_3$Cit] = 0 (that is, in the system "Pb^{2+}–OH$^-$–H$_2$O") reaches its minimum value

$\sim 2 \times 10^{-4}$ M at pH ≈ 10.5. In the region of pH > 10.5 lead hydroxo complexes are formed, so that the total concentration of lead in solution increases.

Increasing concentrations of citrate leads to an increase in the absolute values of the $Pb(OH)_2$ solubility and the solubility minimum is shifted to the region of higher pH values. For example, for the concentration $[Na_3Cit] = 0.05$ M the lead hydroxide solubility in the "Pb^{2+}–Cit^{3-}–OH^-–H_2O" system reaches its minimum value of 3.6×10^{-2} M при at pH ≈ 13, and for the concentration $[Na_3Cit] = 1.0$ M the minimum solubility equal to 0.146 M corresponds to pH ≈ 14 (see Fig. 2.6).

Thus, the calculation with (2.10) and (2.11) allowed us constructing a dependence of the total equilibrium concentration of lead ions in solution on the pH value and find the stability region of lead hydroxide for the system "Pb^{2+}–Cit^{3-}–OH^-–H_2O" (see Fig. 2.6) and its specific section "Pb^{2+}–OH^-–H_2O" (see Fig. 2.5). From the calculation it follows that introduction of citrate ion to the solution increases the equilibrium concentration of lead and solubility of $Pb(OH)_2$ due to the formation of citrate and hydroxo-citrate complexes.

In aqueous solution at 298 K a three-stage dissociation of citric acid occurs with the equilibria:

$$H_3Cit \leftrightarrow H^+ + H_2Cit^-, K_1 = [H^+][H_2Cit^-]/[H_3Cit] = 7.4 \times 10^{-4}, \quad (2.12a)$$

$$H_2Cit^- \leftrightarrow H^+ + HCit^{2-}, K_2 = [H^+][HCit^{2-}][H_2Cit^-] = 1.8 \times 10^{-5}, \quad (2.12b)$$

$$HCit^{2-} \leftrightarrow H^+ + Cit^{3-}, K_3 = [H^+][Cit^{3-}][HCit^{2-}] = 4.0 \times 10^{-7}. \quad (2.12c)$$

In accordance with (2.12), the total analytical concentration of citric acid in solution can be calculated as

$$C_{H_3Cit,\Sigma} = [H_3Cit] + [H_2Cit^-] + [HCit^{2-}] + [Cit^{3-}]. \quad (2.13)$$

With (2.12) and (2.13), the fractional concentrations of molecular and ionized form of citric acid are equal

$$\begin{aligned} \alpha_{H_3Cit} &= [H_3Cit]/C_{H_3Cit,\Sigma} = [H^+]^3/\{[H^+]^3 + K_1[H^+]^2 \\ &+ K_1K_2[H^+] + K_1K_2K_3\}, \end{aligned} \quad (2.14a)$$

$$\begin{aligned} \alpha_{H_2Cit^-} &= [H_2Cit^-]/C_{H_3Cit,\Sigma} = K_1[H^+]^2/\{[H^+]^3 + K_1[H^+]^2 \\ &+ K_1K_2[H^+] + K_1K_2K_3\}, \end{aligned} \quad (2.14b)$$

$$\begin{aligned} \alpha_{HCit^{2-}} &= [HCit^{2-}]/C_{H_3Cit,\Sigma} = K_1K_2[H^+]/\{[H^+]^3 + K_1[H^+]^2 \\ &+ K_1K_2[H^+] + K_1K_2K_3\}, \end{aligned} \quad (2.14c)$$

$$\alpha_{Cit^{3-}} = [Cit^{3-}]/C_{H_3Cit,\Sigma} = K_1K_2K_3/\{[H^+]^3 + K_1[H^+]^2$$
$$+ K_1K_2[H^+] + K_1K_2K_3\}. \qquad (2.14d)$$

Calculations based on (2.14) show that at pH < 2 the citric acid exists in solution in molecular form. In the pH range from 2 to 8 in addition to the molecular form the ions H_2Cit^- and $HCit^{2-}$ appear, and at pH > 8 the solution contains only Cit^{3-} ions. Hence, it is clear that Cit^{3-} ion is involved in complex formation only at pH > 8.

As a final step, we consider the ionic equilibrium in the "Pb^{2+}–Cit^{3-}–OH^-–N_2H_4CS–H_2O" system and taking into account the polynuclear hydroxo complexes refine the field of formation of lead sulfide in an aqueous solution of such sulfidizer as thiocarbamide N_2H_4CS. A spontaneous formation of metal sulfide is possible when the ion product IP_{MS} of considered sulfide MS is higher than the solubility product $K_{sp,MS}$. In the considered case the lead sulfide formation will occur if ion product $IP_{PbS} = [Pb^{2+}] \cdot [S^{2-}]$ is higher than the lead sulfide solubility product, $K_{sp,PbS} = [Pb^{2+}]_{eq} \cdot [S^{2-}]_{eq}$, that is, $IP_{PbS} > K_{sp,PbS}$, where $[Pb^{2+}]$, $[S^{2-}]$, $[Pb^{2+}]_{eq}$ and $[S^{2-}]_{eq}$ are arbitrary (e.g., initial) and equilibrium concentration of ions Pb^{2+} and S^{2-}, respectively. Thus, to discuss the possible formation of metal sulfide it is necessary to know the initial concentration of the ions Pb^{2+} and S^{2-}. The initial concentration of free lead ions was calculated according to (2.5) or (2.7).

The basis for the calculation of the sulfide ion concentration is the concept of the reversible decomposition of thiocarbamide in the aqueous alkaline solution by the reaction

$$N_2H_4CS \leftrightarrow H_2S + H_2NCN, \qquad (2.15)$$

which was first developed in detail and justified in study [16]. According to [16], the initial concentration of the S^{2-} ions is equal to

$$[S^{2-}] = \frac{K_{H_2S}}{[H^+]^2} \left\{ \frac{K_c[N_2H_4CS]\beta_c}{\beta_s} \right\}^{1/2}, \qquad (2.16)$$

where $K_c = 1.6 \times 10^{-23}$ is the N_2H_4CS hydrolysis constant [16]; $\beta_s = [H^+]^2 + K_1[H^+] + K_{H_2S}$ [16]; $\beta_c = [H^+]^2 + K_1'[H^+] + K_{H_2NCN}$ [16]; $K_{H_2S} = 1.1 \times 10^{-20}$ [44] and $K_{H_2NCN} = 7.95 \times 10^{-23}$ [45], $K_1 = 8.9 \times 10^{-8}$ [44] and $K_1^* = 5.25 \times 10^{-11}$ [45] are the complete dissociation constants and the first stage dissociation constants of hydrosulfuric acid and cyanamide, respectively.

Next, from the found initial concentration of ions Pb^{2+} and S^{2-} we determine the value of the ion product and compare it with the solubility product $K_{sp,PbS} = 2.5 \times 10^{-27}$ of lead sulfide PbS [44].

Replacing in (2.5) the concentration $[Pb^{2+}]$ by its value expressed through a solubility product and concentration of sulfide ions $[Pb^{2+}] = K_{sp,PbS}/[S^{2-}]$, we obtain the expression

Fig. 2.7 Regions of
formation of PbS as a sol and
a film upon deposition from
aqueous solutions of N₂H₄CS
with different concentration of
0 and 0.025 mol l⁻¹ [30]. The
total concentration of lead,
$C_{Pb,\Sigma}$, in the solutionьı is
shown in a logarithmic scale

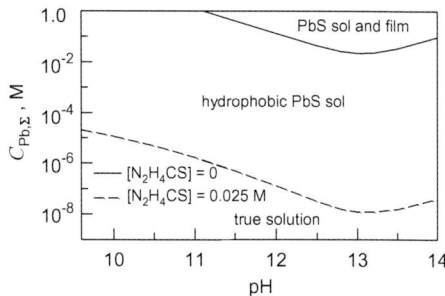

$$C_{Pb,\Sigma}^{cit} = \frac{K_{sp,PbS}}{[S^{2-}]} \left\{ 1 + \frac{\beta_{11}}{[H^+]} + \frac{\beta_{12}}{[H^+]^2} + \frac{\beta_{13}}{[H^+]^3} + \frac{3\beta_{34}K_{sp,PbS}^2}{[H^+]^4[S^{2-}]^2} + \frac{4\beta_{44}K_{sp,PbS}^3}{[H^+]^4[S^{2-}]^3} \right.$$

$$\left. + \frac{6\beta_{68}K_{sp,PbS}^5}{[H^+]^8[S^{2-}]^5} + \frac{2\beta_{21}K_{sp,PbS}}{[H^+][S^{2-}]} + \beta_1[Cit^{3-}] + \beta_{1,2}[Cit^{3-}]^2 + \beta_1'\frac{K_w[Cit^{3-}]}{[H^+]} \right\}$$

(2.17)

Thus, based on (2.5) and (2.16) we obtain the necessary data to construct the
dependence $C_{Pb,\Sigma} = f(pH)$, which characterizes the equilibrium in the system
between the PbS deposit and dissolved thiocarbamide and the lead complex
compounds.

Figure 2.7 shows the boundaries of lead sulfide deposition at different concentrations of N₂H₄CS calculated with (2.17). From a comparison of the boundaries of
the lead hydroxide formation (solid line at [N₂H₄CS] = 0 in Fig. 2.7) and the
boundaries of the lead sulfide formation follows that the lead hydroxide formed at
pH ≥ 11.1 should transform inevitably into the sulfide PbS, as the equilibrium
concentration of lead above the Pb(OH)₂ deposit is much higher than that of the
sulfide. For example, at pH = 12 the lead concentrations above the deposit of Pb
(OH)₂ hydroxide is equal to 0.132 M, and above the sulfide PbS deposit is just
1.3 × 10⁻⁷ M (at [N₂H₄CS] = 0.025 M).

Generally, the calculation procedure proposed in studies [29, 30] is suitable for
predicting the formation of solid lead hydroxide and sulfide phases in the solution,
as well as reliable choosing the compositions of reaction mixtures required for PbS
deposition as a sol (outside of the hydroxide region), quantum dots, core@shell
heterostructures and films.

2.1.3 Deposition of Lead Sulfide Nanopowders

The chemical deposition from aqueous solutions is the most popular method used
for the synthesis of lead sulfide nanopowders. The size of the obtained nanoparticles depends on the chosen reactants and their concentrations and the temperature

of synthesis. Detailed investigations of the conditions of hydrochemical deposition of stable PbS nano-particles using various reagents are reported in studies [20, 21, 46, 47].

Stable nanoparticles of the PbS powder were prepared by deposition from aqueous solutions of $Pb(OAc)_2$ [20, 21, 46, 47] or $Pb(NO_3)_2$ [47]. An aqueous solution of Na_2S served as the source of sulfur ions.

In study [21], lead acetate $Pb(AcO)_2$, sodium sulfide Na_2S and water were used for the one-pot synthesis of PbS nanoparticles in an aqueous solution. The synthesis was carried out under the ambient conditions (298 K, 1 atm). The initial concentrations of aqueous solutions of $Pb(AcO)_2$ and Na_2S salts were varied in the range of 0.0025–0.25 M. The solutions of $Pb(AcO)_2$ and Na_2S were taken in stoichiometric amounts for the each reaction and mixed with vigorous stirring. All PbS nanopowders prepared in work [21] possess a cubic $B1$ type crystal structure. The nanoparticle size D in synthesized PbS powders is equal to 15–20 nm. According to [21], the main drawback of synthesis without complexing agent is that the direct interaction of lead Pb^{2+} and sulfide S^{2-} ions in an aqueous solution is always accompanied by hydrolysis processes, which leads to the formation of poorly soluble impurity phases.

As complexing agents, Na_3Cit or Trilon B were used in studies [20, 46, 47].

In study [20], the $Pb(OAc)_2$ and Na_2S concentrations in the reaction mixtures were varied from 1.25 to 100 mmol l^{-1}, and those of sodium citrate and Trilon B were in the ranges of 0–100 and 0–75 mmol l^{-1}, respectively (Table 2.3). PbS nanoparticles were synthesized at a temperature of 298 K and pH of the solutions during synthesis ranged between 5.5 and 6.5. In study [47], the initial $Pb(NO_3)_2$ and Na_2S (or $Pb(AcO)_2$ and Na_2S) concentrations in all the reaction mixtures were 50 mmol l^{-1}. The concentrations of the complexing agents (sodium citrate and Trilon B) in the reaction mixtures were varied in the ranges from 12.5 to 50 mmol l^{-1}, and from 15 to 65 mmol l^{-1}, respectively (Table 2.4). Synthesis with the use of Trilon B was carried out with the addition of acetic acid CH_3COOH or sodium hydroxide NaOH and without them.

The concentration ratio of lead and sulfur ions was stoichiometric, $[Pb^{2+}]:[S^{2-}] = 1:1$, in all cases.

The solubility product of lead sulfide is low (at 298 K, the K_{sp} value is 2.5×10^{-27} [44] or 8.0×10^{-28} [36]); therefore, at a relatively high content of Na_2S in the reaction mixture, lead sulfide is formed almost immedi-ately, during 1–2 s. As a result, the reaction mixture first turns black and then, in several minutes, PbS particles settle down and the solution becomes transparent. The standard duration of synthesis providing deposition of lead sulfide powder is equal 5 min. The formation of lead sulfide nanoparticles proceeded according to reaction (2.1), while with lead nitrate the reaction was as follows:

$$Pb(NO_3)_2 + Na_2S \longrightarrow Na_3Cit \text{ or } Na_2H_2 - EDTAPbS \downarrow + 2NaNO_3. \qquad (2.18)$$

Table 2.3 Condition of chemical deposition and the size of PbS particles synthesized by reaction scheme (2.1) from aqueous solutions of $Pb(OAc)_2$ with Na_3Cit or Trrilon B [20]

Concentration of reactants in the reaction mixture (mol l^{-1})[a]				t (hour)[b]	D (nm)	Lattice constant a_{B1} (nm)
Pb (CH$_3$COO)$_2$	Na$_2$S	Na$_3$C$_6$H$_5$O$_7$	Trilon B			
0.025	0.025	0	0	0	20–27[c]	0.5933(4)
0.005	0.005	0	0.005	0	24 ± 3	0.5933(1)
0.0125	0.0125	0	0.0125	0	30 ± 3	0.5933(1)
0.050	0.050	0	0.050	0	35 ± 3	0.59337
0.050	0.050	0	0.0125	0	8 ± 1[c]	0.5934(3)
0.050	0.050	0	0.0250	0	10 ± 1	0.5941(3)
0.050	0.050	0	0.0333	0	14 ± 2	0.5936(1)
0.050	0.050	0	0.0375	0	20 ± 3	0.5934(1)
0.050	0.050	0	0.050	0	35 ± 3	0.59337
0.050	0.050	0	0.075	0	50 ± 5[c]	0.59335
0.050	0.050	0	0.050	0	35 ± 3	0.59337
0.050	0.050	0	0.050	120	55 ± 6	0.59335
0.050	0.050	0	0.050	260	55 ± 6	0.59357
0.050	0.050	0	0.050	460	55 ± 6	0.59326
0.050	0.050	0	0.050	770	55 ± 6	0.59340
0.00125	0.00125	0.00125	0	0	No deposition	–
0.00125	0.00125	0.00125	0	1440	No deposition	–
0.0025	0.0025	0.0025	0	0	4.5–5.0[c]	0.5949(7)
0.005	0.005	0.005	0	0	4.5–5.0	0.594(3)
0.005	0.005	0.005	0	70	7–8	0.593(2)
0.0125	0.0125	0.0125	0	0	4.5–5.0	0.594(3)
0.025	0.025	0.025	0	0	4.5–5.0	0.5926(3)
0.050	0.050	0.050	0	0	7–8	0.591(1)
0.100	0.100	0.100	0	0	7–8	0.590(1)
0.025	0.025	0.025	0	0	4.5–5.0	0.5926(3)
0.025	0.025	0.025	0	70	5 ± 1	0.592(2)
0.025	0.025	0.025	0	240	5 ± 1	0.592(3)
0.025	0.025	0.025	0	360	5 ± 1	0.594(3)
0.025	0.025	0.025	0	500	5 ± 1	0.594(1)
0.025	0.025	0.025	0	720	5 ± 1	0.593(3)
0.025	0.025	0.00625	0	0	5 ± 1[c]	0.593(8)
0.025	0.025	0.0125	0	0	5 ± 1	0.593(7)

[a]The deposition time in all the experiments was 5 min
[b]The exposure time of already deposited PbS particles in the matrix solution
[c]The deposit contained impurity phases

The contents of lead, sulfur and oxygen in the annealed single-phase PbS nanopowders determined by Energy Dis-persive X-ray (EDX) analysis were 86 ± 2, 13 ± 1, and 1 ± 0.5 wt%, respectively. According to EDX data (Fig. 2.8), the impurity oxygen present in PbS nano-powders is distributed over the particle surface, the major part being present as adsorbed water and the rest oxygen being in the chemisorbed state.

All the synthesized PbS nanoparticles with any particle size had a cubic $B1$ type lattice with the lattice constant $a_{B1} = 0.5902-0.5949$ nm, which is in line with the results [48, 49].

The XRD pattern of a PbS sample synthesized from a reaction mixture of Pb $(AcO)_2$ and Na_2S with concentrations 25 mmol l^{-1} without adding a complexing agent is presented in Fig. 2.9. According to X-ray phase analysis, the sample contains ~ 45 wt% of impurity of orthorhombic (space group $Pnma$) lead sulfate $PbSO_4$. The average size D of PbS particles ranges from 20 to 27 nm and could not be determined more accurately because of the presence of impurities. A similar result was obtained during synthesis of PbS from a reaction mixture of $Pb(NO_3)_2$ and Na_2S without adding a complexing agent. Thus, synthesis in the absence of a complexing agent fails to yield pure lead sulfide. This is consistent with the data [21].

The XRD patterns of PbS powders synthesized from the reaction mixtures with equal concentrations $[Pb(AcO)_2] = [Na_2S] = [TrB]$ of lead acetate, sodium sulfide, and Trilon B are demonstrated in Fig. 2.10. The inset in Fig. 2.10 shows the dependences between the reduced diffraction reflection broadening $\beta^*(2\theta)$ of the synthesized PbS powders and the scattering vector s. The lead sulfide nanoparticle size $\langle D \rangle$ is calculated to be 24 nm when the concentrations of each reactant are equal to 5 mmol l^{-1}. If the concentrations of these components in the reaction mixture are increased to 12.5 and 50 5 mmol l^{-1}, deposited PbS nanoparticles grow to 30 and 35 nm, respectively (see Table 2.3).

Figure 2.11 displays the SEM images of the nanocrystalline PbS powder with the average particle size $D = 24$ nm as an example. The powder was deposited from a solution of lead acetate, sodium sulfide, and Trilon B having equal con-centrations, 5 mmol l^{-1}. According to SEM data, this nanopowder consists of large irregular agglomerates (Fig. 2.11a) ranging in size from 6 to 10 μm. At a magni-fication of 30000 times, it is seen that these agglomerates have a loose microstructure and are formed by particles smaller than 1 μm (Fig. 2.11b). At a magnification of 100000 times, it is clear that these particles, in turn, consist of nanoparticle with apparent size from 30 to 50 nm (Fig. 2.11c). The same microstructure is typical of other synthesized nanopowders. The results of scanning electron microscopy (SEM) together with the data on coherent scattering region dimensions obtained by the X-ray method are indicative of strong agglomeration of the lead sulfide powders. The agglomeration is likely to be due to water repellence of the surface of the synthesized PbS nanoparticles.

Table 2.4 Conditions of chemical deposition and the size of PbS particles synthesized by reaction scheme (2.18) from aqueous solutions of $Pb(NO_3)_2$ with Na_3Cit or Trilon B [47]

Concentration of reactants in the reaction mixture (mol l^{-1})							t^c	D (nm)
$Pb(NO_3)_2$	$Pb(AcO)^a_2$	Na_2S	Trilon B	Na_3Cit^b	CH_3COOH	NaOH	(hour)	
0.05	0	0.05	0	0	0	0	0	~10–15^d
0.05	0	0.05	0	0.025	0	0	0	6 ± 1
0.05	0	0.05	0	0.0125	0	0	0	5 ± 1
0.05	0	0.05	0	0.02	0	0	120	6 ± 1
0.05	0	0.05	0	0.05	0	0	0	6 ± 1
0.05	0	0.05	0.015	0	0	0	24	11 ± 1
0.05	0	0.05	0.015	0	0	0	72	11 ± 1
0.05	0	0.05	0.025	0	0	0	24	11 ± 1
0.05	0	0.05	0.025	0	0	0	72	11 ± 1
0.05	0	0.05	0.035	0	0	0	24	11 ± 1
0.05	0	0.05	0.035	0	0	0	72	12 ± 2
0.05	0	0.05	0.05	0	0	0	0	21 ± 3
0.05	0	0.05	0.065	0	0	0	120	78 ± 3
0.05	0	0.05	0.065	0	0	0	0	69 ± 3
0	0.05	0.05	0.035	0	0	0	24	38 ± 3
0	0.05	0.05	0.05	0	0	0	0	45 ± 3
0	0.05	0.05	0.05	0	0.05	0	0	56 ± 4
0	0.05	0.05	0.05	0	0.10	0	0	48 ± 3
0	0.05	0.05	0.05	0	0.10	0	24	55 ± 2
0	0.05	0.05	0.05	0	0.10	0	48	56 ± 2
0	0.05	0.05	0.05	0	0.25	0	0	40 ± 2
0	0.05	0.05	0.05	0	0.25	0	24	48 ± 2
0	0.05	0.05	0.05	0	0.25	0	48	49 ± 3
0	0.05	0.05	0.05	0	0.05	0	120	70 ± 5
0	0.05	0.05	0.065	0	0	0	120	86 ± 12
0	0.05	0.05	0.065	0	0.05	0	0	64 ± 3
0	0.05	0.05	0.065	0	0.05	0	72	59 ± 4
0	0.05	0.05	0.065	0	0.05	0	120	88 ± 2
0	0.05	0.05	0.10	0	0.05	0	120	95 ± 10
0	0.05	0.05	0.10	0	0.15	0	120	95 ± 10
0	0.05	0.05	0.05	0	0	0.05	0	27 ± 1
0	0.05	0.05	0.05	0	0	0.10	0	35 ± 3
0	0.05	0.05	0.05	0	0	0.15	0	32 ± 2
0	0.05	0.05	0.05	0	0	0.25	0	33 ± 2

^{a}$Pb(AcO)_2 \equiv Pb(CH_3COO)_2$
^{b}$Na_3Cit \equiv Na_3C_6H_5O_7$
^{c}The exposure time of already deposited PbS nanopowder in the matrix solution
^{d}Deposited PbS powder contains sulfate impurity phase $PbSO_4$, so the particle size of PbS was evaluated only roughly

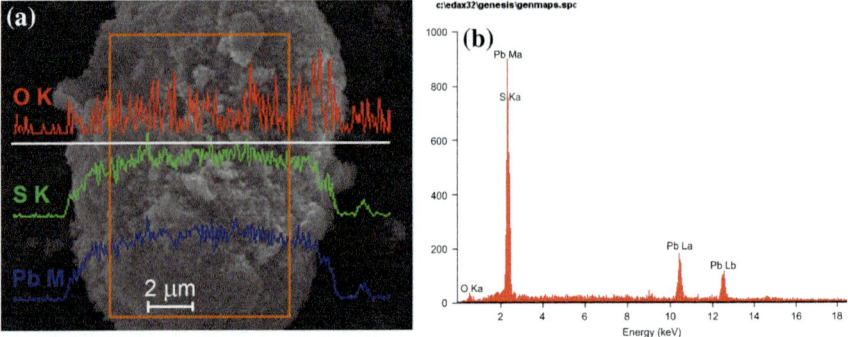

Fig. 2.8 EDX analysis of Pb, S, and O distributions in PbS nanopowders. **a** the EDX distributions of Pb, S, and O and scanning direction (*white horizontal line*) of the agglomerated PbS particle during X-ray energy dispersion analysis. *Red rectangle* shows a part of the agglomerate from which its cumulative EDX pattern **b** is collected. Reprinted from [20] with permission from Elsevier

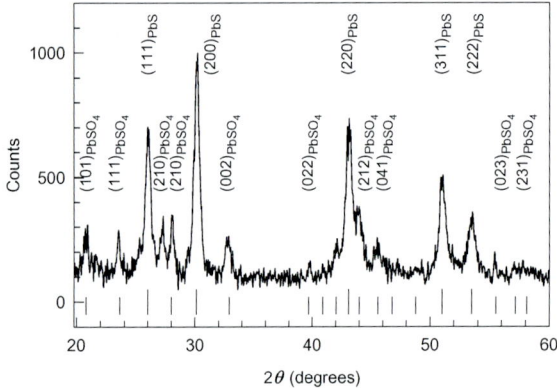

Fig. 2.9 XRD pattern of PbS nanopowder synthesized without a complexing agent from a reaction mixture of lead acetate and sodium sulfide having equal concentrations, 25 mmol l^{-3}. The long and short ticks correspond to diffraction reflections of lead sulfide PbS and lead sulfate $PbSO_4$. The XRD pattern is recorded in $CuK\alpha_{1,2}$ radiation. Reprinted from [20] with permission from Elsevier

The concentration of Trilon B influences both the phase composition of the deposited powders and the nanoparticle size. A powder precipitates from a mixture with the minimum Trilon B concentration of 12.5 mmol l^{-1} contained ~ 15 wt% of lead sulfate impurity (Fig. 2.12). The lead sulfate and sulfur impurity phases in amounts of ~ 10 and ~ 8 wt%, respectively, were found in the precipitates obtained from the reaction mixture with the maximum Trilon B concentration of 75 mmol l^{-1}. Generally, upon the deposition of PbS powder with Trilon B, the

Fig. 2.10 XRD patterns of nanocrystalline PbS powders synthesized in a reaction mixtures with equal concentrations of lead acetate, sodium sulfide, and Trilon B. The *inset* shows the dependences of reduced diffraction reflection broadening $\beta^*(2\theta)$ of synthesized PbS powders on the scattering vector s: *Open square* synthesis from the reaction mixture with reactant concentrations 5 mmol l^{-1}; *Open circle* 12.5 mmol l^{-1}; and *Filled circle* 50 mmol l^{-1}. The *solid lines* denote approximating of β^* (s) dependences. All the XRD patterns are recorded in $CuK\alpha_{1,2}$ radiation. Reprinted from [20] with permission from Elsevier

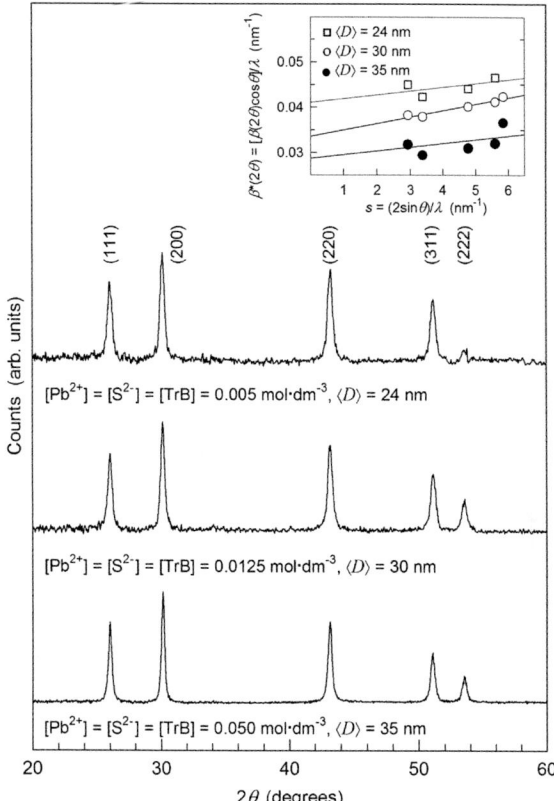

impurity phases are absent only if the $[Pb^{2+}]:[S^{2-}]:[TrB]$ concentration ratio is from 1:1:1 to 1:1:(1/2).

As the Trilon B concentration increases from 12.5 to 75 mmol l^{-1}, nanoparticles measuring 8–50 nm are formed (Fig. 2.13). The increase in the exposure time of the synthesized PbS particles in the matrix solution also induces growth of the nanoparticles.

When sodium citrate is used in a reaction mixture with low (1.25 mmol l^{-1}) concentrations of $Pb(OAc)_2$, Na_2S and Na_3Cit, no lead sulfide precipitate is formed. The deposition of PbS was observed from reaction mixtures containing lead acetate, sodium sulfide and sodium citrate in equal concentrations ranging from 2.5 to 100 mmol l^{-1}. During deposition from a citrate-containing reaction mixture with reactant concentrations of 2.5 mmol l^{-1}, an impurity phase $PbSO_4$ was present in the nanopowder along with lead sulfide. The deposits from the mixtures with large concentrations of reactants did not contain any impurity phases. The XRD patterns of the powders synthesized from citrate-containing mixtures have very broad diffraction reflections (Fig. 2.14). The particle size estimated from diffraction

Fig. 2.11 The microstructure of PbS nanopowder synthesized from a mixture of lead acetate, sodium sulfide, and Trilon B having equal initial concentrations, 5 mmol l^{-1}: **a** at a magnification of 20000 times large agglomerates of size to 10 μm are visible, **b** loose microstructure of the agglomerate at a magnification of 30000 times; **c** magnification of 100000 times. Reprinted from [20] with permission from Elsevier

Fig. 2.12 XRD patterns of nanocrystalline PbS powders synthesized from a mixture of lead acetate and sodium sulfide with initial concentrations 0.050 mol dm^{-3} and Trilon B with initial concentration [TrB] ranging from 12.5 to 75 mmol l^{-1} [20]. The long, medium, and short ticks correspond respectively to diffraction reflections of lead sulfide PbS, lead sulfate PbSO$_4$, and sulfur S. The XRD pattern is recorded in Cu$K\alpha_{1,2}$ radiation. Reprinted from [20] with permission from Elsevier

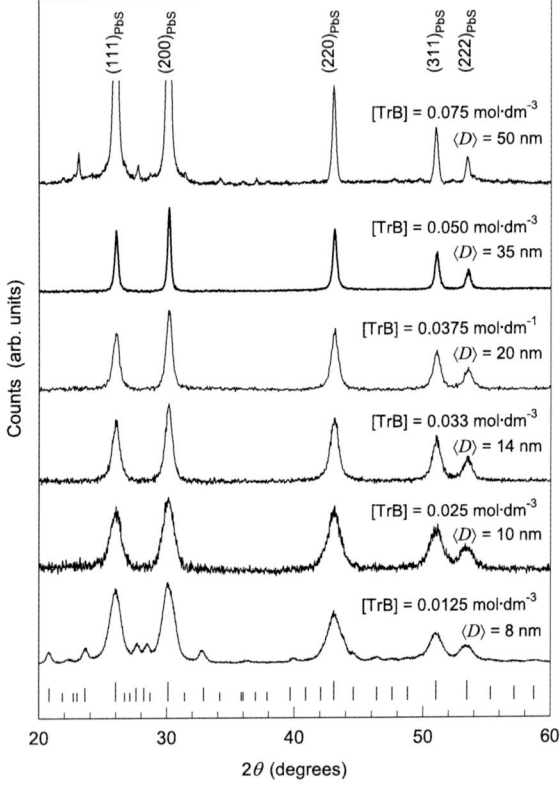

Fig. 2.13 Effect of the
concentration of complexing
agents Trilon B (1) and
sodium citrate (2) on the lead
sulfide nanoparticle size
D upon deposition from the
reaction mixture of lead
acetate and sodium sulfide
having equal reactant
concentrations [20, 47]. The
arrows indicate the increase
in the PbS particle size upon
70 h (for citrate solution) and
120 h (for Trilon B solution)
exposure in the matrix
solution

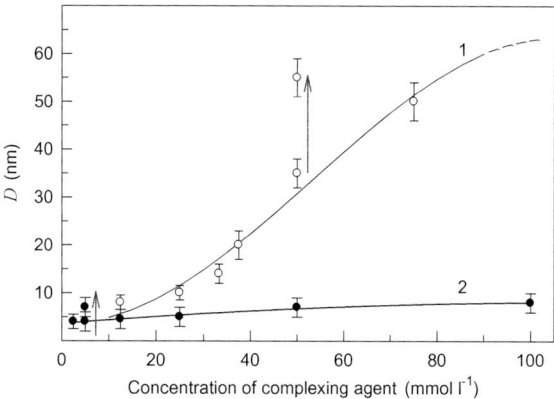

Fig. 2.13 Effect of the concentration of complexing agents Trilon B (1) and sodium citrate (2) on the lead sulfide nanoparticle size *D* upon deposition from the reaction mixture of lead acetate and sodium sulfide having equal reactant concentrations [20, 47]. The *arrows* indicate the increase in the PbS particle size upon 70 h (for citrate solution) and 120 h (for Trilon B solution) exposure in the matrix solution

reflection broadening is from 4–5 to 7–8 nm (see Table 2.3) and increases slightly with increasing concen-tration of Na_3Cit (see Fig. 2.13).

Lead sulfide nanopowders consisting of small particles with average size of 5–15 nm were prepared from reaction mixtures containing $Pb(NO_3)_2$ and Na_2S using sodium citrate as the complexing agent. A change in the Na_3Cit concentration from 12.5 to 50 mmol l^{-1} does not affect the average PbS particle size to within the error of measure-ments. When Na_3Cit is replaced by Trilon B, nanopowders with larger particles with size up to 20 nm are formed in reaction mixtures containing Pb $(NO_3)_2$ and Na_2S.

Thus, according to studies [20, 46, 47], the size of lead sulfide particles obtained by deposition from aqueous solutions depends on the reactant concentrations and the type of complexing agent. By varying these parameters, the PbS particle size can be controlled in the range from 5 to 55 nm at 298 K. For the targeted synthesis of PbS nanoparticles of <10 nm size, it is reasonable to use sodium citrate, while for the preparation of larger particles (up to 50 nm), Trilon B should better be used. Even coarser PbS particles can be obtained by deposition in the temperature range of 330–340 K or using thiourea N_2H_4CS instead of Na_2S [50].

A method for preparing nanocrystalline PbS powder using aqueous solutions of lead nitrate $Pb(NO_3)_2$ is protected by the patent [51].

Uniform cube-shaped PbS nanoparticles with an edge length of 22 nm were obtained by hydrochemical deposition from solutions of lead acetate and sodium thiosulfate in the presence of $C_{17}H_{33}COOK$ as a surfactant [52].

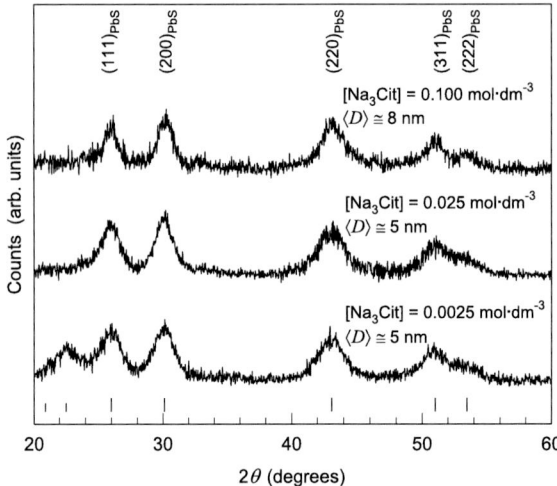

Fig. 2.14 XRD patterns of the nanocrystalline PbS powders synthesized from reaction mixtures of lead acetate, sodium sulfide, and sodium citrate. The XRD pattern of the powder deposited from a mixture with the initial concentrations 5 mmol l^{-1} contains diffraction reflections of impurity PbSO$_4$. The long and short ticks correspond to the diffraction reflections of lead sulfide PbS and lead sulfate PbSO$_4$, respectively. The XRD pattern is recorded in CuK$\alpha_{1,2}$ radiation. Reprinted from [20] with permission from Elsevier

A sort of bath deposition of nanoparticles is electrochemical deposition. In this case, particles are formed as electrical current is passed [53, 54]. Nanocrystalline PbS is synthesized in an aqueous solution of Na$_2$S with metallic lead as the sacrificial anode and polyvinyl alcohol as the stabilizer [55]. Upon application of an electrical potential to the electrodes, PbS nanoparticles of 8–10 nm size are formed in the solution. According to Yang et al. [53], by varying the potential value and scan rate, it is possible to synthesize PbS nanoparticles with a size in a specified narrow range from 3 to 7 nm. An advantage of electrochemical deposition is the possibility of more accurate control over the nanoparticle size. Also, in the case of long deposition time, it is possible to eliminate unstable compounds from the list of reactants.

Lead sulfide nanoparticles can also be synthesized by a sol-gel process [56]; however, this does not ensure the formation of monodisperse particles.

Lead sulfide nanoparticles and nanostructures are synthesized in spatially confined colloidal systems—colloidal nanoreactors. Most often, these are reverse micelles, liquid crystals, adsorption layers, Langmuir-Blodgett films [57] and microemulsions. The confinement of the reaction zone in which the nanophase is formed provides a high degree of monodispersity for the resulting particles. The molecules that confine the reaction zone should be amphiphilic, i.e, they should have a nonpolar tail soluble in nonpolar solvents, or a hydrophobic tail and a polar hydrophilic head. The amphiphilic properties are inherent in surfactant molecules and ions. When they get in water, surfactant molecules cover the water surface by

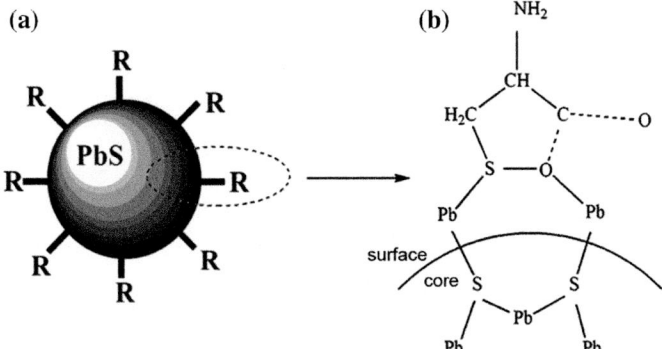

Fig. 2.15 Schematic structure of a **a** PbS nanoparticle with the surface coated by the *L*-cysteine capping agent and **b** a PbS nanoparticle-cysteine coating complex [58]. The radical (substituent) R is a part of *L*-cysteine molecule

forming a monomolecular film. In the film, the polar part of the molecule is immersed in water, while the nonpolar hydrophobic substituent is located in air, and the water surface tension is thus decreased. After the surface has been covered, new surfactant molecules and ions pass to the bulk of water.

In non-polar solvents, colloidal structures are formed in which the hydrocarbon tails are arranged at the outward normal to the surface – reverse micelles and water-in-oil microemulsions. In reverse micelles, hydrophilic groups form a polar core, while the hydrophobic tails point outwards. For the chemical synthesis of nanoparticles, reverse micelles are used most often. In order to stabilize the nanoparticles, compounds that modify and coordinate the surface atoms of the particles are added to the micellar solution [58]. As an example, Fig. 2.15 shows the schematic structure of a PbS nanoparticle with the surface coated by *L*-cysteine $C_3H_7NO_2S$, which serves as the capping agent [58].

Organometallic synthesis of colloid chalcogenide nanoparticles in organic solvents in the presence of trioctylphosphine oxide (TOPO) has been reported by Gerion, Alivisatos et al. [59]. A drawback of this method is that the nanoparticles are soluble only in organic solvents. For using these nanoparticles in biological media, it is necessary to endow them with water solubility by forming hydrophilic groups on the surface. The process of particle stabilization and transfer to the aqueous phase is called solubilization. One solubilization method implies the formation of a polymer layer around the nanoparticle [60]. In this case, the hydrophobicity of the TOPO molecules located on the nanoparticle surface is used for binding the hydrophobic groups of solubilizing polymers (block copolymers, polyethylene glycol (PEG) and so on) (Fig. 2.16). The hydrophilic groups of these polymers point towards the surrounding aqueous solution and form an insulating polar shell. This shell contains active groups for attachment of biological molecules (antibodies, peptides, DNA molecules).

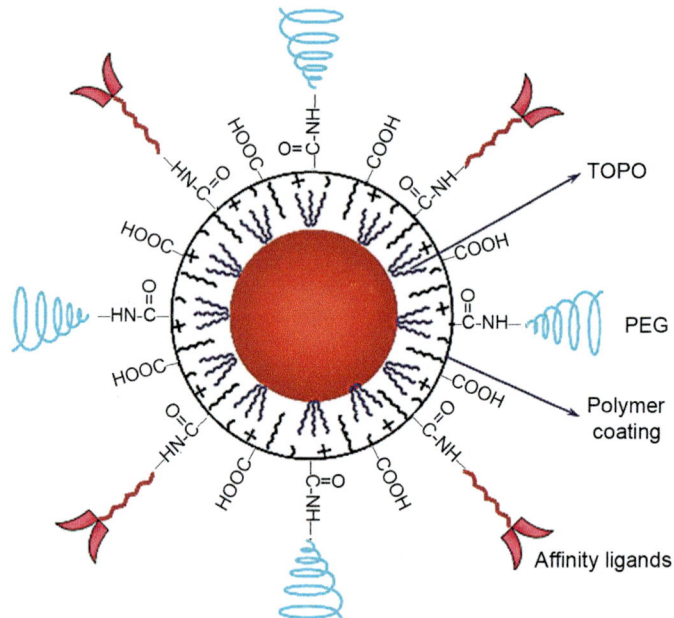

Fig. 2.16 Scheme of the formation of a hydrophilic polymer layer on a quantum dot [60]. The quantum dot includes a TOPO-coated core and a polymeric shell with attached amphiphilic PEG layer and biocompatible ligands

2.1.4 Deposition of Lead Sulfide Particles with Different Morphology

The hydro- and solvothermal synthesis at a temperature of >300 K can be considered as a sort of chemical bath deposition. Note that in some publications, chemical bath deposition followed by heat treatment of the resulting lead sulfide powder is erroneously described as hydro- or solvothermal synthesis.

Lead sulfide nanorods ∼80 nm in diameter and ∼400 nm in length and dendrite nanostructures were obtained by chemical deposition from aqueous solutions of lead acetate and thioacetamide without using ligands or templates [61]. A distinctive feature of this method is the use of solutions of reactants with concentrations differing 5–40 fold. The synthesis was carried out under ambient conditions by adding a highly concentrated solution of one reactant to a solution of the second reactant with a low concentration.

Hydrothermal deposition was used to obtain nanowires and star-shaped PbS crystals [62], and also well-faceted cubic nanostructures [63]. By using non-aqueous solvents (ethyl alcohol, ethylene glycol, ethylenediamine, triethylene glycol, propylene glycol and so on), one can deposit lead sulfide with various particle morphology – dendrite [64–66] and many-arm star-shaped structures [15, 67], separate nanospheres [68], nanotubes with different cross-section shapes [69–71], and flower-like structures [72, 73]. For example, Sun et al. [72] prepared flower-like lead sulfide nanoparticles by thermal decomposition of lead diethyldithiocarbamate dissolved in ethylenediamine (the solution was heated by microwave radiation). The reported micrographs [72] indicate that the flower petals are composed of well-faceted cubic or rectangular PbS particles. Shakouri-Arani and Salavati-Niasari [74] prepared 11 different morphological forms of lead sulfide by solvothermal synthesis from $PbNO_3$ and $C_{13}H_{11}NS$. Propylene glycol served as the solvent; the synthesis was carried out at 433 K; and the duration of the synthesis ranged from 8 to 24 h for the preparation of PbS with different morphology.

The micrographs of lead sulfide nanoparticles with different morphology obtained by hydro- and solvothermal synthesis – nanowires and star-shaped crystals [62], nanotubes [71], many-arm structures [72] and "flowers" [73] – are presented in Fig. 2.17. By ultrasonic treatment of a solution of lead sulfide, it is possible to obtain PbS as separate nanospheres [67], dendrite structures [65] and nanotubes with a square cross-section [71].

It was shown [75] that thermal radiolysis (electron beam irradiation) of a solution containing precursors for the synthesis of nanocrystalline PbS gives rise to particles of equal size, i.e, the final material becomes more uniform. As the reactants, the reseachers [75] used solutions of lead acetate and thioacetamide in polyvinyl alcohol.

An interesting method for the preparation of hollow spherical PbS nanoparticles has been proposed [76]. Aqueous solutions of lead nitrate and sodium thiosulfate

Fig. 2.17 Micrographs of nanoparticles of nanostructured lead sulfide with different morphology obtained by hydro- and solvothermal synthesis. **a** Nanowires [62], **b** dendrites [66], **c** stars [62], **d** many-arm crystals [72], **e** nanotubes with a square cross-section [71], **f** "flowers" [73]

were used, thiosulfate solution being slowly added dropwise to a $PbNO_3$ solution. Then the reaction mixture was maintained for 2 h at a temperature of 403 K. First, the lead thiosulfate complex was formed in the solution

$$Pb^{2+} + 2S_2O_3^{2-} \rightarrow \left[Pb(S_2O_3)_2\right]^{2-}, \qquad (2.19)$$

and on heating, this complex decomposed to give lead sulfide, sulfur and gaseous sulfur dioxide

$$\left[Pb(S_2O_3)_2\right]^{2-} \rightarrow PbS + S + SO_2 \uparrow + SO_4^{2-}. \qquad (2.20)$$

During the synthesis [76], PbS nanoparticles of up to 10 nm size were located on the surface of SO_2 gas bubbles and gradually these structures were transformed into hollow spheres 2–40 nm in diameter. In other words, SO_2 gas functioned as a template and played the key role in the growth of hollow nanospheres from PbS nanoparticles (Fig. 2.18).

Thus, hydro- and solvothermal synthesis can produce a variety of morphological types of lead sulfide. The reasons for the formation of PbS particles of different shapes are not entirely clear, as there are very few experimental or theoretical results on this subject in the literature. Probably, the formation of nanoparticles of different morphology is related to their nucleation and growth kinetics. The nanoparticle nucleation and growth are the subjects of only two comprehensive studies [77, 78] carried out for cadmium sulfide nanoparticles by ultrafast small angle X-ray scattering (SAXS). Data on nucleation and growth of PbS nanoparticle are lacking. The use of elevated temperature for the synthesis promotes the growth of PbS particles; therefore, the characteristic particle size is often beyond the conventional upper boundary (40–60 nm [8, 79–82]) of the range defined for nanocrystalline substances.

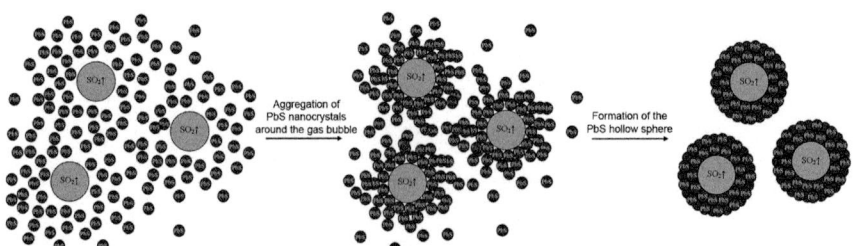

Fig. 2.18 Scheme of formation of hollow spheres from PbS nanoparticles [76]

2.1.5   Lead Sulfide Quantum Dots

The use of semiconductor quantum dots implies the presence of a matrix. Most often, glass, porous ceramics, or substrates are used as QD matrices; therefore, the syntheses of QDs and the matrix are often combined.

Lead sulfide quantum dots are prepared in various media (in glasses [83–86], porous ceramics [87], sol-gel films [88] or polymers [89, 90]) and are deposited on various substrates, for example, GaAs [91].

Okuno et al. [83] obtained PbS quantum dots in a low- melting phosphate glass. A mixture of glass powder and PbS was melted at ~ 1300 K. Then the resulting glass was annealed at ~ 670 K for a period from several minutes to an hour. After annealing, the glass acquired a brown or black color as a result of formation of PbS with an average radius of 2–3 nm, which is much smaller than the lead sulfide exciton radius (18 nm). The PbS quantum dots with a radius of 1.3 nm were also synthesized by passing hydroen sulfide through an aqueous solution of polyvinyl alcohol and lead acetate.

Del Monte et al. [87] grew PbS quantum dots in a hybrid porous organo-inorganic ceramic prepared using 3-mercaptopropyltrimethoxysilane $(MeO)_3Si(CH_2)_3SH$. The growth of the quantum dots is presented schematically in Fig. 2.19. The quantum dots are formed in the pores of the ceramic saturated with a lead acetate solution in acetic acid with diffusion of gaseous hydrogen sulfide through the ceramic.

The growth of lead sulfide QDs is terminated owing to the interaction of the formed PbS particles with 3-mercaptopropyl group giving a 3-mercaptopropyl shell

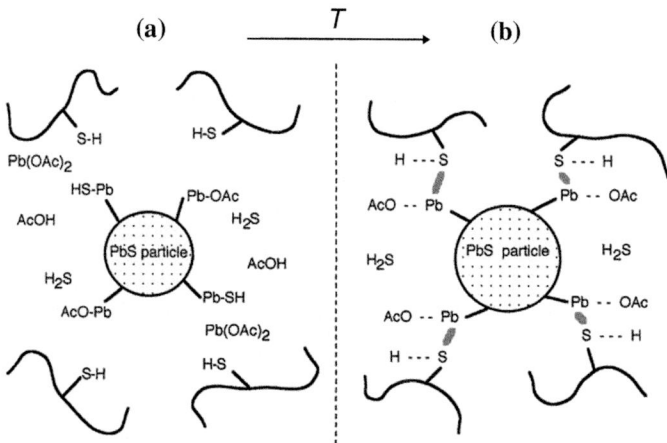

Fig. 2.19 Scheme of PbS nanoparticle growth in the pores of hybrid organo-inorganic ceramics [87]. **a** The formation and growth of PbS particles occur *via* the reaction of lead acetate with gaseous hydrogen sulfide; **b** interaction of PbS particles with 3-mercaptopropyl to give a 3-mercaptopropyl shell on the PbS particle surface and termination of the particle growth

on the particle surface. Sulfidation was carried out at a temperature of 398–418 K for 16–32 h. The size of quantum dots obtained under different conditions was in the range from 11 to 14 nm.

An organically modified SiO_2 gel was used [88] as the matrix for PbS quantum dots. A colloidal dispersion of lead sulfide was prepared by a sol-gel process using complexing agents such as Trilon B and 3-mercaptopropyltrimethoxysilane. Lead sulfide quantum dots with average diameters of 2 and 11 nm, respectively, distributed in a polyvinyl alcohol film, have been studied [89, 90]. The chemical deposition of PbS quantum dots from aqueous solutions of lead acetate and sodium sulfide in the presence of L-cysteine as a complexing and stabilizing agent has been considered [58, 92]. The size of the resulting lead sulfide QDs was 10–20 nm (Fig. 2.20).

The photoluminescence of PbS quantum dots was studied in quite a number of publications [10, 12–14, 58, 70, 83, 91, 92].

Deng et al. [10] proposed a method for solubilization of oleic acid-coated lead sulfide QDs, which provides their transfer into the water-soluble form with high fluorescence. The main idea of the method is that various hydrophilic thiol ligands, including glutathione $C_{10}H_{17}N_3O_6S$ and L-cysteine, are used as capping agents. Glutathione proved to be especially efficient.

The initial oil-soluble PbS quantum dots were synthesized in two stages. First, lead oleate was prepared from solutions of lead acetate, oleic acid and n-decane at 403 K. Then an aqueous solution of sodium sulfide was added dropwise to this solution; as a result, PbS quantum dots coated by oleic acid were formed and grew at the interface between two liquid phases (water and n-decane). In order to switch to water-soluble QDs, oleic acid molecules were replaced by thiol ligands. The

Fig. 2.20 HRTEM image of PbS quantum dots obtained in an aqueous solution of lead acetate, sodium sulfide and L-cysteine [58]

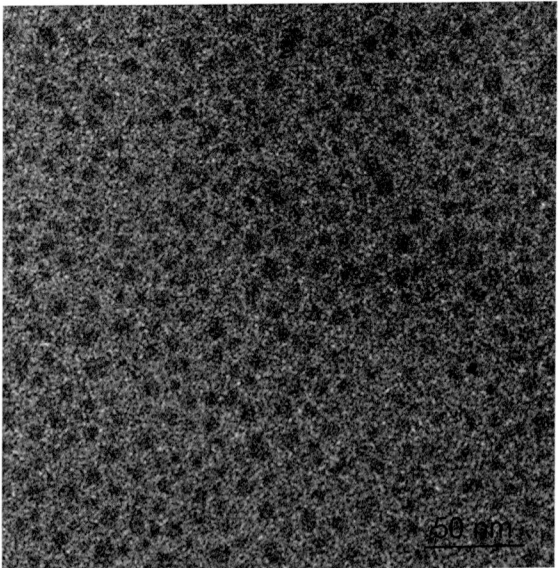

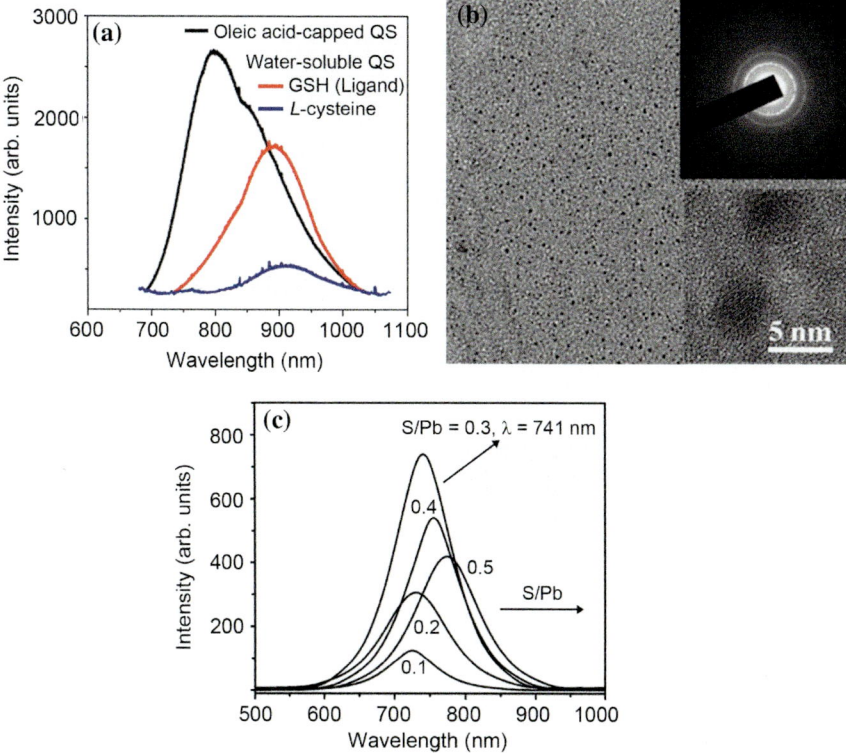

Fig. 2.21 Lead sulfide quantum dots and their properties. **a** Influence of the ligand nature on photoluminescence spectra [10]; **b** uniform distribution of monodisperse PbS QDs with a glutathione shell in an aqueous solution (the *upper* and *lower insets* show the HRTEM image and the electron-diffraction pattern of QDs) [10]; **c** change in intensity and red shift of the luminescence maximum of the L-cysteine-coated PbS QDs depending on the sulfur and lead concentration ratio S/Pb in the solutions [58]

intensity of photoluminescence of the synthesized monodisperse water-soluble PbS quantum dots depended on the sort of the ligand. The highest radiation intensity, which amounted to $\sim 70\%$ of the luminescence intensity of organic QDs, was achieved by using glutathione (Fig. 2.21a). The radiation intensity of PbS quantum dots with an L-cysteine shell amounted to 20% of the initial intensity. Transition from the initial QDs to water-soluble ones is accompanied by a red shift of the luminescence maximum from 800 to 890 and 920 nm in the case of glutathione and L-cysteine, respectively (see Fig. 2.21a). In the opinion of Deng et al. [10], the observed effect is due to the change in the electron density and confinement as a result of formation of Pb-thiol bonds. The formation of these bonds induces also a minor increase in the size of inorganic core of a quantum dot compared with the core of the initial oleic-acid coated PbS quantum dot. Figure 2.21b shows a TEM image of water-soluble glutathione-coated lead sulfide QDs with an emission max imum at 800 nm. The 4–5 nm PbS quantum dots are uniformly distributed in the solution and, according to electron diffraction data (see the lower inset in Fig. 2.21b), they have a cubic structure.

Yu et al. [58] studied the photoluminescence of lead sulfide QDs with a cysteine shell. The 7–8 nm quantum dots were synthesized from aqueous solutions of lead acetate, L-cysteine and sodium sulfide. The highest photoluminescence intensity was found for the QDs synthesized from a solution with the Pb:cysteine:S concentration ratio of 1:2.2:0.3; the intensity maximum was observed at a 741 nm wavelength (Fig. 2.21c). It was demonstrated that photoluminescence of the QDs with cysteine shell depends on the pH of the medium and S:Pb and cysteine/ Pb concentration ratios in the solution (see Fig. 2.21c). An increase in the S/Pb ratio induced a red shift of the emission maximum from 725 to 780 nm. A similar shift of the photoluminescence peak was observed in another study [10].

Lead sulfide QDs on semiconductor substrates (like GaAs) are grown by molecular-beam epitaxy method but in this case, a heteronanostructure is formed.

2.1.6 Nanostructured Lead Sulfide Films

The methods used to obtain nanostructured chalcogenide films can be classified into chemical and physical ones.

The key chemical methods for the fabrication of nanostructured lead sulfide films include deposition from colloidal solutions, chemical vapour deposition, electrochemical deposition, spray pyrolysis, photocatalytic chemical deposition and successive ionic layer adsorption and reaction (SILAR). Among the physical methods for the preparation of PbS films, the major ones are various alternatives of gas phase deposition (vacuum evaporation, molecular-beam epitaxy, glow discharge ion sputtering, ion beam sputtering, ion deposition, reactive ion sputtering, magnetron ion sputtering and high-frequency ion sputtering). Often, combined physical and chemical methods are used to prepare lead sulfide films [93].

Chemical bath deposition of films is distinguished by simplicity and the possibility to deposit a film on a surface of any size and any configuration; there is no need to heat the substrates to high temperature, and the use of toxic or explosive chemicals is avoided.

Three mechanisms of chemical bath deposition of films can be conventionally distinguished: ionic, cluster (hydroxide or colloid) and ionic cluster mechanisms [6, 23, 26, 94]. In the ionic mechanism, the sulfide nucleation occurs upon the reaction between Pb^{2+} and S^{2-} ions directly on the substrate. According to the cluster (colloid) mechanism, it is assumed that the solid phase nucleation and cluster formation take place in a homogeneous solution, while the growth, coagulation and adsorption occur on the substrate [6, 32]. It is more likely that sulfide films are deposited by a mixed mechanism. In this case, the sulfide is formed both upon the direct reaction between the metal and sulfide ions and upon adsorption of the sulfidizer on the surface of hydroxide particles to give intermediate complexes, which decompose to give the sulfide.

The chemical deposition of PbS films on a glass sub-strate has been studied in detail in work [95–100]. Three synthetic procedures were used. According to the first procedure, the PbS-1 film was deposited from aqueous solutions of $Pb(OAc)_2$ and N_2H_4CS in the presence of Na_3Cit and $NaOH$ without stirring of the solution. In the second procedure, the film was deposited under the same conditions as in the first one but without the addition of sodium citrate, and the deposition was carried with stirring (PbS-2 film) or without stirring (PbS-3 film). In the third synthetic procedure, the PbS film was formed on the surface of a Trilon-containing aqueous solution of lead acetate owing to the reaction of the latter with gaseous hydrogen sulfide; the resulting film (PbS-4) was transferred onto a glass substrate. The PbS-1 film was additionally annealed in air at temperatures from 293 to 423 K with a step of 40 K.

The average grain size of PbS-1 film in the substrate surface was ~ 250 nm; however, upon a 20000-fold magnification (Fig. 2.22a), one can see that the grains

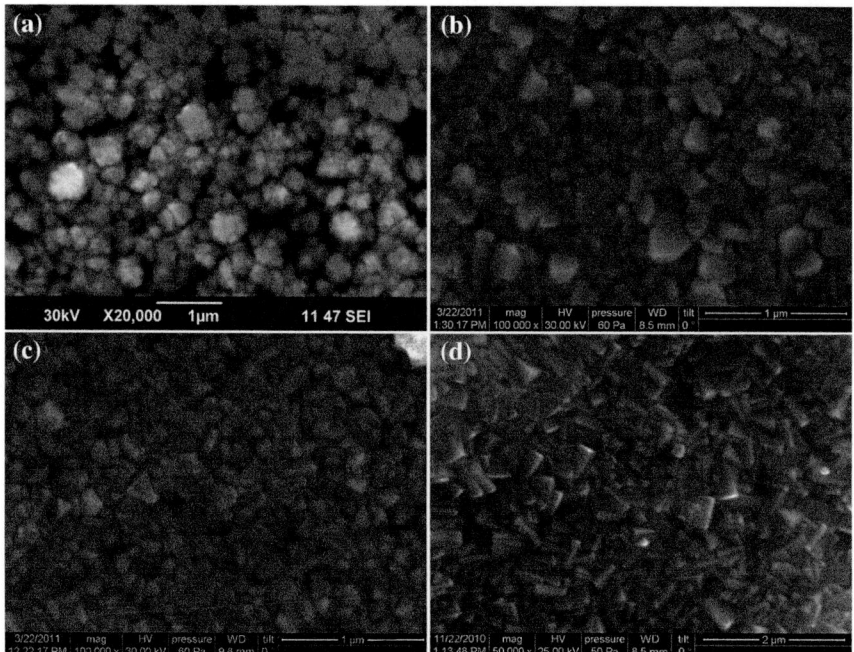

Fig. 2.22 Micrographs of **a** PbS-1, **b** PbS-2, **c** PbS-3, and **d** PbS-4 films [100]. **a** In the PbS-1 film, the nanoparticles smaller than 80–100 nm are assembled into larger aggregates of up to 200–250 nm size; the degree of film continuity does not exceed 80–85%; **b** the PbS-2 film completely coats the substrate; **c** the PbS-3 film microstructure is most uniform in grain size; **d** grains of PbS-4 film are well-faceted; more than 65% of grains have a size less than 120 nm, and $\sim 6\%$ of grains have a size from 250 to 400 nm. Reprinted from [100] with permission from Elsevier

are agglomerates of nanoparticles of the <100–120 nm size. The degree of continuity of PbS-1 films was not more than 80–85%. The PbS-2, PbS-3 and PbS-4 films
completely coated the substrate (Fig. 2.22b–d). The smallest nanoparticle size was
observed in the PbS-3 film (see Fig. 2.22c). In the PbS-4 films (see Fig. 2.22d), all
grains were well-faceted and the size of most grains was <120 nm. The smaller
particle size of PbS-2 and PbS-3 films compared with PbS-1 is due to the fact that
the deposition was carried out without sodium citrate. In the absence of sodium
citrate, the rate of formation of PbS nuclei decreases but the number of the nuclei
increases, which results in a relative decrease in the particle size. The observed size
of PbS nanoparticles in the films varies over a broad range, from 10 to 160 nm, the
average size being 50–70 nm [97–100].

A simple method for the preparation of nanostructured PbS films, ensuring a
specified range of band gap energies, has been proposed and patented by
Sadovnikov and Rempel [101]. Lead sulfide nanofilms active in the near-IR range
were prepared by deposition from aqueous solutions of lead acetate and thiourea in
the presence of sodium citrate and sodium hydroxide; the solution pH was 10–13.
By varying the reactant concentration ratio, PbS films with a controlled average
particle size ranging from 35 to 80 nm were obtained in study [101].

In the case of chemical vapour deposition [102], the substrate is exposed to
vapour of one or several compounds or to reactive gases which are components of
the deposited material. The compound is formed via layer-by-layer condensation of
atoms (molecules), similarly to the preparation of semiconductors using physical
deposition methods (evaporation or ion sputtering). The difference is in the fact that
in the case of chemical vapour deposition, the product arises as a result of
heterogeneous chemical reaction. The advantages of this method include the possibility of high-rate deposition and doping and growing of epitaxial layers with low
impurity contents. However, this method also has disadvantages, including the
complexity of deposition and the necessity to heat the substrate to high temperature,
which restricts the choice of the substrate material.

The spray pyrolysis method [103, 104] implies spraying a solution, most often,
aqueous one, containing salts of components of the deposited compound onto a
heated substrate. As the sprayed solution hits the hot substrate surface, it pyrolyt-
ically decomposes, and the reaction prod-uct forms separate crystallites or groups of
crystals on the surface. This method is suitable for the preparation of films with
strong adhesion to the substrate and high mechanical strength, stable under ambient
conditions and at elevated temperature. A drawback of this method is the absence of
direct correlation between the compositions of the deposited compounds and the
sprayed solution.

The SILAR method for the preparation of thin films proposed in 1985 [105] and
patented in 1987 [106] is based on the heterogeneous reaction taking place at the
liquid/solid interface between the Pb^{2+} cations adsorbed on the substrate and the S^{2-}
anions located in the solution [9, 107–111]. This method is schematically shown
in Fig. 2.23. The purified substrate is placed into a solution containing Pb^{2+} cations
(stage a), which are adsorbed on the substrate surface. In stage b, the surface is
cleaned from non-adsorbed lead cations. A lead sulfide nanolayer is formed in stage

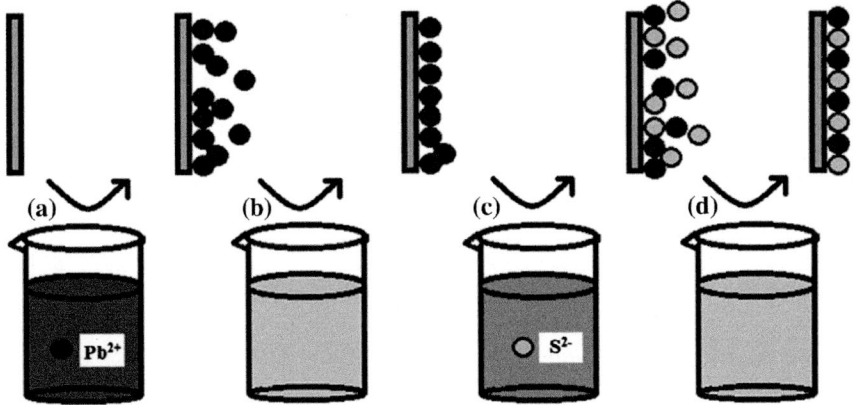

Fig. 2.23 Schematic diagram of the SILAR method [111]. **a, c** cationic and anionic precursoe, respectively; **b, d** double distilled water. For stages **a–d**, see the text also

c, when the substrate bearing the adsorbed Pb^{2+} ions is dipped into a solution containing S^{2-} anions. As a result of reaction between the sulfide anions and lead cations, a thin layer of lead sulfide is formed (stage *d*) (see Fig. 2.23). The deposition process repeated many times provides a layer-by-layer growth of the PbS film of a specified thickness with particles of a desired size on substrates of different type and shape [107–111].

The successive layer ionic deposition as a preparation method for thin-film coatings is upgraded hydrochemical deposition method, which makes it possible to control the morphology and thickness of the obtained films [107–111]. By using various reagents and by varying their concentration and the number of deposition cycles, it is possible to grow polycrystalline PbS films with predominant particle growth along a particular crystallographic direction [108].

An interesting method is photocatalytic chemical deposition of films [112, 113]. Zhukovskiy et al. [113] dipped a TiO_2 nanofilm on a substrate into a lead acetate and sulfur solution in ethyl alcohol. The TiO_2 nanofilm served as the catalyst. When this system was exposed to light, the following processes took place: photocatalytic reduction of sulfur with ethanol, the formation of lead sulfide and deposition of 4–10 nm PbS nanoparticles on the TiO_2 surface. An atomic force microscopy study of the surface of PbS/TiO_2 nanocomposites showed the products of the photocatalytic deposition of lead sulfides to be rod-like 30 nm nanoaggregates. The polycrystalline PbS nanorods are oriented normally to the surface of TiO_2 films and consist of separate nanosized sulfide particles manifesting pronounced quantum confinement effects [113]. Zhukovskiy et al. [113] proposed a mechanism of the formation of PbS nanorods on the surface of nanocrystalline TiO_2 films.

The problems of manufacture of photocatalysts by depositing PbS on TiO_2 are considered in a review [114].

Thin nanostructured films are formed by vacuum evaporation followed by deposition and molecular-beam and gas phase epitaxy. Because of specific features of the deposition process (high vapour supersaturation, specific condensation kinetics, sharp change in the velocity of vaporized atoms on going to the absorbed state), the resulting fine-grained nanofilms are thermodynamically non-equilibrium and contain a large number of "frozen" structural defects. Therefore, this method is more suitable for the preparation of lead sulfide nanocrystals without aggregation on a large surface area.

Well-faceted PbS nanoparticles on an Al_2O_3 substrate were obtained in study [115] by a combined physical and chemical method. A porous membrane with 20 nm holes was used as the template. A metallic lead melt heated to 673 K was hydraulically injected through the membrane nanopores. Sulfidation of lead nanowires was carried out by gaseous hydrogen sulfide. This gave a dense layer of PbS nanoparticles of 100 nm and smaller size (Fig. 2.24). Figures 2.24a, b are the images of a porous alumina membrane in which pore diameter was 20 nm and Pb nanowire arrays formed as the hydraulic force was applied. After the 20 nm Pb nanowire arrays were fabricated, a sulfization procedure was proceeded to obtain PbS nanoparticles, as shown in Fig. 2.24c. The study [115] may give rise to an alternative method of producing PbS nanostructures.

Thus, a number of chemical and physical methods for the synthesis of nanostructured lead sulfide as colloid solutions, nanoparticles, nanopowders and films have now been developed.

2.2   Crystal Structure of Nanoparticles, Nanopowders, and Nanofilms of Lead Sulfide

The bulk lead sulfide is commonly regarded as a stoichiometric compound with a negligibly narrow homogeneity $PbS_{0.9995-1.0005}$ [2]. Under a normal atmospheric pressure and room temperature, lead sulfide crystallizes in a cubic (space group $Fm\bar{3}m$) $B1$ structure with the lattice constant $a = 0.59362$ nm. The lead and sulfur ions occupy sites $4(a)$ with coordinates (0 0 0) and sites $4(b)$ with coordinates (1/2 1/2 1/2), respectively.

Under a pressure of 2.5 GPa, lead sulfide with the $B1$ structure acquires orthorhombic $B16$ ($B33$) structure, i.e, a phase transition of cubic $B1$ structure to orthorhombic (space group $Pnma$) structure of the GeS type takes place [2]. As the particle size decreases, the surface energy increases; therefore, pressure needed to change the crystal structure should increase. Indeed, in a PbS nanopowder, a cubic to orthorhombic phase transition starts at a higher pressure 3.0 GPa [116]. According to the data of Qadri et al. [116], the phase transition is sluggish. The smaller the grain size, the more sluggish the transition. The two phases with the $B1$ and $B16$ ($B33$) structures can coexist up to 8 GPa. As the pressure increases to 25 GPa, a second phase transition of lead sulfide to a CsCl type ($B2$) cubic structure takes place [117].

Fig. 2.24 SEM images of
a Al₂O₃ membrane with pores
of 20 nm diameter, **b** metallic
lead nanofibres obtained by
injection of a lead melt
through membrane pores, and
c PbS nanoparticles obtained
by sulfidation of lead
nanofibres with gaseous
hydrogen sulfide (**c**) [115]

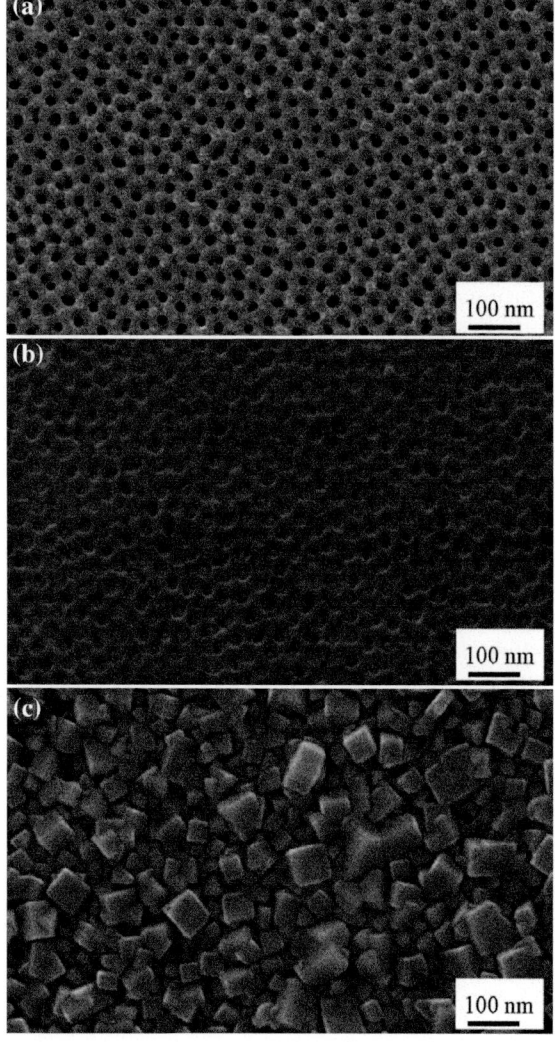

Recent studies showed that the decrease in the grain size of sulfides to several tens of nanometres and below, i.e, transition to the nanostructured state, is accompanied by a noticeable change in their optical properties.

The transition to the nanostate is accompanied by not only transformation of the properties of sulfides associated with size effects but also by structural changes. In particular, cadmium sulfide as single crystals or coarse-grained powders crystallize in the hexagonal wurtzite ($B4$ type) [118] or cubic sphalerite ($B3$ type) type structures [119, 120]. However, CdS thin films or nano- or ultradisperse powders have a structure different from the crystal structure of these two modifications. Recently, it was shown [121, 122] that CdS nanoparticles have a specific disordered structure with random alternation of close-packed atomic planes; the mean lattice of this disordered structure is described by space group $P6/mmm$.

The data on the crystal structure of PbS films are also ambiguous. It is usually assumed that PbS films, like coarse-grained bulk lead sulfide, have a stoichiometric composition and a cubic (space group $Fm\bar{3}m$) $B1$ type structure. On the basis of investigating the structures of PbS films at different temperatures, Quadri et al. [123] suggested that at 375 K, subtle phase transition from the $B1$ to cubic $B3$ structure (space group $F\bar{4}3m$) takes place in the film. The phase with the $B3$ structure has nonstoichiometric composition $PbS_{0.9}$ and can coexist with the $B1$ structure. Meanwhile, it was shown [124] that the lead to sulfur ratio in PbS thin films is stoichiometric (Pb:S = 1:1). The statement of Qadri et al. [123] about the $B3$ phase is a mistake, because for interpreting the diffraction data, they suggested that some lead atoms in the detected phase occupy sites $4(b)$ with the coordinates (1/2 1/2 1/2) but there are no such sites in the $B3$ type structure, however, they exist in space group $F\bar{4}3m$. To conserve the sulfide composition, the occupancies of the 4 (a) and $4(b)$ positions were taken equal to 0.75 and 0.25, respectively. Thus, the structure of PbS films proposed by Quadri et al. [123] is cubic and belongs to space group $F\bar{4}3m$ but does not refer to the $B3$ type structure.

In the thin-layer heterostructures, $PbVS_3$, $PbNbS_3$ and $PbTaS_3$, which represent alternating PbS and transition metal disulfide (VS_2, NbS_2 or TaS_2) layers, lead sulfide can have monoclinic crystal structure (space groups $C2$, $C2/m$, $Cm2a$) [125–128].

Unfortunately, studies into the synthesis and properties of nanostructured PbS with different morphology usually do not include detailed investigation of the crystal structure. The conclusion about the $B1$ type cubic structure of nanostructured PbS is drawn upon visual comparison of the experimental XRD patterns with the standard data for bulk lead sulfide and comparison of the estimated unit cell constant with the unit cell constant of bulk PbS.

Thorough investigations of the crystal structure of PbS nanopowders and nanofilms [95, 96, 98–100, 129] were carried out after 2008.

2.2.1 Crystal Structures of Nanopowders and Quantum Dots

Nanocrystalline PbS powders with an average particle size from 100 to 8 nm were studied by X-ray diffraction [98, 99].

Nanocrystalline PbS powders with the average particle size of 20 nm and smaller were prepared by chemical condensation with the reaction between S^{2-} and Pb^{2+} ions. The synthesis of PbS nanoparticles was performed by mixing stoichiometric amounts of aqueous solutions of lead acetate $Pb(CH_3COO)_2 \equiv Pb(AcO)_2$ and sodium sulfide Na_2S.

The crystal structure and average size of the particles in nanocrystalline PbS powders was studied by XRD method. The measurements were performed by the Bragg-Brentano technique with the use of a Shimadzu XRD-7000 X-ray diffractometer with $CuK\alpha_{1,2}$ radiation in the 2θ angle range 18°–90°, with a step of $\Delta(2\theta) = 0.03°$ and 10 s exposure at each point. The microstructure and crystal structure of PbS nanoparticles were investigated using transmission and scanning electron microscopy and electron diffraction on electron microscopes FEI Quanta_200 and JEOL JEM 200CX.

Figure 2.25 shows the XRD patterns of several finest grained PbS nanopowders with different average particle sizes from 9 to 20 nm. The observed large broadening of the diffraction reflections is associated primarily with the small size of the particles.

The set of reflections was the same for all nanopowders and, as a first approximation, it corresponded to the cubic structure with space group $Fm\overline{3}m$. The final refinement of the structure of the PbS nanopowders was performed with the X'Pert Plus software package [130]. The structure was refined using models corresponding to $B1$, $B2$, $B3$, $B16$ and $D0_3$ type structures mentioned in the literature. The best agreement between the experimental and calculated XRD patterns was achieved for the $B1$ structure.

As an example, Fig. 2.26 shows the experimental, calculated, and difference XRD patterns of the PbS nanopowder with the average particle size of 12 nm. The Rietveld reliability factors are equal to $R_I = 0.0096$, $R_p = 0.0478$, $\omega R_p = 0.0641$. The use of alternative structures such as $B2$, $B3$, $B16$ and $D0_3$ resulted in poorer agreement between the experimental and calculated XRD patterns of nanocrystalline PbS powders. The results of electron microscopy of the particles of PbS nanopowders [98, 99] are consistent with XRD data. Thus, the PbS nanopowders have a cubic (space group $Fm\overline{3}m$) $B1$ structure.

Fig. 2.25 XRD patterns of the nanocrystalline PbS powders with different average sizes D of particles [99]. All powders contain only the cubic (space group $Fm\overline{3}m$) lead sulfide with the $B1$ type structure. The large broadening of the diffraction reflections is due to the small size of PbS particles. XRD patterns are recorded in $CuK\alpha_{1,2}$ radiation

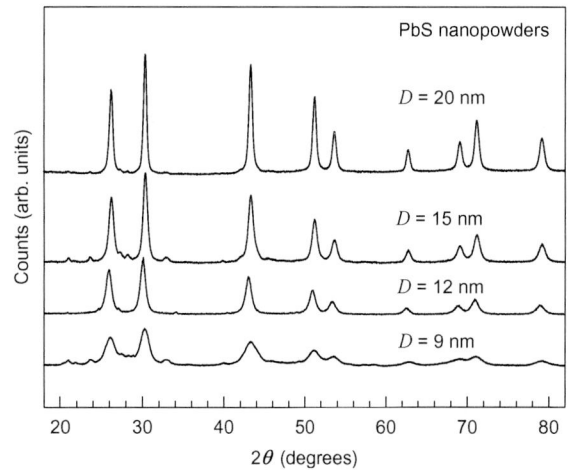

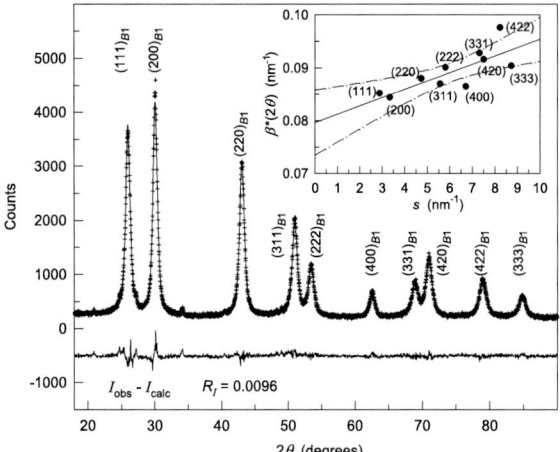

Fig. 2.26 Experimental (*crosses*) and calculated (*solid line*) XRD patterns of the nanocrystalline PbS powder with a cubic (space group $Fm\bar{3}m$) $B1$ type crystal structure [99]. The *bottom curve* shows the difference ($I_{obs} - I_{calc}$) between the experimental and calculated XRD patterns. The *inset* shows the dependence of the reduced broadening $\beta^*(2\theta)$ of diffraction reflections on the magnitude of the scattering vector s and its linear approximation. The *dot-dashed lines* indicate a 95% confidence interval of the determination of $\beta^*(2\theta)$ as a function of s. The average size of particles D is obtained by extrapolating the dependence $\beta^*(s)$ to the value of $s = 0$

Electron diffraction studies of isolated nanoparticles and lead sulfide QDs [10, 11, 58, 92] in combination with high resolution transmission electron microscopy (HRTEM) confirmed their cubic structure. However, a special quantitative refinement of the crystal structure of PbS quantum dots was never carried out; therefore, the assignment of $B1$ type structure to these objects is hypothetical. Indeed, the crystal structure of PbS nanofilms was also considered to be $B1$, but recent studies demonstrated that this opinion is erroneous.

2.2.2 Crystal Structures of Nanofilms

Quantitative determination of the crystal structures of nanostructured PbS films is described in studies [95, 96, 100, 129]. Synthesis of these nanostructured PbS films, their microstructure and designations are described in Sect. 2.1.6.

Lead sulfide films of ~ 100 nm thickness were annealed in air at temperatures from 293 to 423 K with a step of 30–40 K.

XRD examination of the films was carried out in situ on a Philips X'Pert automated diffractometer with $CuK\alpha_{1,2}$ radiation in the 2θ angle range $18°–90°$, with a step of $\Delta(2\theta) = 0.016°$ and large exposure time ~ 500 s in each point. The

Philips X'Pert diffractometer was equipped with a position-sensitive fast X'Celerator sector detector which is an integral device of several parallel detectors [131, 132]. Owing to this, the X'Celerator detector measures the reflection intensity in the wide range 7.2° of the 2θ angles, not in a point, as a common proportional counter-detector does. For example, if the measurement starts from 18°, at the initial moment, the detector captures the angle range 14.4°–21.6° and begins to move to a range of larger angles. By the time when the detector scans the reflection intensity in the angular range from 18.0° to 25.2°, the exposure time in the 18° point with a step of 0.016° is even 450 s. As a result, the duration of measuring the XRD of the film was decreased approximately by a factor of 100 (from 600 to 700 h as a common detector is used to 7–8 h with the X'Celerator detector) without any loss of the resolution quality.

The XRD patterns of the PbS-1 nanostructured film measured in situ at various annealing temperatures are shown in Fig. 2.27. All the observed diffraction reflections are noticeably broadened. The change in the lattice constant on heating shifts the maxima of the XRD diffraction reflections. As an example, the inset to Fig. 2.27 shows the change in the position of the (200) XRD of the film of the nanocrystalline lead sulfide as temperature increases. Thus, the lead sulfide has, at a temperature of 423 K, the lattice constant which is approximately 0.0012 nm larger than that at room temperature. An average particle size in film was estimated from the diffraction reflection broadening. The average particle size increases with

Fig. 2.27 XRD patterns of nanostructured PbS-1 film recorded in situ at different annealing temperatures. *Inset* shows a systematic displacement of the (200) diffraction reflection with increase of annealing temperature. The film are deposited onto the glass substrate, microstructure of as-prepared PbS-1 film is shown in Fig. 2.22a. The XRD patterns are recorded in $CuK\alpha_{1,2}$ radiation. Reprinted from [100] with permission from Elsevier

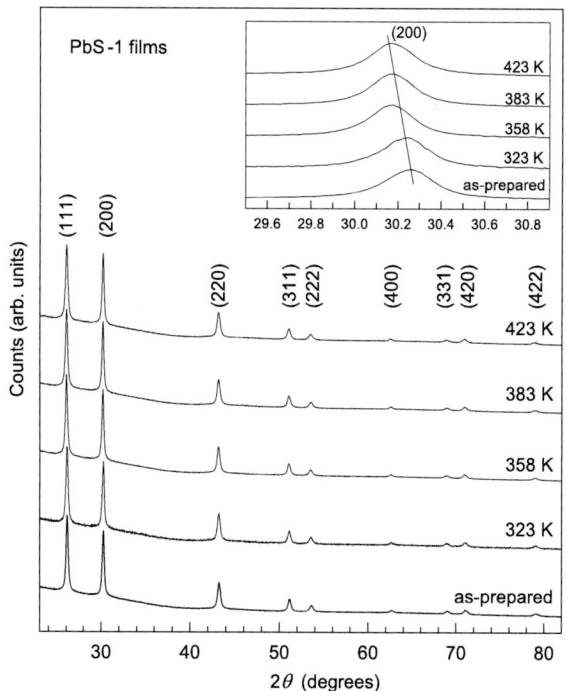

temperature from 70 nm (as-prepared state at 293 K) to 175 nm at a temperature of 423 K.

Since the structural type of PbS films has not been unambiguously determined, the authors [95, 96, 100, 129] carried out quantitative analysis of the measured XRD patterns using various cubic structure models. The following crystal structure models were considered: cubic $B1$ (space group $Fm\overline{3}m$); cubic $B3$ (space group $F\overline{4}3m$); and a two-phase model of a film in which the relative contents of phases with $B1$ and $B3$ type structures are y and $(1 - y)$, respectively.

In the general case, the measured intensity of the ith structure reflection (hkl) is

$$I_i = K \cdot F_{hkl}^2 \cdot P_{hkl} \cdot PLG(\theta) \cdot f_T, \tag{2.21}$$

where K is the instrument constant, F_{hkl}^2 is the structural factor, P_{hkl} is the multiplicity factor, $PLG(\theta)$ is the angular factor of the intensity, and f_T is the temperature factor.

As the $B1$ and $B3$ type structures and the assumed intermediate structure are cubic, their coefficients K, P_{hkl}, $PLG(\theta)$ and f_T are the same at other conditions being identical, and only the factor F_{hkl}^2 is variable quantity.

The structural factor F_{hkl}^2 appearing in relationship (2.21) is the square of the structural amplitude

$$F_{hkl} = \sum_j f_j \exp[-i2\pi(x_j h + y_j k + z_j l)] \tag{2.22}$$

and, in the trigonometric form, has a general expression

$$F_{hkl}^2 = \left[\sum_j f_j \cos[2\pi(x_j h + y_j k + z_j l)] \right]^2$$

$$+ \left[\sum_j f_j \sin[2\pi(x_j h + y_j k + z_j l)] \right]^2, \tag{2.23}$$

where x_j, y_j, z_j are the coordinates of the jth atom, and f_j is the atomic scattering factor. Summation in (2.22) and (2.23) is performed over atoms of the unit cell of the crystal structure under consideration.

The unit cell basis of lead sulfide with the $B1$ structure (space group $Fm\overline{3}m$) contains eight atoms among them four Pb atoms in positions $4(a)$ with the coordinates (0 0 0), (1/2 1/2 0), (1/2 0 1/2), and (0 1/2 1/2), and four S atoms in positions $4(b)$ with the coordinates (1/2 1/2 1/2), (0 0 1/2), (0 1/2 0), and (1/2 0 0). If the atomic scattering factors of lead and sulfur atoms are f_{Pb} and f_S, respectively, the structural factor of the $B1$ type structure calculated by relationship (2.23) has the form

$$F_{B1}^2 = \{f_{Pb}[1 + \cos \pi(h+k) + \cos \pi(h+l) + \cos \pi(k+l)]$$
$$+ f_S[\cos \pi(h+k+l) + \cos \pi h + \cos \pi k + \cos \pi l]\}^2 \quad (2.24)$$

In the unit cell of PbS with the $B3$ structure (space group $F\bar{4}3m$), four lead atoms occupy positions $4(a)$ with the same coordinates which they occupy in the $B1$ structure, namely, (0 0 0), (1/2 1/2 0), (1/2 0 1/2), and (0 1/2 1/2), and four sulfur atoms occupy positions $4(c)$ with the coordinates (1/4 1/4 1/4), (3/4 3/4 1/4), (3/4 1/4 3/4), and (1/4 3/4 3/4). According to the foregoing, the structural factor of lead sulfide with the $B3$ structure is given by

$$F_{B3}^2 = \{f_{Pb}[1 + \cos \pi(h+k) + \cos \pi(h+l) + \cos \pi(k+l)]$$
$$+ f_S\left[\cos \frac{\pi}{2}(h+k+l) + \cos \frac{\pi}{2}(3h+3k+l) + \cos \frac{\pi}{2}(3h+k+3l)\right.$$
$$\left. + \cos \frac{\pi}{2}(3h+k+3l) + \cos \frac{\pi}{2}(h+3k+3l)\right]\}^2 \quad (2.25)$$
$$+ f_S^2\left[\sin \frac{\pi}{2}(h+k+l) + \sin \frac{\pi}{2}(3h+3k+l) + \sin \frac{\pi}{2}(3h+k+3l)\right.$$
$$\left. + \sin \frac{\pi}{2}(h+3k+3l)\right]^2$$

If the film is two-phase and the relative content in the film of the phase with the $B1$ structure is y and the phase with the $B3$ structure is $(1 - y)$, the intensity I_i of arbitrary reflection is a superposition of the intensities of like reflections of the $B1$ and $B3$ structures. In such a two-phase film, the sulfur atoms occupy the non-metallic positions of the $B1$ structure with the probability of y and the nonmetallic positions of the $B3$ structure with the probability $(1 - y)$. In this case, the structural factor is the superposition of the structural factors of the $B1$ and $B3$ phases and has the form

$$F_{B1+B3}^2 = \{f_{Pb}[1 + \cos \pi(h+k) + \cos \pi(h+l) + \cos \pi(k+l)]$$
$$+ yf_S[\cos \pi(h+k+l) + \cos \pi h + \cos \pi k + \cos \pi l] + (1-y)f_S\left[\cos \frac{\pi}{2}(h+k+l)\right.$$
$$\left. + \cos \frac{\pi}{2}(3h+3k+l) + \cos \frac{\pi}{2}(3h+k+3l) + \cos \frac{\pi}{2}(h+3k+3l)\right]\}^2$$
$$+ \left\{(1-y)f_S\left[\sin \frac{\pi}{2}(h+k+l) + \sin \frac{\pi}{2}(3h+3k+l) + \sin \frac{\pi}{2}(3h+k+3l)\right.\right.$$
$$\left.\left. + \sin \frac{\pi}{2}(h+3k+3l)\right]\right\}^2$$

$$(2.26)$$

The determination of the phase composition and crystal lattice parameters of possible cubic phases and final refinement of the structure of the PbS film corresponding to various temperatures from 293 to 423 K were carried out using the X'Pert Plus program [130]. To estimate the validity of the structural models, authors [95, 96, 100] used the Rietveld factor of reliability [133]

$R_I = \sum_{i=1}^{N} |I_{\exp(i)} - I_{\text{calc}(i)}| / \sum_{i=1}^{N} |I_{\exp(i)}|$, where $I_{\exp(i)}$ and $I_{\text{calc}(i)}$ are the experi-

mental and calculated intensities of the ith reflection, respectively. The minimization of the experimental XRD patterns in an approximation of two-phase film gives $y = 0.90 \pm 0.02$ and a better convergence $R_{I(B1+B3)} = 0.04$ than the minimization in an approximation in which the film contains the only phase with either the $B1$ or $B3$ structure ($R_{I(B1)} = 0.05$ и $R_{I(B3)} = 0.12$, respectively). Along with this, it follows from the minimization that the lattice constants of the phases with the $B1$ and $B3$ structures are absolutely equal. Physically, this is unlikely and indicates that the PbS film is single-phase, but its structure is similar to the $B1$ and $B3$ structures but it differs from them.

Sadovnikov, Gusev and Rempel [95] assumed that the real structure of PbS films corresponds to space group $Fm\bar{3}m$ but the S atoms in it occupy not only octahedral (sites $4(b)$) but also tetrahedral interstices (sites $8(c)$) (Fig. 2.28b). In this structure, the probabilities for S atoms to occupy sites $4(b)$ and $8(c)$ are equal to y and $(1 - y)/$ 2, respectively. The $8(c)$ positions of the cubic crystal lattice with space group $Fm\bar{3}m$ have the following coordinates: (1/4 1/4 1/4), (3/4 3/4 1/4), (3/4 ¼ 3/4), (1/4 3/4 3/4), (3/4 3/4 3/4), (3/4 1/4 1/4), (1/4 3/4 1/4), and (1/4 1/4 3/4).

Taking into account the coordinates of the $4(a)$ positions occupied with Pb atoms and the $4(b)$ and $8(c)$ positions occupied with S atoms with the probabilities y and $(1 - y)/2$, respectively, the structure amplitude of the proposed cubic (space group $Fm\bar{3}m$) phase has the form

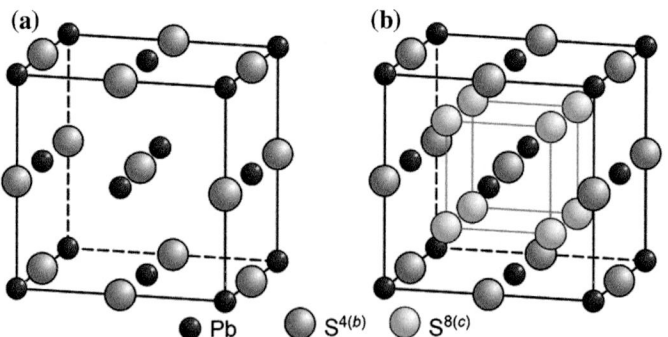

(a)   **(b)**   ● Pb   ⬤ S$^{4(b)}$ ◯ S$^{8(c)}$

Fig. 2.28 Crystal structures of PbS. **a** Unit cell of coarse-grained bulk cubic (space group $Fm\bar{3}m$) lead sulfide with $B1$ type structure; **b** distribution of lead and sulfur atoms in the unit cell of cubic (space group $Fm\bar{3}m$) DO_3-type structure of a $PbS \equiv PbS_y^{4(b)}S_{1-y}^{8(c)}$ nanofilm. *Filled violet circle* atoms Pb в in $4(a)$ positions; *Filled orange (dark) circle* atoms S in $4(b)$ positions; *Filled yellow (light) circle* atoms S in $8(c)$ positions. In the unit cell of the DO_3-type structure, sulfur atoms randomly occupy, with probability y, octahedral sites $4(b)$ and randomly occupy, with probability $(1 - y)$, tetrahedral sites $8(c)$

$$F = f_{Pb}\{1 + \exp[-i\pi(h+k)] + \exp[-i\pi(h+l)] + \exp[-i\pi(k+l)]\}$$
$$+ yf_S\{\exp[-i\pi(h+k+l)] + \exp(-i\pi h) + \exp(-i\pi k) + \exp(-i\pi l)]\}$$
$$+ [(1-y)f_S/2]\{\exp[-i\pi(h+k+l)/2] + \exp[-i\pi(3h+3k+l)/2]$$
$$+ \exp[-i\pi(3h+k+3l)/2] + \exp[-i\pi(h+3k+3l)/2]$$
$$+ \exp[-i\pi(3h+3k+3l)/2] + \exp[-i\pi(h+k+3l)/2]$$
$$+ \exp[-i\pi(h+3k+l)/2] + \exp[-i\pi(3h+k+l)/2]\}. \tag{2.27}$$

With the known structure amplitude (2.27), we can easily obtain the structural factor F^2 of the cubic (space group $Fm\overline{3}m$) phase of the PbS film

$$F^2 = \{f_{Pb}[1 + \cos\pi(h+k) + \cos\pi(h+l) + \cos\pi(k+l)]$$
$$+ yf_S[\cos\pi(h+k+l) + \cos\pi h + \cos\pi k + \cos\pi l]$$
$$+ \frac{1-y}{2}f_S\left[\cos\frac{\pi}{2}(h+k+l) + \cos\frac{\pi}{2}(3h+3k+l) + \cos\frac{\pi}{2}(3h+k+3l)\right.$$
$$+ \cos\frac{\pi}{2}(h+3k+3) + \cos\frac{\pi}{2}(3h+3k+3l) + \cos\frac{\pi}{2}(3h+k+l)$$
$$+ \left.\cos\frac{\pi}{2}(h+3k+l) + \cos\frac{\pi}{2}(h+k+3l)\right]\Big\}^2 \tag{2.28}$$
$$+ \left\{\frac{1-y}{2}f_S\left[\sin\frac{\pi}{2}(h+k+l) + \sin\frac{\pi}{2}(3h+3k+l) + \sin\frac{\pi}{2}(3h+k+3l)\right.\right.$$
$$+ \sin\frac{\pi}{2}(h+3k+3l) + \sin\frac{\pi}{2}(3h+3k+3l) + \sin\frac{\pi}{2}(3h+k+l)$$
$$+ \left.\left.\sin\frac{\pi}{2}(h+3k+3l) + \sin\frac{\pi}{2}(h+k+3l)\right]\right\}^2$$

Figure 2.29 shows, as an example, the experimental XRD pattern of the PbS-1 film measured in situ at temperatures of 293 K, and the XRD pattern calculated in the approximation of the new cubic structure of the film, in which sulfur atoms are placed not only in the octahedral positions 4(b), but also in the tetrahedral positions 8(c).

Minimization of the experimental XRD patterns of PbS films demonstrated that at any temperature from 293 to 423 K, sites 4(b) and 8(c) are occupied by sulfur atoms with the probabilities of ~ 0.84 and ~ 0.08 (Table 2.5).

The reliability factor R_I for all XRD patterns did not exceed 0.017. Thus, the crystal lattice of PbS films is characterized by a latent nonstoichiometric distribution of sulfur atoms among sites 4(b) and 8(c). The new cubic structure (space group $Fm\overline{3}m$) found in the lead sulfide nanofilms corresponds to the $D0_3$ structural type with partially disordered (random) distribution of sulfur atoms over the sites of two types (see Fig. 2.28b). Considering the crystal structure of nanofilms and the occupancies of sites 4(b) and 8(c), the formula of the lead sulfide can be written as $PbS_{0.84}^{4(b)}S_{0.16}^{8(c)} \equiv Pb(S_{0.84}\square_{0.16})^{4(b)}(S_{0.16}\square_{0.84})^{8(c)}$, where $\square$ is a structural vacancy.

Fig. 2.29 Experimental (*crosses*) and calculated (*solid line*) XRD patterns of the as-prepared nanostructured PbS-1 film with a cubic (space group $Fm\bar{3}m$) DO_3-type crystal structure (each third experimental point is shown only) [100]. The *bottom curve* shows the difference ($I_{obs} - I_{calc}$) between the experimental and calculated XRD patterns

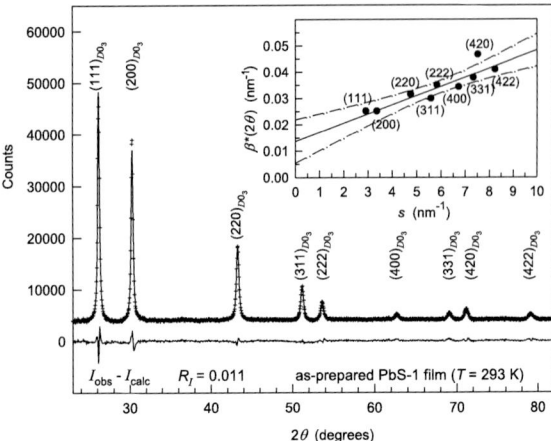

Table 2.5 Cubic (space group $Fm\bar{3}m$) DO_3 type structure of nanostructured PbS films [95, 100]: $a = 0.59395$ nm

Atom	Position and multiplicity	Atomic coordinates			Occupancy
		x/a	y/a	z/a	
Pb	$4(a)$	0	0	0	1
S 1	$4(b)$	0.5	0.5	0.5	0.84
S 2	$8(c)$	0.25	0.25	0.25	0.08

Previously, there was no experimental evidence for the occupation of tetrahedral sites in cubic sulfides by sulfur atoms, and the "latent" nonstoichiometry of lead sulfide was found for the first time in the studies considered above [95, 100, 129].

In the cubic structure (space group $Fm\bar{3}m$) of PbS, the radii of the octahedral and tetrahedral interstices are equal $r_{octa} = a/2 - r_{Pb^{2+}}$ and $r_{tetra} = a\sqrt{3}/4 - r_{Pb^{2+}}$, respectively.

The lattice constant a of the PbS films is 0.5940 nm, and the radii of the Pb^{2+} and S^{2-} ions are 0.121 and 0.184 nm, respectively. Therefore, the radii of the octahedral and tetrahedral interstices are ~ 0.176 and ~ 0.136 nm, respectively. Since $r_{S^{2-}} > r_{tetra}$, the location of the S^{2-} anion in the tetrahedral interstice would result in considerable displacements of the nearest lead atoms, which was actually observed in experiments: the microstrains in the films were 0.20–0.30% at any annealing temperatures up to 423 K.

The occupation of sites $4(b)$ and $8(c)$ by sulfur atoms with the probabilities of ~ 0.84 and ~ 0.08, respectively, implies that out of every twelve octahedral interstices, approximately ten are occupied by sulfur atoms and two are vacant.

The cubic structure with space group $Fm\bar{3}m$ has twice more tetrahedral interstices than octahedral ones. Therefore, there are 24 tetrahedral interstices per 12 octahedral ones; of these, two are occupied by sulfur atoms and the other are vacant. The absence of superstructural reflections means that the arrangement of sulfur atoms in each type of sites is disordered (random). This is illustrated in Fig. 2.30, which shows the model of the cubic structure of a PbS film in comparison with the $B1$ structure. The arrangement of some sulfur atoms in tetrahedral interstices entails some increase in the lattice constant in comparison with the lattice constant of lead sulfide with the $B1$ structure and gives rise to microstrains. As can be seen in Fig. 2.30, when a sulfur atom occupies a tetrahedral interstice, at least one neighbouring octahedral interstice is vacant, i.e, the crystal lattice of the cubic phase in question has some short-range order.

Sadovnikov and Rempel [134] studied the distribution of the sulfur atoms and vacancies in the square (001) and hexagonal (111) planes formed by sites 4(b) and 8 (c) in the nonmetal sublattice of the nanocrystalline PbS films with cubic $D0_3$ structure and latent nonstoichiometry.

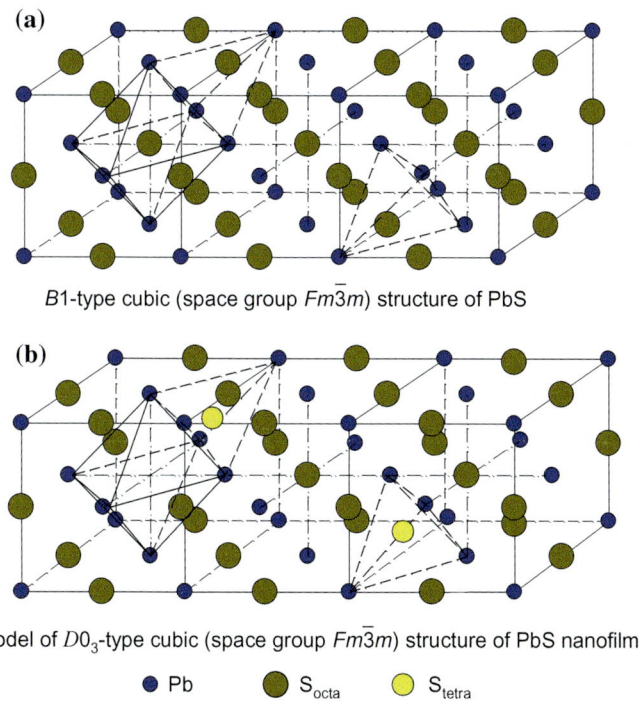

(a)

$B1$-type cubic (space group $Fm\bar{3}m$) structure of PbS

(b)

Model of $D0_3$-type cubic (space group $Fm\bar{3}m$) structure of PbS nanofilm

● Pb   ● S_{octa} ○ S_{tetra}

Fig. 2.30 **a** Cubic (space group $Fm\bar{3}m$) $B1$ type structure of lead sulfide and **b** model of the cubic (space group $Fm\bar{3}m$) $D0_3$ type structure of a PbS nanofilm with random distribution of sulfur atoms in the octahedral and tetrahedral interstices [95, 100]

2.2.3 Correlations of Sulfur Atoms S in PbS Nanostructured Films

In studies [95, 96, 100, 129], it was demonstrated that PbS nanostructured films with nanoparticle sizes less than 80 nm have a cubic (space group $Fm\overline{3}m$) structure with a disordered arrangement of sulfur atoms S not only in octahedral positions 4 (b) but also in tetrahedral positions 8(c) (see Sect. 2.2.2). The newly revealed structure of PbS nanostructured films belongs to the DO_3 structure type. In this structure, the nonmetal lattice is separated into two nonmetal sublattices, one of which is formed by sites occupying the 4(b) crystallographic positions and the other sublattice is formed by sites occupying the 8(c) crystallographic positions.

In the case of chemical deposition of PbS films from aqueous solutions, films with different crystallographic orientations can be formed on the substrate surface. For the cubic DO_3 structure, the formation of PbS nanofilms corresponding to the (001) and (111) planes or equivalent to them is most probable. In the (001) planes, the 8(c) sites of the nonmetal sublattice form a planar square (plane group $p4mm$) Kepler net of the 4^4 type (Fig. 2.31a), and, in the (111) plane, the 4(b) sites of the nonmetal sublattice form a planar hexagonal (plane group $p6mm$) net of the 3^6 type (Fig. 2.31b) [134]. In the revealed structure of PbS nanofilms [95, 100], the S atoms occupy the octahedral sites 4(b) and tetrahedral sites 8(c) of the nonmetal sublattice with the probabilities $P_{\text{S-octa}} = 0.84$ and $P_{\text{S-tetra}} = 0.08$, respectively. This means that the nonmetal lattice of the PbS nanofilm with the DO_3 structure is characterized by a latent nonstoichiometry [129]. Therefore, in the (001) and (111) nonmetal planes of the lead sulfide nanofilm, a number of sites are occupied by S atoms and the other sites are vacant (Fig. 2.31). In planar square (Fig. 2.31c) and hexagonal (Fig. 2.31d) lattices, S atoms and vacant sites $\square$ form an $S_y\square_{1-y}$ substitutional solid solution. Nonequivalent configurations of figures (site, bond, and square cluster) used for describing the $S_y\square_{1-y}$ solid solution with planar square lattice are shown in Fig. 2.31e. Nonequivalent configurations of figures (site, bond, and triangular cluster) used for describing the $S_y\square_{1-y}$ solid solution with planar hexagonal lattice are shown in Fig. 2.31f. With allowance made for the results obtained in works [95, 96, 100, 129], the difference between the properties of PbS nanostructured films and coarse-grained (bulk) lead sulfide can be caused, with an equal probability, by the size effects and also the change in the structure and specific features in the mutual arrangement of atoms S and vacancies $\square$ in defect nonmetal planes (001) and (111).

The distribution of atoms of different types (or atoms and structural vacancies) in a crystal is characterized by the long- and short-range orders [135]. Authors [95, 96, 100, 129] did not reveal superstructure reflections in the XRD patterns of lead sulfide nanofilms with the DO_3 structure, which indicates the absence of ordering in

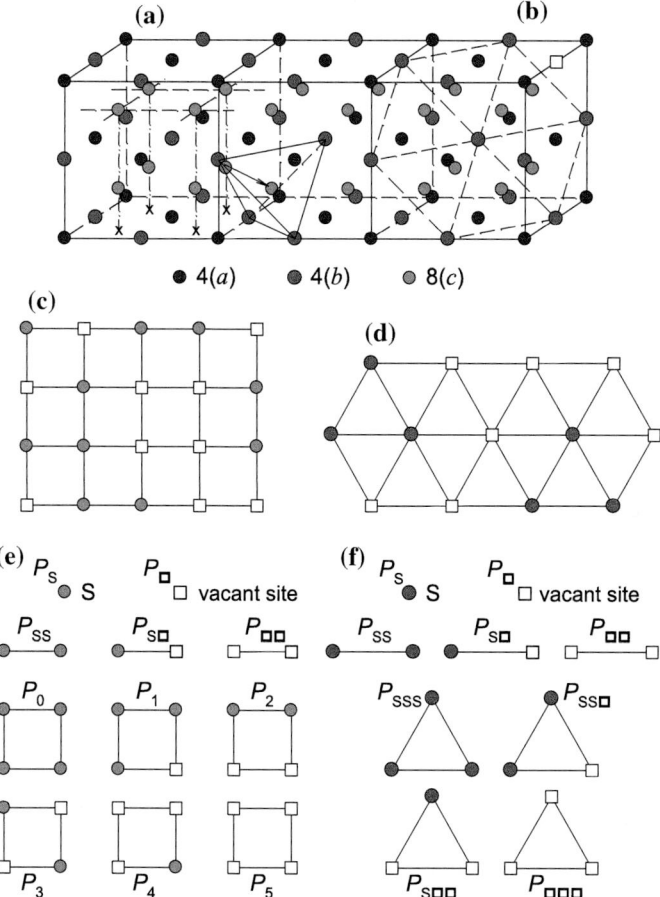

Fig. 2.31 Positions of nonmetal **a** square (001) and **b** hexagonal (111) planes in the unit cell of the cubic (space group $Fm\bar{3}m$) DO_3 structure of a lead sulfide nanostructured film, and **c** square (plane group $p4mm$) and **d** hexagonal (plane group $p6mm$) lattices simulating these planes [134]. The (001) and (111) nonmetal planes pass through the $8(c)$ and $4(b)$ sites, respectively. In the square and hexagonal lattices, sulfur atoms S and vacant sites □ form an $S_y\square_{1-y}$ substitutional solid solution. Nonequivalent configurations (**e** and **f**) of figures (site, bond, square and triangular clusters) used for describing the $S_y\square_{1-y}$ solid solution with planar square and hexagonal lattices are shown at the bottom

the nonmetal lattice. However, the possible presence of correlations (short-range order) in the mutual arrangement of sulfur atoms S and vacancies □ was not ruled out. Therefore, the structure and properties of lead sulfide nanofilms should be described taking into account the short-range order. This is especially important because the relative number of vacancies in each nonmetal sublattice is large and sufficient for the appearance of correlations between sulfur atoms S and vacancies □.

The radius r_{tetra} of the tetrahedral interstitial site $8(c)$ is smaller than the radius $r_{S^{2-}}$ of the S^{2-} ion. Since $r_{S^{2-}} > r_{tetra}$, the transition of the S atom from the $4(b)$ site to the $8(c)$ site leads to local displacements of the nearest Pb atoms from the 8 (c) site occupied by the S atom. This results in the appearance of microstrains in the PbS nanofilm: their value at room temperature is approximately equal to 0.20% [95, 96, 100, 129]. In the PbS sulfide with the DO_3 structure, any $8(c)$ site is a center of the tetrahedral interstitial site formed by Pb atoms occupying nearest $4(a)$ sites, on the one hand, and is a center of tetrahedral interstitial site of the four nearest 4 (b) sites of the nonmetal lattice, on the other hand (Fig. 2.32).

Let us evaluate the possibility of transferring the S atom from the $4(b)$ site to the $8(c)$ site. This transition leads to the formation of a pair consisting of a vacancy in the (111) nonmetal plane and a sulfur atom in the (001) nonmetal plane (see Fig. 2.32), i.e., to the appearance of a vacancy in one of the four nearest sites $4(b)$. This additionally increases the size of the tetrahedral interstitial site. However, the (111) nonmetal planes already contain vacant sites and their relative concentration is $(1 - P_{S-octa}) \approx 0.16$. Therefore, in the environment of the S atom located in the (001) nonmetal plane and occupying the $8(c)$ tetrahedral interstitial site, the tetrahedron will be formed in two adjacent planes (111). Two sites of this tetrahedron are occupied by S atoms and two sites are vacant.

In the case of a disordered arrangement of sulfur atoms and vacancies over the 4 (b) sites, the probability of the formation of a tetrahedron formed by four neighboring sites $4(b)$ has the form

$$\lambda_i P_i \equiv C_4^{n_i} P_i = C_4^{n_i} P_{S-octa}^{4-n_i} (1 - P_{S-octa})^{n_i}, \qquad (2.29)$$

where n_i is the number of vacancies in the i-configuration of the tetrahedron, λ_i is the multiplicity of the i-configuration (the number of configurations coinciding with each other after application of symmetry operations), and P_{S-octa} is the probability

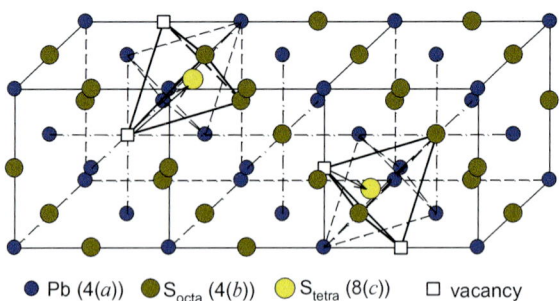

● Pb $(4(a))$ ● S_{octa} $(4(b))$ ○ S_{tetra} $(8(c))$ □ vacancy

Fig. 2.32 Model of the cubic (space group $Fm\bar{3}m$) DO_3 structure of a PbS nanofilm with a statistical distribution of sulfur atoms S with probabilities of 0.84 and 0.08 over the octahedral and tetrahedral sites $4(b)$ and $8(c)$. *Arrows* indicate possible transitions of S atoms from the $4(b)$ sites to the $8(c)$ sites. *Solid lines* represent tetrahedral clusters formed by the $4(b)$ sites of the nonmetal lattice. *Dashed lines* show tetrahedral clusters formed by the $4(a)$ sites of the metal lattice, i.e., by Pb atoms

of the occupation of the 4(b) octahedral positions by S atoms (in the case under consideration, $P_{\text{S-octa}} = 0.84$). The total number of nonequivalent configurations is equal to five: the tetrahedron P_{SSSS} with all sites occupied by S atoms; tetrahedra $P_{\text{SSS}\square}$, $P_{\text{SS}\square\square}$, and $P_{\text{S}\square\square\square}$ containing one, two, and three vacant sites, respectively; and tetrahedron $P_{\square\square\square\square}$, in which all sites are vacant.

The occupation of the octahedral and tetrahedral positions 4(b) and 8(c) with the probabilities $P_{\text{S-octa}} = 0.84$ and $P_{\text{S-tetra}} = 0.08$ means that, among every twelve octahedral interstitial sites, ten sites are occupied by S atoms and two sites are vacant. With due regard for the probability $P_{\text{S-octa}}$ and relationship (2.29), the probability of the formation of a tetrahedral cluster containing two vacant sites is $P_{\text{SS}\square\square} = C_4^2 P_{\text{S-octa}}^2 (1 - P_{\text{S-octa}})^2 = 0.108$. Since $P_{\text{SS}\square\square} > P_{\text{S-tetra}}$, the transition of sulfur atoms to the 8(c) tetrahedral interstitial sites with occupancies of ~ 0.08 is mathematically quite probable.

The short-range order describes the distribution of atoms around some lattice site and, specifically, determines existing two-particle atomic or atom–vacancy correlations in a particular coordination sphere [135]. The short-range order is characterized by the short-range order parameters α_j and the correlation parameters ε_j in the jth coordination sphere.

Up to now, it has not been revealed how many coordination spheres are covered by the short-range order appeared in the first coordination sphere. In study [134], this problem was solved using the computer simulation of the dependence of the correlation parameters in the jth coordination spheres ($j \geq 2$) on the correlation parameter ε_1 of the first coordination sphere. The object of the simulation was a $S_y\square_{1-y}$ substitutional solid solution with atoms located in sites of defect planar square and hexagonal lattices: a number of sites are occupied by sulfur atoms, and the other sites are vacant. Sulfur atoms S and vacancies $\square$ are mutually substituted elements of the solid solution.

The pair correlation parameter ε_j is defined as the difference between the probability of the occurrence of a pair bond formed by like elements in the jth coordination sphere of the crystal lattice with a short-range order and the probability of the occurrence of the same bond in a disordered lattice. In an infinite disordered lattice, the probability of the occurrence of pair bonds is governed by the binomial distribution [135]. As a result, we have

$$\varepsilon_{\text{SS}}(R_j) = P_{\text{SS}}^{(j)} - P_{\text{SS}}^{\text{dis}} = P_{\text{SS}}^{(j)} - P_{\text{SS}}^{\text{bin}} = P_{\text{SS}}^{(j)} - y^2, \tag{2.30}$$

$$\varepsilon_{\square\square}(R_j) = P_{\square\square}^{(j)} - P_{\square\square}^{\text{dis}} = P_{\square\square}^{(j)} - P_{\square\square}^{\text{bin}} = P_{\square\square}^{(j)} - (1 - y)^2, \tag{2.31}$$

where $P_{\text{SS}}^{(j)}$ and $P_{\square\square}^{(j)}$ are the probabilities of the occurrence of like pair bonds in the lattice with the short-range order. In the infinite disordered lattice, these probabilities are as follows: $P_{\text{SS}}^{\text{bin}} = y^2$ and $P_{\square\square}^{\text{bin}} = (1 - y)^2$. According to [135], in the presence only of a short-range order, the probabilities of the occurrence of pair bonds are given by the relationships

$$P_{SS}^{(j)} = P_{SS}^{bin} + y(1-y)\alpha_j = y^2 + y(1-y)\alpha_j, \tag{2.32a}$$

$$P_{\square\square}^{(j)} = P_{\square\square}^{bin} + y(1-y)\alpha_j = (1-y)^2 + y(1-y)\alpha_j, \tag{2.32b}$$

$$P_{S\square}^{(j)} = (1-\alpha_j)P_{S\square}^{bin} = (1-\alpha_j)y(1-y), \tag{2.32c}$$

where α_j is the short-range order parameter in the jth coordination sphere.

It follows from relationships (2.30)–(2.32) that, in the absence of a long-range order, the correlations between sulfur atoms (or vacancies) are equal to each other; that is,

$$\varepsilon_{SS}(R_j) \equiv \varepsilon_{\square\square}(R_j) \equiv \varepsilon_j = y(1-y)\alpha_j. \tag{2.33}$$

The parameters of unlike pair correlation of sulfur atoms and vacancies are equal in magnitude to the pair correlation parameters of sulfur atoms (vacancies) but opposite in sign: $\varepsilon_{S\square}(R_j) \equiv \varepsilon_{S\square}(R_j) \equiv -\varepsilon_j$. In the case of a disordered distribution of atoms in an infinite lattice, the correlation parameter is exactly equal to zero. In study [134], all probabilities were calculated for the finite lattice model without long-range order. In this case, the probabilities of the occurrence of bonds in the disordered lattice are close to the binominal probabilities $P_{SS}^{bin} = y^2$ and $P_{\square\square}^{bin} = (1-y)^2$ but are not equal to them due to the limited size of the lattice and the contribution of bonds lying at the boundary of the two-dimensional crystal.

The probabilities of the occurrence of the pair bonds S–S, S–$\square$, and $\square$–$\square$ in the $S_y\square_{1-y}$ solid solution with the atoms located at the sites of the square and hexagonal lattice were calculated for nine coordination spheres. This is enough to judge the influence of the short-range order in the first coordination sphere on the pair correlation parameters in distant coordination spheres and a decrease in the short-range order parameters with an increase in the radius R_j of the coordination sphere. Furthermore, the simulation showed that the correlations in the 9th and 10th coordination spheres decrease to values comparable to the error in the calculations and almost decay.

In the disordered infinite lattice, the correlation parameter ε_j in all coordination spheres is exactly equal to zero. In the case of the short-range order when the nearest coordination sphere of the S atom predominantly contains vacancies $\square$, the correlation parameter ε_1 is negative. For the short-range decomposition, when the environment of the S atom predominantly involves S atoms (or the environment of the vacancy $\square$ predominantly contains vacancies), the correlation is positive, i.e., $\varepsilon_j > 0$.

In order to reveal the dependences between the correlation parameters in different coordination spheres, authors [134] considered two-dimensional square and hexagonal lattices of fixed size with a specified vacancy content. The calculations were performed for 23 × 23 and 32 × 32 square lattices containing 529 and 1024 sites, respectively, and for 33 × 33 hexagonal lattice containing 1089 sites. Some number of sites was filled with atoms, and the other sites were vacant. In the simulation, the boundaries of the solid solution lattice were assumed to be perfectly rigid. As a consequence, the lattice boundaries were fixed in the space and the number of atoms and vacancies in the lattice always remained unchanged.

At the first stage of the simulation of the lattice with a specified size, the computer synthesis of an $S_y\square_{1-y}$ disordered solid solution with a given quantity y, i.e., with the known number of atoms and vacancies, was performed. The composition of the $S_y\square_{1-y}$ solid solution was determined by the number n of sulfur atoms in the crystal lattice containing N sites. In this case, the probability of the occupation of a site in the infinite disordered crystal lattice by an atom is equal to $y = n/N$ and the relative number of vacancies is defined as $(1 - y)$. The composition was varied from 0.1 to 0.9 with a step of 0.1. The synthesis of the disordered solution was carried out after the choice of the lattice size and the solution composition. The number of sulfur atoms necessary for providing the specified composition $S_y\square_{1-y}$ was introduced into the lattice with the use of a random-number generator. As a result of statistical occupation of sites, the atoms and vacancies were distributed in the lattice in a disordered manner. The disordered state of the synthesized solution was initial state for the subsequent simulation.

In the simulation of the solid solution with the hexagonal lattice, there arises a problem associated with the fractional irrational coordinates of sites in the rectangular coordinate system. The fractional atomic coordinates lead to the error in the determination of interatomic distances. In the description of correlations in distant coordination spheres, when the increment of the radius upon changing over from the nth coordination sphere to the $(n + 1)$th coordination sphere is equal to a relatively small value comparable to the error in the atomic arrangement, there arises a probability that the site belonging to the nth coordination sphere appears to be in the $(n + 1)$th coordination sphere. In order to exclude this error, the hexagonal lattice was simulated using a special algorithm for transforming the coordinates of the sites of the hexagonal lattice into the coordinates of a hypothetical square lattice, which was described in work [136].

Then, we carried out correlation walks of atoms over the lattice sites. For this purpose, energy fluctuations were introduced into arbitrarily chosen sites, as a result of which the atom can pass to vacant sites. Only energy fluctuations sufficient for the "jump" of the atom from its site of the crystal lattice into the neighboring vacant site were taken into account.

The virtual annealing was performed after the synthesis of the disordered lattice. An arbitrary lattice site was chosen using the Monte Carlo method. The probability of choice of an occupied or vacant site depends directly on the solid solution composition. The annealing process involved displacements of atoms depending on the initial value of the correlation parameter ε_1 in the first coordination sphere. If the site was occupied by the atom, it diffused to the neighboring vacant site. The direction of the displacement was stochastically chosen using a random-number generator. Therefore, the probabilities of atomic displacements to each side of the simulated lattice were equal to each other.

The calculations showed that there are two radically different processes: the decomposition of the $S_y\square_{1-y}$ solid solution when the pair correlation parameter ε_1 in the first coordination shell is positive ($\varepsilon_1 > 0$) and ordering at $\varepsilon_1 < 0$ (Fig. 2.33).

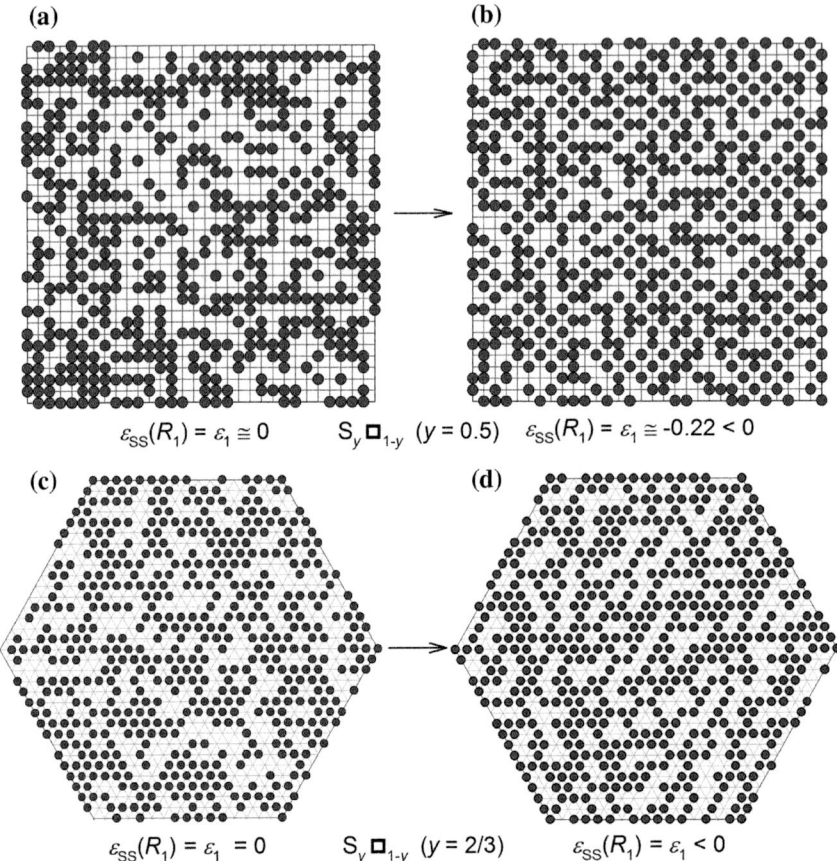

Fig. 2.33 Model distributions of sulfur atoms in **a, b** the square (plane group $p4mm$) lattice of the $S_y\square_{1-y}$ ($y = 0.5$) solid solution and **c, d** the hexagonal (plane group $p6mm$) lattice of the $S_y\square_{1-y}$ ($y = 0.66$) solid solution [134]. **a, c** disordered solid solutions. **b, d** short-range order formed after virtual annealing of the solid solution at a negative pair correlation parameter $\varepsilon_1 < 0$ in the first coordination sphere

Figure 2.34 shows the variation in the pair correlation parameters ε_j as a function of the relative radius R_j/a_{sq} or R_j/a_{hex} of the jth coordination sphere for the $S_y\square_{1-y}$ solid solutions with the square and hexagonal lattices. It can be seen from Fig. 2.34 that the correlations occurring in the first coordination sphere of the $S_y\square_{1-y}$ ($y = 1/2$) solid solution with the square lattice or the $S_y\square_{1-y}$ ($y = 1/3$) solid solution with the hexagonal lattice propagate (gradually decaying) up to the 9th coordination sphere; i.e., they extend over a distance no shorter than $4a_{sq}$ or $4a_{hex}$, where a_{sq} and a_{hex} are the lattice constants of these lattices. For the cubic DO_3 structure of the lead sulfide, these lattice constants are as follows: $a_{sq} = a_{cub}/2$ and $a_{hex} = (\sqrt{2}/2)a_{cub}$, where a_{cub} is the lattice constant of the cubic (space group $Fm\overline{3}m$) lead sulfide. In the case of the short-range order, the correlation parameters ε_j oscillate, change sign, and asymptotically tend to zero in magnitude: at $\varepsilon_{SS}(R_1) \equiv \varepsilon_1 < 0$, we have $|\varepsilon_j| \to 0$ при $j \to \infty$. For the short-range decomposition at $\varepsilon_{SS}(R_1) \equiv \varepsilon_1 > 0$, the correlation parameters ε_j are positive in all coordination spheres, decrease with an increase in the radius of the coordination sphere, and tend to zero.

The simulation of correlations in the mutual arrangement of sulfur atoms with due regard only for the nearest pair interactions has revealed that these interactions lead to correlations in distant coordination spheres of the square planes (001) and hexagonal planes (111). These correlations decay only at the coordination sphere radius equal to $4a_{sq}$ or $4a_{hex}$. Moreover, the results of the simulation [134] have demonstrated that, in the hexagonal planes (111), the pair correlations result in the appearance of three-particle correlations.

2.3   Properties of Nanostructured Lead Sulfide

Lead sulfide is used in fire alarm sensors and flame sensors, IR detectors and heat source detection systems, photodetectors operating in the range from infrared to ultraviolet, converters of solar energy to electrical energy and other thermoelectric transducers and optical switches. These applications are dependent on the optical and thermal properties of lead sulfide. These properties are considered below.

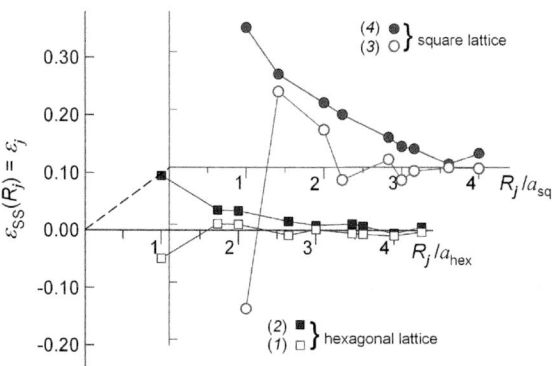

Fig. 2.34 Variation in the pair correlation parameters ε_j as a function of the relative radius of the jth coordination sphere [134]. Parameters ε_j for the $S_{1/3}\square_{2/3}$ ($y = 0.33$) solid solution with the hexagonal lattice for (*1*) the short-range order with $\varepsilon_1 < 0$ and (*2*) the short-range decomposition with $\varepsilon_1 > 0$. Parameters ε_j for the $S_{0.5}\square_{0.5}$ ($y = 0.5$) solid solution with the square lattice for (*3*) the short-range order with $\varepsilon_1 < 0$ and (*4*) the short-range decomposition with $\varepsilon_1 > 0$. Designations a_{hex} and a_{sq} are the lattice constants of the hexagonal and square lattices, respectively

2.3.1 Optical Properties and Band Gap

Of most interest are the properties of nanostructured lead sulfide related to the band gap. There are quite a few publications on determination of the band gap of PbS as isolated nanoparticles and films.

The first measurements of light transmission and reflection and studies of the absorption edge for PbS films were carried out in the period from 1948 to the 1970s. According to the results of these studies, which were surveyed in a number of publications [4, 137–139], at a temperature of 300 K, the direct band gaps in the single crystalline PbS and in polycrystalline PbS films are 0.41 and 0.42 eV, respectively.

Light scattering and absorption by nanoparticles have specific features and differ from those of coarse-grained materials. For this reason, investigations of size effects in the optical and luminescent properties of semiconductors have attracted considerable attention in recent years.

Electronic excitation of semiconductor crystals gives rise to loosely bound electron-hole pair – an exciton. The exciton delocalization area can be much larger than the crystal lattice constant of the semiconductor; therefore, the decrease in the semiconductor crystal size to a value comparable with the exciton radius affects the properties of the crystal [79–82, 140]. For example, as the PbS particle size decreases to 5 nm, the band gap in the electronic spectrum can increase from 0.41 to 1.92 eV [53, 141] or to 2.30 eV [142]. For films based on PbS nanoparticles, the band gap can increase to 2.81 eV [143]. A decrease in the semiconductor particle size should be accompanied by a shift of the absorption band to higher frequency, i.e., so-called blue shift [144, 145]. In the case of lead sulfide films, blue shift of the absorption band is observed at a film thickness less than 150 nm [146–148].

The recombination of charges generated under the action of light (photons with energy exceeding the band gap) gives rise to luminescence. A blue shift following a decrease in the particle size was also found in the luminescence spectra of PbS nanoparticles [115, 143, 149, 150].

Systematic investigations of optical transmission of PbS nanofilms have been reported [97, 100, 151]. The optical transmission spectra were measured for PbS-1, PbS-2, PbS-3 and PbS-4 films and for the PbS-1 film annealed at temperatures of 473 and 523 K. Microstructure of these films are shown in Fig. 2.22.

The particle size distributions (Fig. 2.35) for the PbS films studied are determined from SEM-images with the use of the Altami Studio 2.0.0 software package. Unimodal distribution is observed only in the as-prepared PbS-1 film: the peak of the distribution corresponds to grains, whose size is 70 nm, with about half of all grains having sizes in the range 60–80 nm (Fig. 2.35a). In the size distribution of nanograins in the PbS-2 film, we observe two peaks, one at 50 nm and the other at 80 nm. The average grain size D is 73 nm, and the sizes of $\sim 55\%$ of the grains are in the range 60–90 nm (Fig. 2.35b). In the PbS-3 film, the two size distribution peaks correspond to grains, whose sizes are 40 and 70 nm, the average size D is 59 nm, and more than 43% of grains have a size from 30 to 50 nm (Fig. 2.35c). In

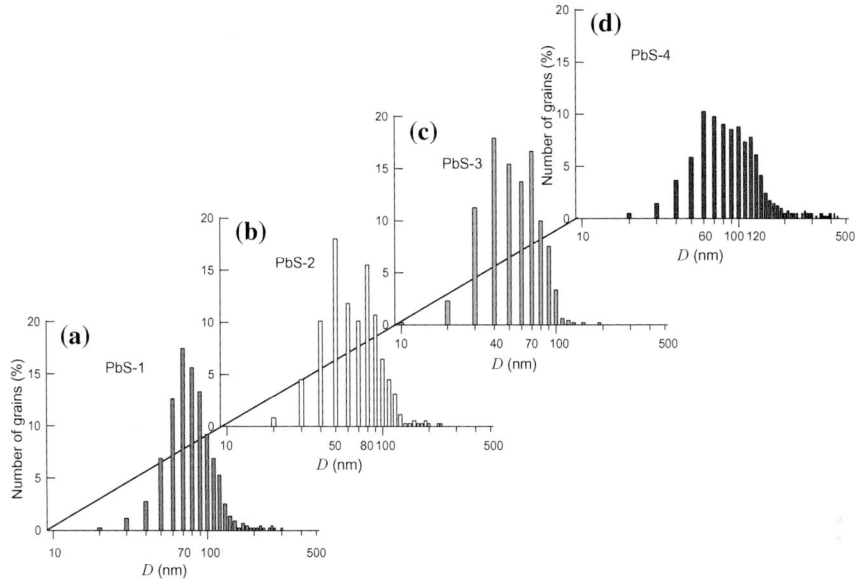

Fig. 2.35 Particle size distributions in the lead sulfide films [97, 100]: **a** In the as-prepared PbS-1 film, about 50% particles are from 60 to 80 nm in size. **b** The PbS-2 film has a bimodal size distribution of nanoparticles with peaks at 50 and 80 nm. **c** In the PbS-3 film, two peaks of the size distribution correspond to particles 40 and 70 nm in size. **d** In the PbS-4 film, a broad peak in the size distribution is observed at 60 nm, the size of about 50% of particles are in the range defined by the inequality 60 nm $< D \leq$ 120 nm. Reprinted from [100] with permission from Elsevier

the microstructure of the PbS-4 film, we observe a broad 60 nm size distribution peak, $\sim 20\%$ of grains are $\leq$ 60 nm in size, about 50% of grains have a size in the range from 60 to 120 nm, and the average grain size D is equal 110 nm (Fig. 2.35d).

For determining the band gap, most useful is the part of the optical spectra in which the transmittance changes noticeably as a function of wavelength λ, i.e., the region from 700–800 to 2500–600 nm (Fig. 2.36) corresponding to the pho-ton energies from ~ 1.6 to ~ 0.5 eV. Analysis of the optical spectra at $\lambda > 600$ nm allows one to exclude errors that can appear in determining E_g due to the absorption band edge of the substrate at $\lambda \approx 320$ nm. Studies of the microstructure (see Fig. 2.22) show that the PbS-1 film covers no more than 80–85% of the substrate surface since PbS-4 film covers the substrate completely. As a result, the trans-mittance $\tilde{T}$ of the PbS-1 film at $\lambda > 800$ nm is nearly twice as high as the trans-mittance of the PbS-4 film (Fig. 2.36). The highest transmittance $\tilde{T}$ is inherent to the PbS-2 and PbS-3 films (Fig. 2.36). It should be noted that the transmission spectra of the PbS-2 and PbS-3 films exhibit several evident inflection points (see Fig. 2.36), in contrast to the smoother transmission spectra of the PbS-1and PbS-4 films.

Fig. 2.36 Transmission
spectra $\tilde{T}(\lambda)$ (*top*) of the
PbS-1 films as-prepared and
annealed at 473 and 573 K,
and also (*bottom*) of PbS-2,
PbS-3, and PbS-4 films [97,
100]. The thickness of PbS-1,
PbS-2, PbS-3, and PbS-4
films is equal to 120 ± 20,
200 ± 20, 300 ± 20 and
400 ± 20 nm, respectively.
Reprinted from [100] with
permission from Elsevier

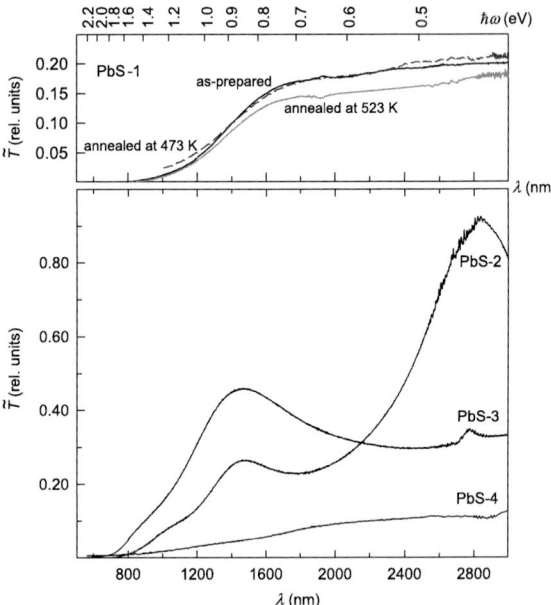

In order to quantitatively estimate E_g, absorption spectra rather than transmission
spectra are considered. The absorption coefficient σ is the absorbance for a 1 cm
thick material layer, $\sigma = (-\lg \tilde{T})/s$, where $\tilde{T} = I/I_0$ is transmittance in relative
units, s is the material thickness in cm (in some publications, the coefficient σ is
designated by α).

The spectral dependence of the absorption coefficient $\sigma(\omega)$ on the semiconductor
band gap E_g (or the Tauc function) has the form [139, 152–154]

$$[\hbar\omega\sigma(\omega)]^{1/n} = A_n(\hbar\omega - E_g),\qquad(2.34)$$

where $\omega = 2\pi c/\lambda$ is the incident radiation frequency, $\hbar\omega = 2\pi\hbar c/\lambda = hc/\lambda$ is the
photon energy, $A_n = const$ is a coefficient independent of frequency ω but
depending on the type of transition. The magnitude n in an exponent is determined
by the transition nature ($n = 1/2$ for the direct allowed transition; $n = 3/2$ for the
direct forbidden transition; $n = 2$ for the indirect allowed transition and $n = 3$ for
the indirect forbidden transition. When reflection spectra are measured, the $\sigma(\omega)$
value in relation (2.34) is replaced by the reflection function $F(R_\infty) = (1 - R)^2/
(2R)$ defined via the reflection coefficient R.

For the direct allowed transitions with $n = 1/2$, relation (2.34) has the form

$$[\sigma(\omega)\hbar\omega]^2 = A_d^2(\hbar\omega - E_g) \equiv A(\hbar\omega - E_g),\qquad(2.35)$$

where $A_d = 2e^2(2\mu_{ex})^{3/2}|P_{if}|^2/(\tilde{n}m_0^2\,c\hbar^2)$, $\mu_{ex} = m_e m_h/(m_e + m_h)$ is the reduced exciton mass; m_e and m_h are the electron and hole effective masses; P_{if} is the matrix element of the $i \rightarrow f$ transition between the ith and fth energy levels; $\tilde{n}$ is the refractive index; and e and m_0 are the charge and the free electron rest mass, respectively. This equation is more convenient for the quantitative treatment of experimental optical absorption data and band gap determination. In an ideal case, $\sigma(\omega) \geq 0$ only when $\hbar\omega \geq E_g$ and the experimental results in the "$[\sigma(\omega)\hbar\omega]^2 \leftrightarrow \hbar\omega$" coordinates are described by a linear plot. In a real experiment, since the absorption band is diffuse, the dependence $[\sigma(\omega)\hbar\omega]^2 = f(\hbar\omega)$ is non-linear near the absorption edge; therefore, the band gap E_g is determined as the intercept on the $\hbar\omega$ axis tangential to the linear part of the experimental absorption curve.

Relation (2.34) was derived without taking account of the electron—hole interaction, which may be significant if the difference $(\hbar\omega - E_g)$ is not too high. The spectral dependence of optical absorption derived taking account of the electron-hole interaction [155] has the form

$$\sigma(\omega) = B\{1 - \exp[-C(\hbar\omega - E_g)^{-1/2}]\}^{-1}, \qquad (2.36)$$

where B and C are constants. In lead sulfide, the electron-hole interaction is markedly shielded because of the high dielectric constant ε_d (at 300 K, the ε_{PbS} value varies from 17.2 to 17.9 [138]; therefore, the optical absorption of PbS at $n = 1/2$ is better described by relation (2.30).

Usually, the films are not monodisperse. Considering this fact, Sadovnikov et al. [97] proposed a model for the absorption spectra of polydisperse films. The absorption spectrum of monodisperse nanoparticles is a smooth function of the wavelength λ or energy $E = hc/\lambda$. In the case of polydisperse films, the absorption spectra represent superpositions of the spectra corresponding to nanoparticles of different size,

$$\sigma(\omega) = \sum_i a_i\sigma_i(\omega) = \sum_i a_i A(\hbar\omega - E_{g_i})^{1/2}/\hbar\omega \equiv \sum_i a_i A(E - E_{g_i})^{1/2}/E$$

$$(2.37)$$

where c_i and E_{g_i} are the relative content of nanoparticles of a definite size in the film and the band gap corresponding to these nanoparticles.

For illustration, Fig. 2.37a shows the model absorption spectrum calculated as a superposition of the absorption spectra of four groups of nanoparticles with the band gap $E_{g_i} = 0.42, 0.8, 1.0$, and 1.5 eV, and the relative content $c_i = 30, 30, 15$, and 25%. In Fig. 2.37b, the same model absorption spectrum is shown in the $[\sigma(\omega)\hbar\omega]^2 \leftrightarrow \hbar\omega$ coordinates. In this spectrum, we can see four clearly pronounced regions, which can be used to estimate the band gap corresponding to nanoparticles of different sizes. The observable band gap $E_{g\Sigma}$ can be found by minimizing the

Fig. 2.37 **a** *Solid line* shows the model absorption spectrum calculated as a superposition of (*dashed lines*) the absorption spectra of four groups of nanoparticles (formula (2.37)) with the band gap = 0.42, 0.8, 1.0, and 1.5 eV, and the relative content c_i = 30, 30, 15, and 25%, respectively [97]. **b** The same absorption spectrum plotted in the energy scale: contributions to the absorption in regions *1–4* are made by particles with E_{g_i} = 0.42, 0.8, 1.0, and 1.5 eV (region *1*), 0.42, 0.8, and 1.0 eV (region *2*), 0.42 and 0.8 eV (region *3*), and 0.42 eV (region *4*)

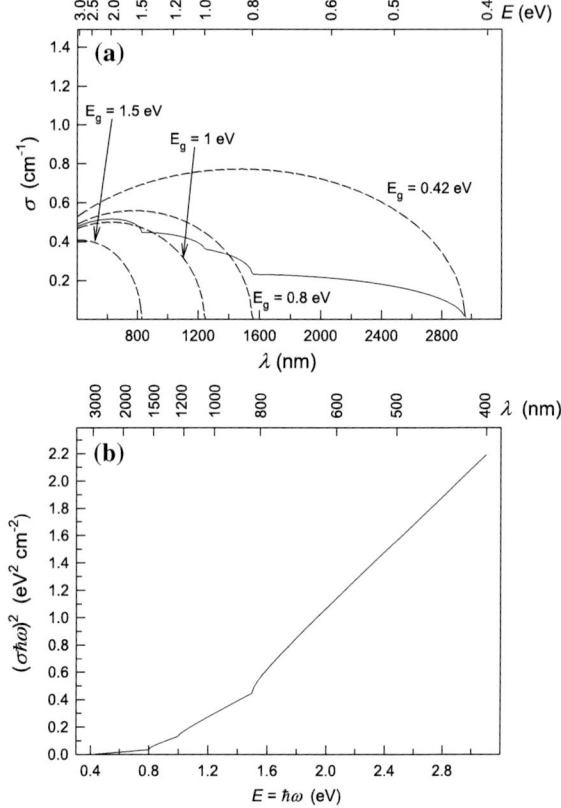

quantity of the experimental data with formula (2.35) and determining the portion intercepted on the energy axis $\hbar\omega$ by any linear portion of the curve $[\sigma(\omega)\hbar\omega]^2$ extrapolated to the axis. In polydisperse films, any section of the dependence $[\sigma(\omega)\hbar\omega]^2 \equiv [\sigma E]^2$ is a superposition of $[\sigma_i(\omega)\hbar\omega]^2 \equiv [\sigma_i E]$ values; therefore, the observed band gap $E_{g\Sigma}$ is an additive function of the energies E_{g_i} corresponding to nanoparticles of different size. Considering the identity $[\sigma E]^2 \equiv A^2(E-E_{g\Sigma})$, it was approximately shown [97] that

$$E_{g\Sigma} \sim \sum_i c_i^2 E_{g_i} + 2\sum_{i\neq j} c_i c_j (E_{g_i} E_{g_j})^{1/2} + |E_{g_i} - E_{g_j}|/3. \qquad (2.38)$$

For a monodisperse film, i.e., when $c_1 = 1$ and $c_{i\neq 1} = 0$, relation (2.38) is transformed to the obvious boundary conditions $E_{g\Sigma} \equiv E_{g1}$.

An important feature of continuous films is the existence of developed grain boundaries. This results in diffuse positions of the conduction and valence bands in the electronic structures of particular grains and, as a consequence, in some uncertainty of the band gap and in additional smearing of the optical spectra of the films. This effect is of the same order as the presence of different size nanoparticles in the film, i.e., the size inhomogeneity of the film, as indicated by Mittleman et al. [156].

Thus, the intricate pattern of the absorption spectra of semiconductor films is a consequence of polydispersity.

The absorption spectra of lead sulfide nanofilms plotted in the $[\sigma(\omega)\hbar\omega]^2 \leftrightarrow \hbar\omega$ coordinates are shown in Fig. 2.38. Two (PbS-1 and PbS-4) or three (PbS-2 and PbS-3) linear segments can be distinguished in the spectra. More detailed analysis of the plots for $[\sigma(\omega)\hbar\omega]^2$ dependences revealed a non-monotonic variation of the absorption of PbS-2 and PbS-3 films (see insets in Fig. 2.38) in the energy region $\hbar\omega = 0.5$–1.1 eV; no specific behaviour of absorption was found in this region for PbS-1 and PbS-4 films.

The low-energy absorption edge of all spectra is smeared. The "tails" observed near the absorption edge, i.e, in the region $\hbar\omega < E_\mathrm{g} = 0.41$ eV, are mainly caused by the dispersion (deviation from the mean value) of the size of the film

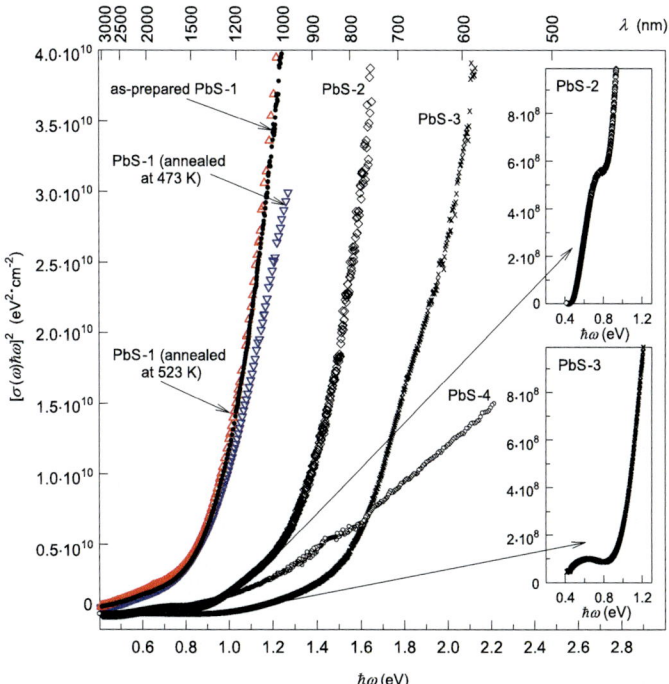

Fig. 2.38 Absorption spectra of the PbS-1 films (*filled circle*) as-prepared and annealed at (*inverted open triangle*) 473 and (*open triangle*) 573 K, and also of (*open diamond*) PbS-2, (*multiplication sign*) PbS-3 (*open circle*), and PbS-4 films on the energy scale [97, 100]. For simplicity, each 5th experimental point is shown only). Insets show the nonmonotonic behavior of an absorption coefficient of the PbS-2 and PbS-3 films in the energy interval from 0.4 to 1.1 eV. Reprinted from [100] with permission from Elsevier

nanoparticles, i.e, inhomogeneous broadening [156]. Other possible causes for absorption edge smearing are structure defects of the semiconductor and/or indirect interband transitions. The portions of the absorption curves $\sigma(\omega)$ at $\hbar\omega < E_g$ are described by the Urbach rule [157]: $\sigma(\omega) \sim \exp[\beta(\hbar\omega - E_g)]$.

The quantitative minimization of the experimental data with respect to absorption coefficient $\sigma(\omega)$ of the PbS-1, PbS-2, PbS-3 and PbS-4 films by function (2.35) made it possible to find E_g for these films. The calculation showed that the band gap of the synthesized PbS-1 film is 0.92 eV, while the band gaps of the same film after annealing at 473 and 523 K are 0.87 and 0.89 eV, respectively. The average nanoparticle size in the annealed PbS-1 films is 80 nm. For the synthesized and annealed PbS-1 films, the spectral regions with $E < 0.7$ eV are characterized by $E_g \approx 0.42$ eV.

The $[\sigma(\omega)\hbar\omega]^2 = f(\hbar\omega)$ plots for the PbS-2 and PbS-3 films (see Fig. 2.38) have clear-cut singular points, and between these points, linear segments can be distinguished. For the PbS-2 film, these segments are characterized by E_g values 0.43, 0.84, and 1.30 eV. For the PbS-3 film, these segments are characterized by E_g values 0.98 and 1.50 eV, and for the PbS-4 film E_g values equal to 0.41 and 1.05 eV.

A correlation of the E_g values with the particle size distribution in the PbS films (see Fig. 2.35) demonstrated that all films contain up to 50% particles with the size from 100–120 to 300 nm and more. Apparently, the band gap of these large particles is equal to $E_g = 0.41$ eV of the bulk lead sulfide.

A considerable contribution to the absorption spectra is made by the groups of nanoparticles corresponding to the maxima of size distributions and the adjacent smaller nanoparticles [97, 100]. In the PbS-1 film, the distribution maximum corresponds to 70 nm particles, and 40% of all particles in the film have a size from 10 to 70 nm (Fig. 2.35a). The average nanoparticle size of the PbS-1 film in this range is 60 nm (Table 2.6). This average size can be correlated to the E_g value, which was found to be 0.92 eV. In the annealed PbS-1 films, the nanoparticles have an average size of 80 nm and are characterized by $E_g = 0.87$–0.89 eV. Size distribution maxima corresponding to nanoparticles of 40 and 70 nm average sizes are inherent in the PbS-2 film (see Fig. 2.35b). This nanoparticle size can be correlated to the E_g values of 1.30 and 0.84 eV, respectively. The size distribution of particles of the PbS-3 film has two maxima, the average particle sizes at these maxima being 35 and 60 nm (see Fig. 2.35c). The nanoparticles of these sizes are characterized by E_g values found to be 1.50 and 0.98 eV, respectively. For the PbS-4 film, the weakly pronounced size distribution maximum corresponds to an average particle size of 50 nm (see Fig. 2.35d) for which $E_g = 1.05$ eV (Table 2.6).

The established correlation between the nanoparticle size in the films and the band gap (Table 2.6) can serve to elucidate the influence of the PbS nanoparticle size on the E_g value. The E_g value tends to increase as the nanoparticle size decreases (Fig. 2.39). The band gap of coarse-grained lead sulfide E_b is 0.41–0.42 eV [4, 138], and the E_g of the considered films [97, 100] increases from $\sim 0.8 \pm 0.1$ to 1.5 ± 0.1 eV as the average nanoparticle size decreases from 80 to

Table 2.6 Thickness s, nanoparticle size D, and the band gap E_g of nanostructured PbS films [97, 100]

Film	Thickness $s \pm 20$ (nm)	Nanoparticle size			Band gap $E_g \pm 0.1$ (eV)
		XRD D (nm)	SEM $D \pm 10$ (nm)[a]	D (nm)[b]	
PbS-1 (as-prepared)	120	70 ± 10	70	60 (10–70)	0.92
PbS-1 (annealed at 473 K)	120	80 ± 10	–	–	0.87
PbS-1 (annealed at 523 K)	120	80 ± 10	–	–	0.89
PbS-2	200	75 ± 15	50	40 (10–50)	1.30
			80	70 (60–80)	0.84
PbS-3	300	65 ± 15	40	35 (10–40)	1.50
			70	60 (50–70)	0.98
PbS-4	400	90 ± 15	60	50 (10–60)	1.05

[a]Nanoparticle sizes correspond to the maxima of the size distribution (see Fig. 2.35)

[b]Average size of nanoparticles in the range specified in brackets

35 nm. In view of this, one can consider that a blue shift of the optical absorption band was observed for PbS nanofilms in studies [97, 100]. The value $E_g = 2.0 \pm 0.1$ eV, similar to the cited results [97, 100], was found in another study [158] for PbS films with an average particle size of 40 nm deposited on a glass substrate.

According to the results of Brus [144], for monodisperse semiconductor nanoparticles, the dependence of the exciton energy E_g on the particle radius $R = D/2$ is described as

$$E_g = E_b - \mu_{ex} e^4 / 2n^2 \hbar^2 + (n^2 \pi^2 \hbar^2 / 2\mu_{ex} R^2) - (1.78 e^2 / \varepsilon_d R), \qquad (2.39)$$

where $n = 1, 2, 3, \dots$. In relation (2.39), the first and second terms are the band gap of the bulk crystal and the electron and hole binding energy (effective Rydberg energy). For coarse-grained lead sulfide, $E_b = 0.41$–0.42 eV. The third term $n^2 \pi^2 \hbar^2/(2\mu_{ex} R^2)$ is the exciton kinetic energy, and the last term $1.78e^2/\varepsilon_d R$ takes into account the Coulomb interaction between an electron and a hole. Thus, interband absorption should give rise to a series of discrete lines for different n values and the absorption threshold is determined by transition to the ground state with $n = 1$. The sum of the first, third and fourth terms of (2.39) is the effective band gap; for the ground state ($n = 1$), it is expressed as

$$E_g(R) = E_b + (\pi^2 \hbar^2 / 2\mu_{ex} R^2) - (1.78 e^2 / \varepsilon_d R). \qquad (2.40)$$

For lead sulfide with the particle size $D \geq 10$ nm, the correction for Coulomb interaction is lower than ~ 0.029 eV ($\sim 7\%$ of the value $E_b = 0.41$–0.42 eV typical

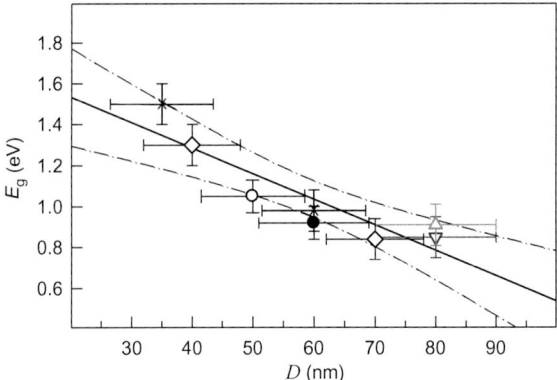

Fig. 2.39 Variation in the observed band gap E_g depending on the nanoparticle size D in the nanostructured PbS films [97, 100]. *Filled circle* as-prepared PbS-1 (*inverted open triangle*), PbS-1 annealed at 473 K, *Open triangle* PbS-1 annealed at 523 K, *Open diamond* PbS-2, (*multiplication sign*) PbS-3, and *Open circle* PbS-4 films. The dashes restrict the 95% confidence interval of determination of the band gap E_g as a function of particle size D. Reprinted from [100] with permission from Elsevier

of bulk PbS); therefore, in the first approximation, this correction in (2.40) can be neglected. Then the dependence of E_g on the particle radius R is described by the relation $E_g = E_b + \pi^2 \hbar^2/(2\mu_{ex}R^2)$, откуда следует, что $E_g(R) \propto 1/R^2$ or $E_g(D) \propto 1/D^2$. Using the data obtained in studies [97, 100] and reported previously literature data [141, 143, 158, 159], we plotted the band gap of PbS as a function of $1/D$ where D is the PbS particle size (Fig. 2.40). In should be noted that, in studies [141, 143, 159] the band gap E_g was determined for isolated (separated) PbS particles dispersed in a film polymer matrix. It is seen from Fig. 2.40 that the dependences $E_g(D^{-1})$ measured on PbS films in works [97, 100] and in study [158], substantially differ from those measured on isolated PbS nanoparticles by authors [141, 143, 159].

In the optical properties of semiconductors, the size effects are manifested most clearly if the particle size is commensurable with (or less than) the characteristic size $R_{ex} = (n^2 \varepsilon m_0/\mu_{ex})a_B$ of the Wannier-Mott exciton where $a_B = \hbar^2/m_0 e^2 = 0.053$ nm is the Bohr radius. A pronounced localization is best achievable for narrow-gap semiconductors with a high dielectric constant; this weakens considerably the electron-hole electrostatic interaction and, as a result, the exciton radius increases (the exciton size may far exceed the crystal lattice constant)., respectively, For lead sulfide at 4 K, the electron and hole effective masses are equal $m_e = (0.092 \pm 0.012)m_0$ and $m_h = (0.090 \pm 0.015)m_0$ [160] therefore, the exciton mass is $\mu_{ex} = (0.045 \pm 0.006)m_0$. For this μ_{ex} value and dielectric constant $\varepsilon_{PbS} \approx 18$ for lead sulfide [138, 160], the radius R_{ex} of the first exciton state with $n = 1$ is $(22 \pm 3)\varepsilon a_B$ or 18–24 nm.

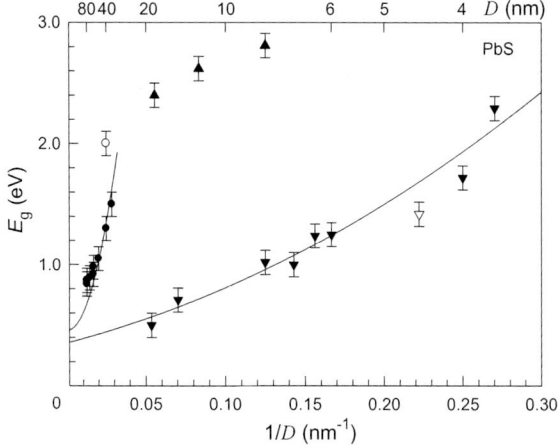

Fig. 2.40 Dependence of the band gap E_g on the PbS nanoparticle size D, plotted as the function $E_g(D) \propto 1/D^2$. *Filled circle* the results [97, 100] obtained on the PbS films, *Open circle* data reported in [158] for the PbS film, *inverted open triangle, filled triangle,* and *inverted filled triangle* data obtained for isolated PbS nanoparticles in [141, 143, 159], respectively. In [159], the average size of particles is estimated with methodical errors. For reduction of these errors all data [159] on the particle size D on this figure are increased in 1.5 times. *Solid lines* show the parabolic approximations of the dependences $E_g(D^{-1})$, obtained in the present study and in [159]. The extremely large values of E_g determined in [143] are doubtful. Reprinted from [100] with permission from Elsevier

Despite the fact that the average PbS particle size in the films studied by Sadovnikov et al. [97, 100] and Kumara et al. [158] is 2–4 times as large as the exciton size, the band gap is much wider that that in the bulk PbS. This is due to the considerable dispersion of nanoparticle sizes: The presence of particles smaller than ~ 10–30 nm in the films accounts for the increase in the band gap, while the presence of much larger particles leads to the average particle size increasing to ~ 60–80 nm [97, 100].

The band gap E_g found for the isolated PbS particles which were dispersed in a polymeric film matrix (Fig. 2.41) is, most often, greater than E_g of nanofilms (see Fig. 2.38). According to Jana et al. [143], the E_g values for 8, 12 and 18 nm nanoparticles are 2.81, 2.62 and 2.40 eV, respectively. The isolated PbS nanoparticles of ~ 13 nm size distributed in a copolymer matrix have $E_g = 3.36$ eV [161]. For isolated PbS particles of 1.5–12.5 nm size dispersed in a polymeric film matrix, E_g was found to be from 2.4 to 0.5 eV [159]. However, the size of PbS nanoparticles in the cited study [159] is apparently underestimated. In the case of 0.8–2.8 nm PbS nanoparticles dispersed in a polyacrylamide film, the blue shift of absorption band to 2.6–2.8 eV is observed [162]. An E_g value of 3.76 eV was reported for isolated PbS nanoparticles not exceeding 15 nm size coated by thio-glycerol and dispersed in polyvinyl alcohol [163] (Fig. 2.42). According to Chakraborty and Moulik [12], colloid PbS nanoparticles of 2.4–3.0 nm size synthesized in an aqueous heptane microemulsion are characterized by E_g ranging from 3.45 to 3.30 eV.

Yucel et al. [111] determined the band gap of nanostruc-tured PbS films prepared on a glass substrate by the SILAR process. Deposition was performed from a mixture of 1 M aqueous solutions of lead acetate and sodium sulfide at room temperature. The E_g value was found as a function of conditions of synthesis (pH and the number n and duration t of cycles of substrate dipping into the reaction medium) (Fig. 2.43).

According to the study [111], a PbS film with the greatest band gap of 2.24 eV can be synthesized by the SILAR method at pH = 9.1 using ten dipping cycles, 10 s each.

2.3.2 Thermal Properties

Data on the thermal expansion of lead sulfide, including films, are very scarce, although these data are useful for the application of films at elevated temperature and for selection of substrates with similar thermal expansion coefficients. It was found [164, 165] that the thermal expansion coefficient of polycrystalline PbS at 300 K is $(19–20) \times 10^{-6}$ K^{-1}. According to study [165] the Grüneisen constant γ of polycrystalline PbS monotonically decreases from ~ 2.2 to ~ 1.9 as the temperature is raised from 20 to 340 K, while according to the other study [164] the constant γ, conversely, increases from 1.98 to 2.26 with temperature rise from 320

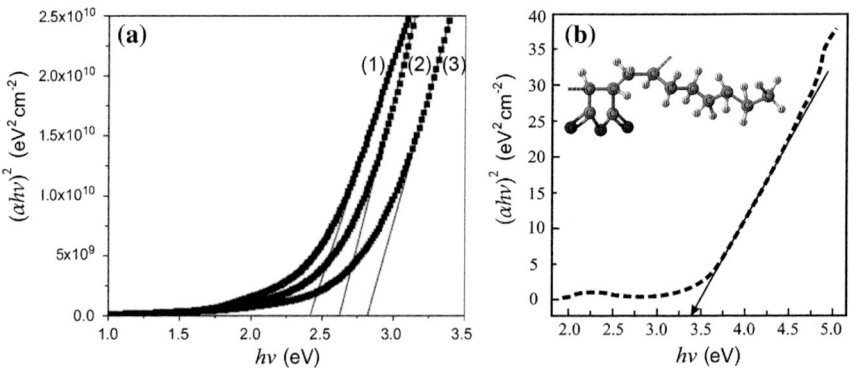

Fig. 2.41 Absorption spectra and band gap of isolated PbS nanoparticles. **a** Polyvinyl alcohol matrix; the deposition temperatures for (*1*) 18, (*2*) 12, and (*3*) 8 nm particles are 303, 293 and 283 K, respectively [143]. **b** Copolymer matrix with PbS nanoparticles of ~ 13 nm size; the inset shows the presumptive molecular structure of the copolymer [161]

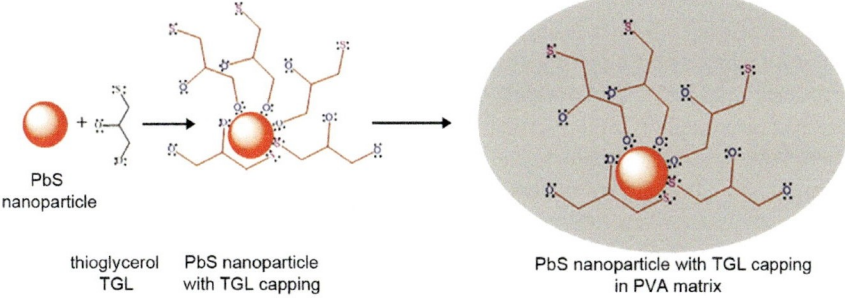

PbS
nanoparticle

thioglycerol   PbS nanoparticle   PbS nanoparticle with TGL capping
    TGL   with TGL capping   in PVA matrix

Fig. 2.42 Formation mechanism of thioglycerol-coated PbS nanoparticles in a polyvinyl alcohol matrix [163]

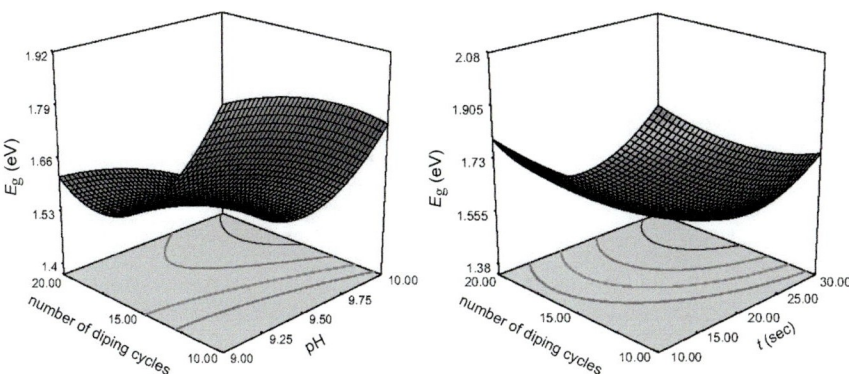

Fig. 2.43 Band gap of PbS films obtained by the SILAR process versus pH of the reaction mixture, duration t and number n of dipping cycles of the substrate into the reaction mixture [111]

to 670 K. The lattice properties of PbS were calculated by *ab initio* density functional theory in the quasiharmonic approximation [166]. At 300 K, the thermal expansion coefficient α and the Grüneisen constant γ of lead sulfide were found to be 29.8×10^{-6} K^{-1} and 2.50–2.52, respectively [166].

The thermal expansion coefficient of a nanostructured PbS film has been measured by Sadovnikov et al. [96, 167]. According to study [96], as the PbS film is cooled down from 423 to 293 K, the lattice constant a decreases from 0.59637 to 0.59326 nm, and this change in the lattice constant is correlated with the thermal expansion coefficient $\alpha(423$ K$) = \sim40 \times 10^{-6}$ K^{-1}. Upon the repeated heating of the PbS film from 293 to 393 K, the lattice constant increases from 0.59326 to 0.59492 nm. This change implies that $\alpha(393$ K$) = \sim28 \times 10^{-6}$ K^{-1}.

The first study of the thermal expansion of the PbS nanofilm in situ at temperatures of up to 473 K was reported by Sadovnikov et al. [167]. The measured linear thermal expansion coefficient of the PbS nanofilm was (37–39) $\times$ 10^{-6} K^{-1}, which is close to the α value determined in another publication [96]. Note that the α value of nanofilms [96, 167] is approximately twice as high as α of the bulk PbS. The authors of study [167] stated, without discussing the physical grounds of the observed effect, that the great difference between the coefficients a is attributable to the small particle size of PbS in the films and to different crystal structures of the films and coarse-grained PbS.

The precision diffraction measurements of the PbS films [167] were carried out in situ at temperatures from 293 to 473 K on a Philips X'Pert automated diffractometer equipped with a position-sensitive fast X'Celerator sector detector. The error of determination of the lattice constant did not exceed ± 0.00005 nm.

The structure refinement of the PbS film demonstrated [167] that reflection positions and intensities correspond to cubic (space group $Fm\overline{3}m$) lead sulfide with DO_3 type structure, which is in line with the results of other publications [95, 96, 100]. The crystal lattice constant a_{cub} of the PbS film is 0.59335 ± 0.00005 nm. All of the observed diffraction reflections were substantially broadened. Estimation showed that the average coherent scattering region (CSR) in the film is ~ 40 nm.

According to the results of measurements [167], the dependence of the lattice constant a on the annealing temperature T is non-linear and in the range of 293–473 K, it is described by a polynomial

$$a(T) = a_0 + a_1 T + a_2 T^2 + a_3 T^3, \tag{2.41}$$

with $a_0 = 0.59057$ nm, $a_1 = -8.1896 \times 10^{-6}$ nm K^{-1}, $a_2 = 7.9452 \times 10^{-8}$ nm K^{-2}, and $a_3 = -6.5455 \times 10^{-11} T^3$ nm K^{-3}. The nonlinear character of the $a(T)$ dependence means that the thermal expansion coefficient α is not constant and it also changes with the temperature.

Figure 2.44 shows the found temperature dependence of the thermal expansion coefficient $\alpha(T)$ for the nanostructured PbS film.

The experimental linear thermal expansion coefficient α is determined as the average expansion coefficient in the temperature interval between the initial temperature 293 K and measured temperature T, i.e.

$$\alpha(T) = \frac{1}{a_{293K}} \frac{\Delta a}{\Delta T} = \frac{a(T) - a_{293K}}{a_{293K}(T - 293)}, \tag{2.42}$$

where $a(T)$ and $a_{293\ K}$ are the crystal lattice constants of PbS film measured at temperature T and at the initial temperature 293 K, respectively. As is seen from Fig. 2.44a, during slow heating of PbS film from 323 to 473 K the thermal expansion coefficient α increases nonlinearly from $\sim 37 \times 10^{-6}$ to $\sim 39 \times 10^{-6}$ K^{-1}. The dependence of the thermal expansion coefficient α on the annealing temperature T in the range 323–473 K can be represented through coefficients of the polynomial (2.41) as

Fig. 2.44 The linear thermal expansion coefficient α of PbS nanofilm versus **a** the temperature and **b** duration of annealing: *Filled circle* the coefficient α during temperature increase from 293 to 423 K, *Inverted filled triangle* the coefficient α during cooling from 423 to 293 K, *Open circle* the coefficient α as a function of the annealing duration of the film at 423 K [167, 168]. Reprinted from [168] with permission from Elsevier

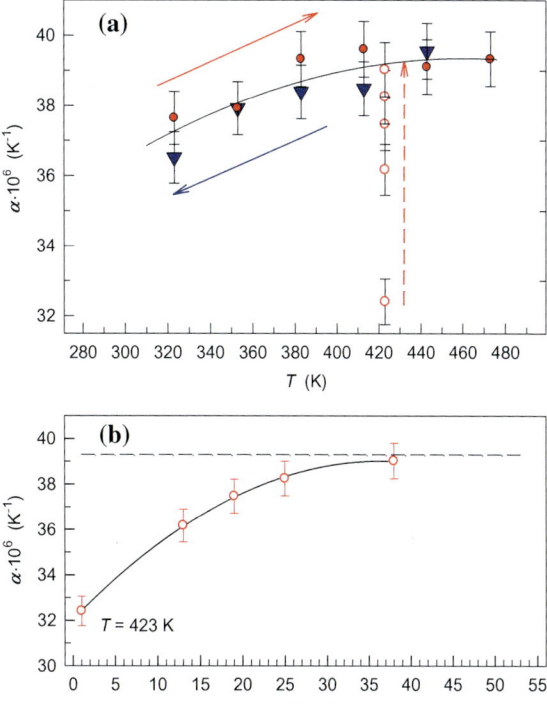

$$\alpha(T) = [(a_1 + a_2 T_0 + a_3 T_0^2) + (a_2 + a_3 T_0)T + a_3 T^2]/a_{293\,\text{K}}$$
$$= 9.4705 \times 10^{-6} + 6.0273 \times 10^{-8} T - 6.5455 \times 10^{-11} T^2. \tag{2.43}$$

According to [167], on the scale of the performed experiments the relaxation processes are very slow and equilibrium values of the lattice constant a and thermal expansion coefficient α are achieved only after many-hour annealing. The measured thermal expansion coefficient α of PbS nanofilms is much greater than that reported in the literature for bulk lead sulfide [164–166].

In the most general case, the difference in the coefficients α of nanofilm and coarse-grained PbS can be caused by four reasons [168, 169]: (1) the substrate influence, (2) the method of measurement, (3) different crystal structure of the nanofilm and coarse-grained PbS, and (4) the small size of the particles in nanofilm which leads on to change of phonon spectrum and its edges. The thermal expansion coefficient α of the glass substrate in the temperature range between 293 and 473 K is measured in study [167] by the dilatometry method and is equal to 12.5×10^{-6} K^{-1}, which is a third of that for the PbS nanofilm. Therefore, the effect of the substrate on α of PbS can be excluded. Different methods of measurements of the thermal expansion coefficient used in work [167] and in studies [164, 165] can lead to quantitative differences in the values of the coefficient α, but no more than

5%. Thus, the main reason of the difference in the coefficients α for the PbS film and the coarse-grained PbS are different crystal structure (DO_3 type for the PbS nanofilm and $B1$ type for the bulk coarse-grained PbS) and the small size of particles in the nanostructured film.

It is known from the literature that a decrease in the size of particles (grains, crystallites) to the nanometer scale is accompanied by an increase in the thermal expansion coefficient as compared with conventional polycrystals [81, 82]. It is exactly this effect that is observed in study [167]. This effect was explained by Sadovnikov and Gusev in studies [168, 169].

Thermal expansion of a solid state is due to anharmonicity of vibration of atoms. In case of small deviations of r of atoms from the equilibrium position r_0, the potential energy $U(r)$ of a system of interacting atoms can be expanded into a Taylor series according to the displacement value $u = (r - r_0)$. In the equilibrium position $r = r_0$, the energy is minimal; therefore $(\partial U/\partial r)_{r=r_0} = 0$. By limiting the expansion to the third-order terms, which is sufficient for qualitative description of thermal expansion of solids, the energy U, being a function of the displacement value u, can be written in the first approximation as

$$U(u) = U_0 + A_h u^2/2 - \beta_{anh} u^3/3 , \qquad (2.44)$$

where $A_h = (\partial^2 U/\partial r^2)_{r=r_0}$ is a some constant and $\beta_{anh} = \frac{1}{2}(\partial^3 U/\partial r^3)_{r=r_0}$ is a coefficient allowing for the anharmonicity of vibration of atoms.

According to Boltzmann distribution, the probability $f(u)$ of deviation of an atom from the equilibrium position by the value u is equal to $f(u) = (A_h/2\pi k_B T)^{1/2}\exp[-U(u)/k_B T]$, therefore the time-average deviation $\langle u \rangle$ of an atom from the equilibrium position is

$$\langle u \rangle = \left[\int_{-\infty}^{+\infty} u f(u) du \right] \Big/ \left[\int_{-\infty}^{+\infty} f(u) du \right] = \frac{k_B T \beta_{anh}}{A_h^2}. \qquad (2.45)$$

The linear expansion coefficient has the form

$$\alpha(T) = (1/a_{293K})d\langle u \rangle/dT = k_B \beta_{anh}(A_h^2 a_{293K}), \qquad (2.46)$$

where $a_{293\,K}$ is the crystal lattice constant at 293 K. Thermal expansion (or compression) of solid can be explained only with allowance for anharmonicity of atomic vibrations, i.e. at $\beta \neq 0$, since in harmonic approximation, when $\beta = 0$, the linear thermal expansion coefficient is equal to 0 and thermal expansion is lacking. From (2.46) it is clear that the observed enhancement of the thermal expansion coefficient during transition from bulk to nanocrystalline lead sulfide can occur only when the coefficient β increases, i.e. when anharmonicity of atomic vibrations takes place.

The linear thermal expansion coefficient $\alpha(T)$ is related to specific heat capacity $C_{sp} = C_V/v_m$ per unit volume of substance and measured in J m^{-3} K^{-1} by [170]:

$$\alpha(T) = \gamma C_{sp}(T)/3B \equiv \frac{\gamma}{3B} \frac{C_V(T)}{v_m}, \tag{2.47}$$

where B is the bulk modulus, $\gamma = \left[\sum_k \gamma_k C_V(\mathbf{k})\right]/\left[\sum_k C_V(\mathbf{k})\right]$ is the Grüneisen constant, $\gamma_k = -(V/\omega_k)[\partial(\omega_k)/\partial V]\} = -\partial(\ln\omega_k)\partial(\ln V)$ is the Grüneisen parameter for the vibrational mode with wave vector $\mathbf{k}$ and angular frequency $\omega_k = ck$, c is the velocity of propagation of elastic vibrations in the lattice, $k = |\mathbf{k}|$, and v_m is the molar volume.

If the temperature dependences of the Grüneisen constant and the compression modulus are neglected, then in the zeroth approximation the thermal expansion coefficient $\alpha(T)$ in the low-temperature region should exhibit the same temperature dependence as the specific heat capacity C_{sp}. However, in a real solid the values of γ_k are not similar for all normal modes, therefore the Grüneisen constant γ, which can be considered as a measure of anharmonicity, depends on temperature.

The electronic contribution to heat capacity of dielectrics and semiconductors at $10\text{ K} < T \le 0.4T_{melt}$ (T_{melt} is the melting temperature)   is small, and in this temperature region lattice heat capacity plays the leading role.

If the lattice heat capacity C_V of monoatomic substance is represented in terms of the average energy $\varepsilon(\omega, T) = (\hbar\omega/2)\coth(\hbar\omega/2k_BT)$ of a linear oscillator and the frequency distribution function $g(\omega)$ as $C_V(T) = \int \frac{\partial\varepsilon(\omega,T)}{\partial T} g(\omega)d\omega$, then the expression (2.47) takes the form

$$\alpha(T) = \frac{\gamma}{3v_m B} \int \frac{\partial\varepsilon(\omega, T)}{\partial T} g(\omega)d\omega. \tag{2.48}$$

From (2.48) it is clear that the temperature dependence of the thermal expansion coefficient depends directly on the spectral distribution of frequencies $g(\omega)$.

Thus, the main reason of variation of the lattice properties of nanocrystals in comparison with bulk substances is variation of the shape and boundaries of the phonon spectrum, i.e. change of the frequency distribution function of atomic vibrations.

The main idea of the majority of models for modification of the phonon spectrum of small particles and/or nanostructured systems consists in the appearance of low-frequency modes in the phonon spectrum, which are absent in the spectrum of the bulk crystal [80–82]. According to [81, 171], waves can occur in the nanoparticles, whose length does not exceed the doubled maximum size of the particle D, i.e. $\lambda \le 2D$; so, on the side of low-frequency vibrations the phonon spectrum is limited by a certain minimal frequency $\omega_{min} \ge 2\pi\frac{c_t}{2D}$, where c_t is the velocity of propagation of transverse elastic vibrations (i.e. transverse velocity of sound). In bulk crystals, there is no limitation like this. Besides, the phonon spectrum is limited on the side of high frequencies. Taking into account the low-

and high-frequency restrictions of the phonon spectrum, the molar heat capacity of a substance, whose molecule contains n atoms, is equal to

$$C_V(T) = n \int_{\omega_{min}}^{\omega_{max}} \frac{\partial \varepsilon(\omega, T)}{\partial T} g(\omega) d\omega. \qquad (2.49)$$

The distribution of natural vibrations in the presence of low-frequency restrictions was discussed in studies [172–174]. In work [174], expressions for the frequency distribution function $g(\omega)$ and the upper boundary ω_{max} of the phonon spectrum of small rectangular particle have been obtained:

$$g(\omega) = V\omega^2/2\pi^2 c_3 + S\omega/8\pi c_2 + L/16\pi c_1, \qquad (2.50)$$

$$\omega_{max} \approx \left(\frac{18\pi^2 N c_3}{V}\right)^{1/3}\left[1 - \frac{S}{144\pi c_2 N^{1/3}}\left(\frac{18\pi^2 c_3}{V}\right)^{2/3} + \Delta(N^{-2/3})\right], \qquad (2.51)$$

where V, S and L are the volume, surface area and the total edge length of small particle, N is the number of atoms in the particle, and $\Delta(N^{-2/3})$ is the correction term of the order $N^{-2/3}$. The quantities c_1, c_2 and c_3 are effective propagation velocities of elastic vibrations determined through the velocities of longitudinal and transverse vibrations, c_l and c_t: $c_1^{-1} = c_\ell^{-1} + 2c_t^{-1}$, $c_2^{-1} = \frac{2c_t^4 - 3c_t^2 c_\ell^2 + 3c_\ell^4}{c_t^2 c_\ell^2 (c_\ell^2 - c_t^2)}$ and $c_3^{-1} = c_\ell^{-3} + 2c_t^{-3}$.

Considering (2.50) and (2.51), it was shown in studies [168, 169] that the molar heat capacity (2.49) of a nanocrystalline substance with particles of a right-angled shape can be written upon mathematical transformations as a function not only of the temperature T, but also of the size D of the small particle:

$$C_V(T, D) = n\left[\frac{12\pi^4 N_A k_B}{5}\left(\frac{k_B T}{\hbar \omega_{max}}\right)^3 + k_1 L_\Sigma T + k_2 S_\Sigma T^2\right] \qquad (2.52)$$

$$= C_V^{bulk}(T) + n(k_1 L_\Sigma T + k_2 S_\Sigma T^2),$$

where the first summand represents the Debye heat capacity of bulk crystal in the low-temperature region, L_Σ and S_Σ are the total edge length and the total surface area of small particles. If we assume that small particles have a cubic shape with edges in length D, then the number of such particles in the volume v_m is equal to $n_p = v_m/D^3$. In this case $L_\Sigma = 12n_p D = 12v_m/D^2$ and $S_\Sigma = 6n_p D^2 = 6v_m/D$. The values $k_1 = (k_B^2 c_1^{-1}/8\pi\hbar)I_2$ and $k_2 = (k_B^3 c_2^{-1}/2\pi\hbar^2)I_3$ are positive constants, $I_m = (4m!/2^{m+1})\sum_{N=1}^{\infty} N^{-m} \equiv (4\, m!/2^{m+1})\zeta(m)$, $\zeta(m)$ is the Riemann zeta function ($I_3 = 1.8031$; $I_2 = \pi^2/6$). It is clear from (2.52) that the heat capacity contribution $\Delta C_V(T, D)$ caused by the small size of particles, is equal

$$\Delta C_V(T, D) = n(k_1 L_\Sigma T + k_2 S_\Sigma T^2). \tag{2.53}$$

With allowance for (2.48), (2.52) and expressions for L_Σ и S_Σ, the thermal expansion coefficient of nanocrystalline substance was be presented in studies [168, 169] as

$$\begin{aligned}
\alpha(T, D) &= \frac{\gamma C_{V(\text{bulk})}(T)}{3v_m B} + n\frac{\gamma}{3v_m B}(k_1 L_\Sigma T + k_2 S_\Sigma T^2) \\
&= \alpha_{\text{bulk}}(T) + n\frac{\gamma}{3B}\left(\frac{12k_1 T}{D^2} + \frac{6k_2 T^2}{D}\right),
\end{aligned} \tag{2.54}$$

where contribution $\Delta\alpha(T, D)$ in the thermal expansion coefficient, which is caused by small size of particles, is equal to

$$\Delta\alpha(T, D) = n\frac{\gamma}{3B}\left(\frac{12k_1 T}{D^2} + \frac{6k_2 T^2}{D}\right). \tag{2.55}$$

Let us determine the theoretical value of the contribution $\Delta\alpha(T, D)$ for nanocrystalline PbS with cubic structure. For lead sulfide, $n = 2$, the average particle size D in PbS nanofilm at 300 K is ~ 50 nm, and the density of cubic lead sulfide is assumed to be equal to 7.5 g cm^{-3}. Sadovnikov and Gusev [168, 169] calculated the coefficients k_1 and k_2, necessary for quantitative estimation of $\Delta\alpha(T, D)$, using the propagation velocities of longitudinal and transverse elastic vibrations c_l and c_t or elastic constants c_{11}, c_{12}, and c_{44} for lead sulfide [166, 175–178].

In high-symmetry crystals, completely longitudinal or completely transverse atomic vibrations propagate only in those directions that correspond to synchronously moving atomic planes. For cubic crystals, such crystallographic directions are [100, 110, 111] and their equivalents. According to calculation [168, 169] with using data [175], at ~ 300 K c_l and c_t are ~ 3975 and ~ 1881 m s^{-1}. These values are in good agreement with experimental data [179], according to which the velocities c_l and c_t at room temperature are equal to 4080 and 1840 m s^{-1}. Figure 2.45 demonstrates the calculated changes in the velocities of transverse c_t and longitudinal c_l elastic vibrations versus the crystallographic direction for a single crystal of cubic lead sulfide PbS at 300 K. The distances from the figure center (Fig. 2.45) are proportional to the corresponding velocity value ($c_t^{[100]} = c_t^{[010]} = c_t^{[001]} = 1.9 \times 10^3$ m s^{-1} and $c_l^{[100]} = c_l^{[010]} = c_l^{[001]} = 4.0 \times 10^3$ m s^{-1}).

According to estimations performed with taking into account these values of c_l and c_t, the low and upper boundaries $\omega_{\min}$ and $\omega_{\max}$ of the phonon spectrum of a small cubic particle of PbS with edges in length $D = 50$ nm are equal to $\sim 1.18 \times 10^{11}$ and $\sim 2.76 \times 10^{13}$ rad s^{-1}, respectively [168, 169]. At 300 K, the constants k_1 and k_2 are equal to $\sim 1.55 \times 10^{-16}$ J m^{-1} K^{-2} and $\sim 6.00 \times 10^{-8}$ J m^{-2} K^{-3}, i.e. $k_2 \gg k_1$. With increasing temperature to 500 K, the constants k_1 and k_2 increase weakly up to $\sim 1.67 \times 10^{-16}$ J m^{-1} K^{-2} and

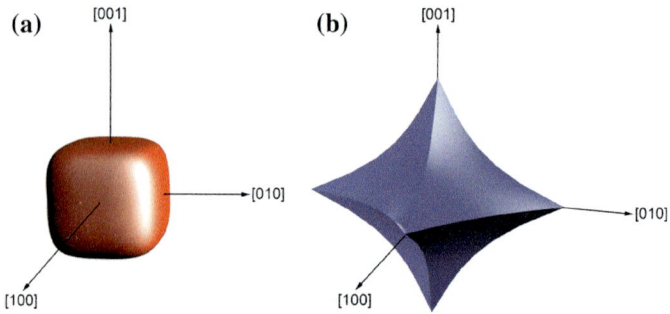

Fig. 2.45 The calculated velocities of transverse c_t (**a**) and longitudinal c_l (**b**) elastic vibrations versus the crystallographic direction for cubic lead sulfide PbS at 300 K [168]. The distance from the figure center is proportional to the velocity value ($c_t^{[100]} = c_t^{[010]} = c_t^{[001]} = 1.9 \times 10^3$ m s^{-1} and $c_\ell^{[100]} = c_\ell^{[010]} = c_\ell^{[001]} = 4.0 \times 10^3$ m s^{-1}). Reprinted from [168] with permission from Elsevier

$\sim 7.06 \times 10^{-8}$ J m^{-2} K^{-3}, respectively [168, 169]. Since $k_2 \geq k_1$, from (2.49) it follows that in the case of small particles, essentially, an additional contribution appears in the thermal expansion coefficient, which is proportional to $S_\Sigma T^2$ and is due to large surface area of small particles. Therefore reduction of the particle size to the nanometer scale should be accompanied by the thermal expansion coefficient growth.

Exactly this effect, namely, ~ 1.7 time greater coefficient α of nanostructured PbS film as compared with bulk PbS, has been found in studies [167–169].

Figure 2.46 shows the dependence of the thermal expansion coefficient of PbS nanofilm in the temperature region between 300 and 500 K calculated as $\alpha_{nano}^{calc}(T, D) = \alpha_{bulk}(T) + \Delta\alpha(T, D)$ with allowance for the derived coefficients k_1 and k_2. The experimental results [162] were used as the $\alpha_{bulk}(T)$ dependence. The variation of the equilibrium particle size in PbS nanofilm versus the temperature was taken into account in the calculation (see Fig. 2.44). It was also considered that with an increase in temperature a small linear growth of the Grüneisen constant γ is observed [164] and the compression modulus B calculated from data [175] decreases. The experimental linear thermal expansion coefficient α of PbS nanofilms [167], as well as the literature data [164, 165] on the coefficient α for bulk coarse-grained lead sulfide and the data [96] for the nanofilm are given in Fig. 2.46 for comparison.

The calculated dependence $\alpha_{nano}^{calc}(T, D)$ for a PbS nanofilm is slightly lower than the experimental dependence $\alpha_{nano}^{exp}(T, D)$: in the examined temperature region, 300–473 K, the difference in the coefficients α is $(5–6) \times 10^{-6}$ K^{-1}. Apparently, this difference is due to the approximations of the model [174], which is the basis of the calculation [168, 169]. Besides, the experimental data [175] used for the estimation of effective propagation velocities of elastic vibrations c_1^{-1}, c_2^{-1}, c_3^{-1} and for subsequent calculation of the $\Delta\alpha(T, D)$ value, as well as the data [164] taken as α_{bulk}

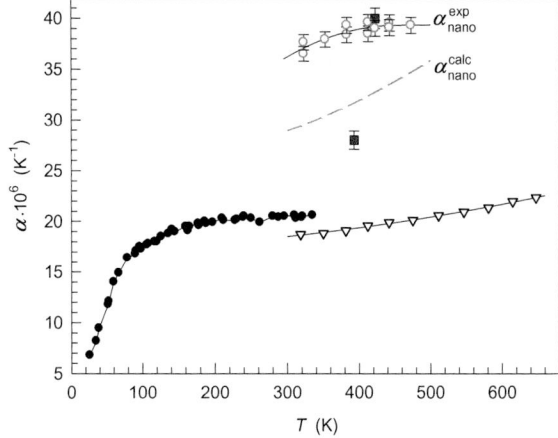

Fig. 2.46 The linear thermal expansion coefficient α of lead sulfide: *Filled square* PbS film [96], *Inverted open triangle* bulk coarse-grained polycrystalline PbS [164], *Filled circle* bulk coarse-grained polycrystalline PbS [165], *Open circle* experimental data of studies [167, 168] for nanostructured PbS film. The *dotted line* shows calculated dependence $\alpha_{nano}^{calc}(T,D) = \alpha_{bulk}(T) + \Delta(T, D)$ of thermal expansion coefficient for PbS nanofilm. Reprinted from [168] with permission from Elsevier

(T) of lead sulfide can be not quite accurate. In particular, the compression modulus B calculated from the data [175] is half as much again the modulus B measured in study [180]. Besides, the nonstoichiometric distribution of sulfur atoms S in the nanofilm and the difference in the crystal structures ($D0_3$ type for the PbS nanofilm and $B1$ type for the bulk coarse-grained PbS) can make a small, about a few percents, additional contribution to the thermal expansion coefficient of the nanofilm. Therefore the calculated dependence of the thermal expansion coefficient of PbS nanofilm agrees rather well with the experimental dependence $\alpha_{nano}^{exp}(T,D)$ (Fig. 2.46).

The thermal expansion coefficient α of PbS film at 423 K estimated from data [96] on lattice constant variation during cooling of the film from 423 to 293 K is $\sim 40 \times 10^{-6}$ K^{-1} and agrees well with the results obtained in studies [167, 168]. Estimation of α from data [96] obtained during cooling of the film from room temperature to 393 K yields a smaller value, $\sim 28 \times 10^{-6}$ K^{-1}. This can be caused by the circumstance that in work [96] the annealing duration of PbS film at 393 K was insufficient for attaining the equilibrium value of the crystal lattice constant. Indeed, the equilibrium values of the lattice constant and thermal expansion coefficient (see Fig. 2.44) are achieved only after long-term 40 h exposure of films.

Since $\beta_{anh} = A_h^2 a_{293K} \alpha/k_B$, then with allowance for (2.49) the dependence between the coefficient of anharmonicity of atomic vibrations and the particle size can be expressed as

$$\beta_{\text{anh}}(T, D) = \frac{A_h^2 a_{293\text{K}}}{k_{\text{B}}} \left[\alpha_{\text{bulk}}(T) + n \frac{\gamma}{3 v_{\text{m}} B} (k_1 L_\Sigma T + k_2 S_\Sigma T^2) \right]. \qquad (2.56)$$

According to (2.56), reduction of the particle size that leads to the growth of their surface area S_Σ should be accompanied by an increase in the anharmonicity of atomic vibrations.

According to [49, 95–97, 100, 123, 181], the lattice constant of lead sulfide remains unchanged when its particle size decreases. If we assume that the coefficient A_h also remains constant when the particle size decreases, then the ratio of anharmonicity coefficients of nanofilm and bulk PbS is $\beta_{\text{anh-nano}}/\beta_{\text{anh}} = \alpha_{\text{nano}}/\alpha$, hence $\beta_{\text{anh-nano}} = (\alpha_{\text{nano}}/\alpha)\beta_{\text{anh}}$. At 293 K, the coefficient α of bulk PbS is $\sim 21 \times 10^{-6}$ K^{-1} [160, 165], while the measured thermal expansion coefficient of nanofilm is $\alpha_{\text{nano}} \cong 36 \times 10^{-6}$ K^{-1}. Therefore $\beta_{\text{anh-nano}} \approx 1.7 \beta_{\text{anh}}$.

Thus, the linear thermal expansion coefficient α of a PbS nanofilm in the temperature range of 293–443 K is approximately 1.5–2 times as high as the a of bulk PbS. The observed difference between the coefficients α is associated with the small PbS particle size in the film, resulting in a change of the distribution function of the atomic vibration frequencies and increase in the vibration anharmonicity.

2.3.3 Stability of Nanostructured Lead Sulfide

The most widely used devices based on lead sulfide are fire alarm sensors, flame sensors and heat source detection systems; therefore, the thermal and oxidation stabilities of the nanostate largely determine the application of nanostructured PbS in this type of devices. However, nanoparticles and quantum dots are metastable materials and are prone to recrystallization and oxidative degradation, which is promoted by their large surface area. Therefore, extensive use of nanostructured PbS requires the knowledge of temperature range in which the nanoparticles do not grow (no transition from the nano- to the coarse-grained state) and the temperature of the onset of oxidation.

The thermal stability of the structure of PbS nanofilms obtained by laser deposition on heated quartz substrates has been studied [123]. The annealing of films in air demonstrated that upon temperature rise to 375 K, the intensity ratio of some diffraction reflections changes but no additional peaks are observed up to 475 K; hence, the PbS nanofilm is not oxidized. The change in the diffraction reflection intensities was attributed [123] to the fact that a phase transition from the $B1$ structure to the high-temperature cubic $B3$ structure (space group $F\bar{4}3m$) takes place in the PbS film at 375 K.

From the XRD patterns of the vacuum-annealed nanostructured PbS powders and films, the particle size, lattice microstrain, and lattice constant of the powders and films were determined as functions of temperature.

The vacuum annealing of PbS nanopowders at a temperature of up to 700 K leads to minimization of microstrains and a slight growth of nanoparticles (Fig. 2.47a); hence, this temperature range can be regarded as the thermal stability region of lead sulfide nanopowders.

The temperature range from 700 to 800 K, in which the particle size increases 5–10-fold, corresponds to the temperature of collective recrystallization of PbS nanopowder. The vacuum annealing of PbS nanofilms at a temperature of up to 673 K is also accompanied by a minor growth of nanocrystallites, while at higher annealing temperature, the particle size increases jumpwise (Fig. 2.47b). The recrystallization in nanofilms starts at a lower temperature than the recrystallization of nanopowders. This is a consequence of tighter contact of crystallites in nanofilms compared with the particle contact in nanopowders.

According to [82, 185], collective recrystallization in nanocrystalline materials starts, most often, at the temperatures $T_{recr} \approx (0.30–0.35)T_{melt}$. In other studies [182, 183], the recrystallization temperature of PbS nanopowders and nanofilms was found to be ~ 700 K, which is half the melting point of bulk PbS (1391 K). Thus, PbS nanopowders and nanofilms have higher relative recrystallization temperature $T_{recr} \approx 0.5T_{melt}$, i.e, they differ from other nanomaterials by a higher thermal stability of particle (crystallite) size.

Analysis of phase composition of the annealed PbS powders shows that annealing at 930 K is accompanied by changes in phase composition: the formation of metallic lead Pb in addition to PbS.

The oxidation of PbS nanopowders with atmospheric oxygen was studied on samples with a particle size of 10–20 nm [183]. The XRD patterns of the initial PbS

Fig. 2.47 Effect of the vacuum annealing temperature on the particle (crystallite) size in nanostructured PbS **a** powders and **b** films [182, 183]. The temperature of the onset of recrystallization is marked by the *dashed line*

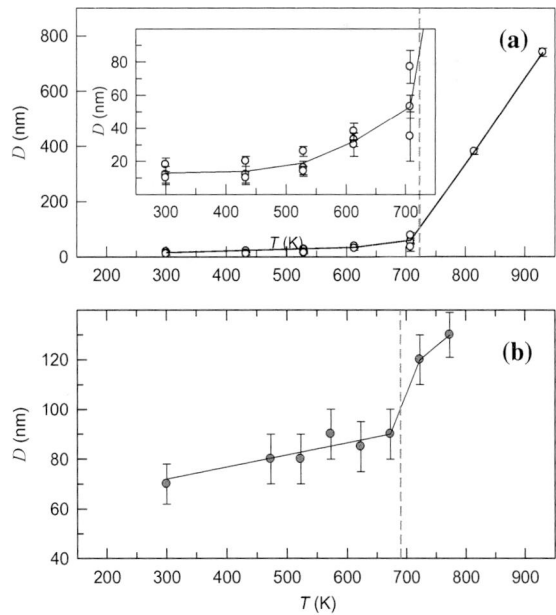

nanopowder with a particle size of 15 nm and of the same powder after annealing at a temperature from 473 to 623 K are presented in Fig. 2.48. The XRD pattern of the initial PbS nanopowder shows only reflections from cubic lead sulfide. Small increasing anneling temperature up to 423 K initiates the reaction between PbS and O_2, as is indicated by the appearance of strong diffraction reflections from oxygen-containing phases on XRD patterns.

Annealing of PbS nanopowder at 473–523 K does not cause any narrowing of the diffraction reflections (Fig. 2.48). Therefore, the lead sulfide particles retain their small size in this temperature range. After annealing of the nanopowder at $T > 523$ K, weakening of an intensity of all diffraction reflections from cubic lead sulfide is observed (Fig. 2.48). It is a consequence of decreasing content of PbS phase in annealed nanopowder.

Most of the nanopowders annealed in air at a temperature from 423 to 523 K contain lead sulfite $PbSO_3$. The absence of the $PbSO_4$ lead sulfate phase in the oxidation products does not mean that this phase does not form at all. It is likely that, under the given conditions, the $PbSO_4$ disappearance rate is higher than the rate of formation of this phase and, accordingly, this phase does not accumulate in the reaction products. As the temperature is raised to 573 K, the $PbO \cdot PbSO_4$ phase appears along with lead sulfate. When the temperature increases to 623 K, $PbO \cdot PbSO_4$ phase disappears and $PbSO_4$ lead sulfate starts to accumule.

Fig. 2.48 XRD patterns of the initial PbS nanopowder with a particle size of 15 nm and of the same nanopowder after annealing in air at a temperature from 423 to 623 K. Long and short ticks indicate reflections from PbS and $PbSO_4$, respectively [183]. The XRD patterns are recorded in $CuK\alpha_{1,2}$ radiation

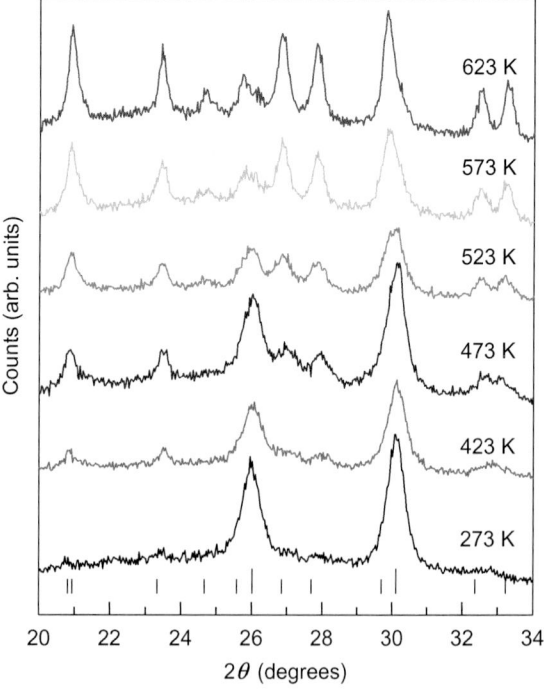

Figure 2.49 presents the XRD patterns of the initial nanostructured PbS film with particle size of 70 nm and of the same film after annealing in air at a temperature from 473 to 773 K [183].

The XRD pattern of the initial PbS film shows only reflections from cubic lead sulfide. Annealing of the film at 473–573 K leads to a narrowing of the diffraction reflections of cubic lead sulfide because of the growth of the PbS nanoparticles from 70 to 90 nm.

Annealing of the PbS film at $T < 623$ K does not change its phase composition; that is, no oxidation takes place. The film annealed at 623 K is partially oxidized, and its XRD pattern shows reflections from the surface oxide-sulfate $PbO \cdot PbSO_4$ phase (Fig. 2.49). According to XRD data, the size of the PbS particles decreases from 90 to 50 nm in the temperature range from 573 to 673 K because of their surface oxidation. The formation of the loose $PbO \cdot PbSO_4$ phase on the surface increases the film thickness to 420 nm. As determined by XRD, the particle size of the surface $PbO \cdot PbSO_4$ phase is about 12 nm, and the particle size remains invariable as the temperature is further raised up to 773 K [182].

Fig. 2.49 XRD patterns of the initial PbS film with thick of ∼ 100 nm and particle size of 70–80 nm on a glass substrate and of the same film annealed in air at a temperature from 473 to 773 K. Long and short ticks indicate reflections from cubic PbS and the oxide-sulfate $PbO \cdot PbSO_4$ phase, respectively [183]. The XRD patterns are recorded in $CuK\alpha_{1,2}$ radiation

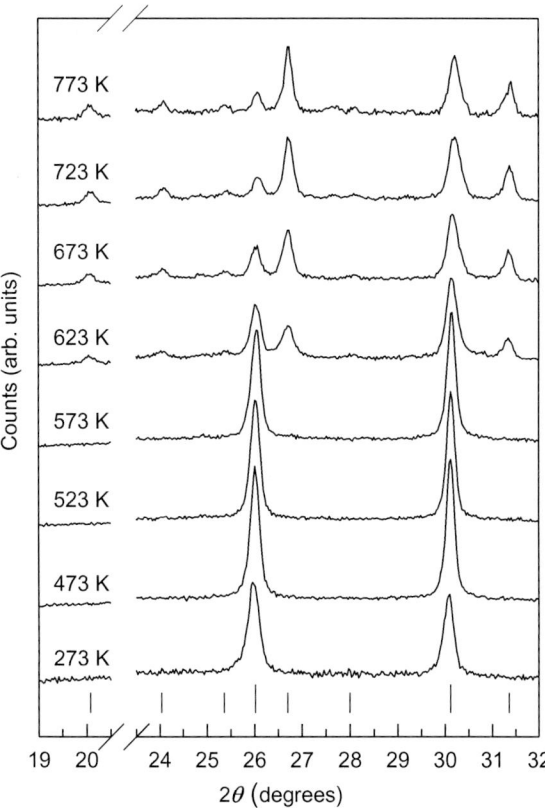

According to XRD data, the content of the oxide-sulfate phase in the film increases as the annealing temperature is increased, and no other oxidation products appear in the film up to 773 K.

Thus, the nanostructured lead sulfide film is stable in air up to 573 K.

Comparative analysis of the oxidative activities of lead sulfide demonstrated that the temperatures of the onset of all oxidative processes are much lower for PbS nanopowders than for the bulk coarse-grained sulfide, these processes being facilitated by oxygen-containing impurities [183, 184]. A decrease in the particle size is accompanied by a shift of the onset of oxidation to lower temperatures, while the ox idation of bulk lead sulfide starts at about 800 K [7].

The sequence of transformations involved in the oxidation of lead sulfide is illustrated in Fig. 2.50 in relation to PbS nanopowder with a particle size of 15 nm [183].

The presence of a small amount (~ 10 wt%) of $PbSO_3$ lead sulfite in the initial nanopowder is due to its formation at the PbS synthesis stage. As the temperature is raised from 423 to 473 K, the amount of $PbSO_3$ first increases and then (on further temperature rise) decreases, and at 573 K, this phase disappears. The absence of PbO in the temperature range of 423–623 K indicates that the reactions yielding this oxide as the final product proceed at higher temperature > 700 K. The temperature of the onset of oxidation of PbS nanopowders increases with increase in the nanoparticle size.

When PbS nanofilms are annealed in air at temperatures below 623 K, no oxidation takes place. This is consistent with the data of study [123]. Annealing at $T > 623$ K leads to partial film oxidation to give the PbO PbSO$_4$ oxide sulfate phase on the film surface. Thus, the nanostructured film of lead sulfide is stable in air up to 573 K and is more stable against oxidation than nanopowders [183].

The results obtained by Mikhlin et al. [186] are useful for the understanding of the oxidation pattern of nanostructured PbS. According to [186] s, the surface of a single-crystalline sulfide is oxidized in air only an increased relative humidity

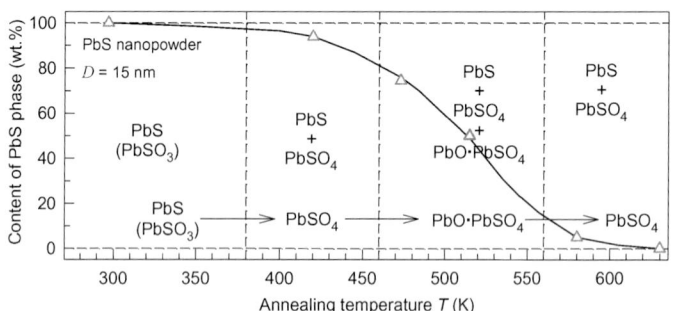

Fig. 2.50 Content of the PbS phase in a PbS nanopowder with a particle size of 15 nm as a function of annealing temperature in air [183]. A minor amount of $PbSO_3$ is present in the synthesized nanopowder as an impurity. The *vertical dashed lines* mark the conventional temperatures of transitions from one phase to the other

reaching $\sim$90%. The oxidation products form $\sim$20-nm protrusions on the PbS surface. Water condensed on the protrusions promote the oxidation of PbS. Over a period of three to five days, the protrusions uniformly cover the surface and them merge together to form a loose surface oxide layer. At normal or low air humidity, the lead sulfide surface is not oxidized. The oxidation of nanostructured PbS also starts on the surface but at a lower humidity.

According to Ihly et al. [187], keeping PbS quantum dots in air is accompanied by a blue shift of the first exciton peak in the optical spectra. This shift is caused by partial oxidation of the QDs. Nanostructured films were prepared by layer-by-layer deposition of lead sulfide accomplished by dipping the substrate into a colloid solution of QDs. The oxidation of PbS nanofilms in air was found to induce broadening of the first exciton peak. In order to prevent the oxidative degradation of nanostructured lead sulfide films, it was proposed to cover the films by an amorphous Al_2O_3 layer [187]. A 18-nm thick Al_2O_3 layer completely prevents the oxidation of PbS nanofilms and prevents the growth of nanoparticles in the films.

2.4 Applications of Nanostructured Lead Sulfide

Traditional applications of lead sulfide, in particular nanostructured one, in devices such as broad-band photodetectors, high-efficiency solar cells, thermoelectric transducers, optical switches, fire alarm sensors and photovoltaic, photonic and optoelectronic cells are known and have been repeatedly described [188–196]. The application of nanostructured lead sulfide is exemplified in Fig. 2.51, which shows a photograph of a flexible photodetector based on paper covered by a layer of PbS quantum dots [192]. Nanostructured PbS has been used as the key working part in systems of detection of extraterrestrial objects by their thermal emission and in the operated and autonomous (unaided) target guidance systems.

Less widely known is the use of nanostructured PbS as a solid-phase sensor for detection of gases [CO, CO_2, SO_2, N_2, NO and especially NO_2 and NH_3 [197–201],

Fig. 2.51 Photograph of flexible photodetector made of PbS quantum dots on a paper substrate [192]

for monitoring of heavy metals, including lead, in aqueous media [202, 203]. Polymeric matrices with PbS nanoparticles are used for protection from γ-radiation [204, 205].

As a result of active integration of the achievements of chemistry, biology and medicine, a background for biomedical applications and development of more perfect tools for medical diagnosis and therapy [206] using nanostructured lead sulfide has been created in the last decade. An example is provided by hybrid PbS nanoparticles coated by L-cysteine as a capping agent [58]. Of interest is the drug delivery to injured organs and tumour cells, which is a prerequisite for more successful therapy. For this purpose, it is especially promising to use core-shell sulfide heteronanostructures [207].

Enhancement of the luminescence intensity of the hybrid sulfide heteronanostructures based on PbS is attained owing to combination of luminescent sulfide nanoparticles with metal (especially silver) nanoparticles possessing the surface plasmon resonance. The injured areas can be visualized by using luminescent PbS quantum dots combined into a single nanostructure with organic drugs. Hybrid nanomaterials of this type based on PbS find use and are often indispensable in biosensorics, medicine, green chemistry and other areas [207–210]. Furthermore, the design of hybrid nanomaterials involving lead sulfide promotes elaboration of new analytical methods for the detection and identification of impurities, preconcentration and separation of compounds and visualization of biological action [211–213]. For example, in recent years, many reports have appeared about the successful synthesis and application in vivo of multifuctional hybrid nanostructures for the targeted drug delivery to damaged cells. Apart from other compounds, lead, cadmium and zinc sulfides are used in these structures as luminescent markers.

Lead sulfide nanoparticles embedded in multi-walled carbon nanotubes have been proposed as highly sensitive genomic sensors for the detection of pathogenic bacteria [214].

In parallel with the studies devoted to biomedical applications of PbS nanoparticles, the toxicity of nanostructured lead sulfide was studied [215, 216]; the results confirmed the applicability of these particles in living organisms provided that the PbS particles are coated by protective biologically harmless shells.

Controllable self-assembly of star-shaped PbS nanocrystals into large-scale ordered close-packed arrays was demonstrated in study [217]. The obtained arrays of star-shaped PbS nanocrystals can be used as anisotropic nanostructured materials for photoelectronic and photonic nanodevices.

The key principles and approaches developed for the synthesis and investigations of various forms of nanostructured lead sulfide are used in relation to other nanocrystalline compounds ranging from sulfides to oxides and carbides. For example, the effect of the small particle size on the thermal expansion and non-stoichiometry of nanocrystalline Ag_2S has been considered in studies [218–220] using an approach proposed previously [168, 169] for nanostructured PbS. Modification of optical and electronic properties of indium and gallium oxide nanofilms has been considered by Chen et al. [221], who took into account data of a study of PbS films [100].

Nanostructured PbS acquire desirable properties when synthesized by processes that ensure the controlled size of structure elements and passivation of the surface. Among numerous methods of synthesis of nanostructured PbS, these conditions are met by deposition from aqueous solutions, which is implemented in many ways. Synthesis in water makes it possible to avoid the influence of toxic organic ligands or toxic organic solvents on biological objects when PbS quantum dots are utilized as biolabels. It can be expected that further development of the methods for the synthesis of PbS nanoparticles of specified size and shape would extend the scope of their applicability, in particular, in optoelectronics, biology and medicine.

The conversion and storage of clean and sustainable energy are major challenges that we are facing today under the background of the energy crisis. As the performance of energy conversion and storage devices depend significantly upon the advancement of their materials, many new energy materials including nanostructured PbS have been developed to meet these challenges. Based on its unique physical and chemical properties, nanostructured lead sulfide and modified lead sulfide nanomaterials promise many advanced applications in the development of technologies of conversion and storage of energy. Moreover, lead sulfide nanomaterials possess such advantage as low cost, which is of great importance for their practical applications.

References

1. Lin, J.C., Sharma, R.C., Chang, Y.A.: Pb-S (Lead-Sulfur). In: Massalski, T.B. (ed.) Binary Alloy Phase Diagrams, 2nd edn, vol. 3, pp. 3005–3009. Materials Park, Ohio (1990) (ASM Intern. Publ.)
2. Lin, J.C., Sharma, R.C., Chang, Y.A.: The Pb-S (lead-sulfur) system. Bull. Alloy Phase Diagr. **7**(4), 374–381 (1986)
3. Scanlon, W.W.: Recent advances in the optical and electronic properties of PbS, PbSe, PbTe and their alloys. J. Phys. Chem. Solids. **8**(1), 423–428 (1959)
4. Schoolar, R.B., Dixon, J.R.: Optical constants of lead sulfide in the fundamental ab-sorption edge region. Phys. Rev. **137**(2A), 667–670 (1965)
5. Zemmel, J.N., Jensen, J.D., Schoolar, R.B.: Electrical and optical properties of epitaxial films of PbS, PbSe, PbTe and SnTe. Phys. Rev. **140**(1A), 330–342 (1965)
6. Hodes, G.: Chemical Solution Deposition of Semiconductor Films, p. 377. Dekker, New York (2002)
7. Sadovnikov, S.I., Gusev, A.I., Rempel, A.A.: Nanostructured lead sulfide: Synthesis, structure, and properties. Russ. Chem. Rev. **85**(7), 731–758 (2016)
8. Rempel, A.A.: Nanotechnologies. Properties and applications of nanostructured materials. Russ. Chem. Rev. **76**(5), 435–461 (2007)
9. Pawar, S.M, Pawar, B.S., Kim, J.H., Joo, O.-S., Lokhande, C.D.: Recent status of chemical bath deposited metal chalcogenide and metal oxide thin films. Curr. Appl. Phys. **11**(2), 117–161 (2011)
10. Deng, D., Xia, J., Cao, J., Qu, L., Tian, J., Qian, Z.: Forming highly fluorescent near-infrared emitting PbS quantum dots in water using glutathione as surface-modifying molecule. J. Coll. Interf. Sci. **367**, 234–240 (2012)

11. Zhang, B., Li, G., Zhang, J., Zhang, Y., Zhang, L.: Synthesis and characterization of PbS nanocrystals in water/C12E9/cyclohexane microemulsions. Nanotechnology **14**(3), 443–446 (2003)
12. Chakraborty, I., Moulik, S.P.: On PbS nanoparticles formed in the compartments of water/AOT/n-heptane microemulsion. J. Nanoparticle Res. **7**(2–3), 237–247 (2005)
13. Jiao, Y., Gao, X., Lu, J., Chen, Y., Zhou, J., Li, X.: A novel method for PbS quantum dot synthesis. Mater. Lett. **72**, 116–118 (2012)
14. Karim, M.R., Zaman, M.D.A., Zaman, M.D.B.: A conventional synthesis approach to prepare lead sulfide (PbS) nanoparticles via solvothermal method. Chalcogen. Lett. **11**(10), 531–539 (2014)
15. Li, F., Huang, X., Kong, T., Liu, X., Qin, Q., Li, Z.: Synthesis and characterization of PbS crystals via a solvothermal route. J. Alloys Comp. **485**(1–2), 554–560 (2009)
16. Kitaev, G.A., Bol'shchikova, T.P., Fofanov, G.M., Yatlova, L.E., Goryukhina, N.M.: Thermodynamic justification of metal sulfide deposition conditions from aqueous solutions by thiourea. In: Kinetika i Mekhanizm Obrazovaniya Tverdoi Fazy (Kinetics and Formation Mechanism of the Solid Phase), pp. 113–126. Ural Pedagogical Institute, Sverdlovsk (1968) (in Russian)
17. Gaiduk, A.P., Gaiduk, P.I., Larsen, A.N.: Chemical bath deposition of PbS nanocrystals: Effect of substrate. Thin Solid Films. **516**, 3791–3795 (2008)
18. Wang, J., Tang, S.H., Wang, B.Y., Li, Y.Q.: In-situ interaction of nano-PbS with gelatin. Sci. China Chem. **56**(11), 1593–1600 (2013)
19. Sadovnikov, S.I., Kuznetsova, Yu., V., Rempel, A.A.: Synthesis of a stable colloidal solution of PbS nanoparticles. Neorg. Mater. **50**(10), 1049–1056 (2014) (in Russian). (Engl. Transl.: Inorg. Mater. **50**(10), 969–975 (2014))
20. Sadovnikov, S.I., Gusev, A.I.: Chemical deposition of nanocrystalline lead sulfide powders with controllable particle size. J. Alloys Comp. **586**, 105–112 (2014)
21. Kozhevnikova, N.S., Sadovnikov, S.I., Rempel, A.A.: One-pot synthesis of lead sulfide nanoparticles. Zh. Obshch. Khim. **81**(10), 1608–1613 (2011) (in Russian). (Engl. Transl.: Russ. J. Gen. Chem. **81**, 2062–2066 (2011))
22. Sadovnikov, S.I., Kuznetsova Yu, V., Rempel, A.A.: A method of producing a colloidal solution of lead sulfide nanoparticles. Patent No. 2567326 of Russian Federation, pp. 1–5 (2015)
23. Froment, M., Lincot, D.: Phase formation processes in solution at the atomic level: Metal chalcogenide semiconductors. Electrochem. Acta. **40**(10), 1293–1303 (1995)
24. Yusupov, R.A., Abzalov, R.F., Smerdova, S.G., Gafarov, M.R.: Sophisticated heterophase equilibria in the system "Pb (II)—water—KOH". Chem. Comput. Simul. Butlerov Commun. **3**, 29–36 (2000) (in Russian)
25. O'Brien, P., Saeed, T.: Deposition and characterization of cadmium sulfide thin films by chemical bath deposition. J. Cryst. Growth. **158**(4), 497–504 (1996)
26. O'Brien, P., McAleese, J.: Developing an understanding of the processes controlling the chemical bath deposition of ZnS and CdS. J. Mater. Chem. **8**(11), 2309–2314 (1998)
27. Osherov, A., Ezersky, V., Golan, Y.: The role of solution composition in chemical bath deposition of epitaxial thin films of PbS on GaAs(100). J. Cryst. Growth. **308**(2), 334–339 (2007)
28. De Farias, P.M.A., Saegesser, Santos D., de Menezes, F.D., de Carvalho Ferreira, R., de Lourdes Barjas-Castro, M., Castro, V., Moura Lima, P.R., Fonte, A., Cesar, C.L.: Core-shell CdS/Cd(OH)2 quantum dots: synthesis ana bioconjugation to target res cells antigens. J. Microscopy. **219**(3), 103–108 (2005)
29. Kozhevnikova, N.S., Sadovnikov, S.I., Uritskaya, A.A., Gusev, A.I.: Lead homogeneous and heterogeneous ion equilibria in water solutions. Izv. VUZov. Khimiya i Khim. Technologiya—Chem. Chem. Technol. **55**(3), 13–18 (2012). (in Russian)

30. Kozhevnikova, N.S., Sadovnikov, S.I., Uritskaya, A.A., Gusev, A.I.: Considering the polynuclear complexes in the ionic equilibria of the Pb^{2+}–H_2O system. Zh. Obshch. Khim. **82**(4), 538–547 (2012) (in Russian) (Engl. Transl.: Russ. J. Gen. Chem. **82**(4), 626–634 (2012))

31. Powell, K.J., Brown, P.L., Byrne, R.H., Gajda, T., Hefter, G., Leuz, A.-K., Sjöberg, S., Wanner, H.: Chemical speciation of environmentally significant metals with inorganic ligands. Part 3: The Pb^{2+} + OH^-, Cl^-, CO_3^{2-}, SO_4^{2-}, and PO_4^{3-} systems (IUPAC Technical Report). Pure Appl. Chem. **81**(12), 2425–2476 (2009)

32. Markov, V.F., Maskaeva, L.N., Ivanov, P.N.: Calculation of the conditions of formation of the solid phase of metal chalcogenides by hydrochemical deposition. Condens. Media Interph. Bound. **6**(4), 374–380 (2004). (in Russian)

33. Wang, Y., Chai, L., Chang, H., Peng, X., Shu, Y.: Equilibrium of hydroxyl complex ions in Pb^{2+}–H_2O system. Trans. Nonferrous Met. Soc. China. **19**(2), 458–462 (2009)

34. Semenov, V.N., Ovechkina, N.M., Khoviv, D.A.: Influence ща hydroxo ycomplexes on the process of deposition and phase composition of the SnS and PbS films. Vestn. Voronezhsk. Univ., Ser. Khimiya, Biologiya, Farmatsiya. **2**, 50–55 (2007) (in Russian)

35. Dean, J.N. (ed.): Lange's Handbook of Chemistry, 15th edn., p. 1424. McGraw-Hill, New York (1998)

36. Patnaik, P.: Dean's Analytical Chemistry Handbook, 2nd edn., pp. 1280. McGraw-Hill, New York (2004) (Table 4.2)

37. Kawai, T.: Ishiguro Shin–ichi, Ohtaki H. A thermodynamic study on hydrolytic reactions of lead(II) ion in an aqueous solution and dioxane-water mixtures. I. Potentiometric study. Bull. Chem. Soc. Jpn. **53**(8), 2221–2227 (1980)

38. Sylva, R.N., Brown, P.L.: The hydrolysis of metal ions. Part 3. Lead(II). J. Chem. Soc., Dalton Trans. **9**(9), 1577–1581 (1980)

39. Cruywagen J.J., van de Water R.F: The hydrolysis of lead(II). A potentiometric and enthalpimetric study. Talanta **40**(7), 1091–1095 (1993)

40. Perera, W.N., Hefter, G., Sipos, P.M.: An investigation of the lead(II) - hydroxide system. Inorg. Chem. **40**(16), 3974–3978 (2001)

41. Pettit, L.D., Powell, K.J.: IUPAC Stability Constants Database (SC-Database, Release 5.8 for Windows). Academic Software and K.J. Powell, Ottley. www.acadsoft.co.uk (2009)

42. Tikhonov, A.S.: Study of complex lead citrate compounds depending on the pH of the aqueous medium. Sb. Trudov Voronezhsk. Gos. Univ. **49**, 23–24 (1958) (in Russian)

43. SigmaPlot 2001 for Windows, Version 7.0 © 1986–2001, SPSS Inc., USA

44. Lur'e Yu, Yu.: Handbook on Analytical Chemistry, p. 448. Khimiya, Moscow (1967) (in Russian)

45. Kitaev, G.A., Bol'shchikova, T.P., Yatlova, L.E.: On the question of solubility of salts of cyanamide with some metals. Zh. Neorg. Khim. **16**(12), 3173–3175 (1971). (in Russian)

46. Sadovnikov, S.I., Gusev, A.I.: Preparation of nanocrystalline lead sulfide powder with controlled particles size. Zh. Obshch. Khim. **84**(2), 177–184 (2014) (in Russian). (Engl. Transl.: Russ. J. Gen. Chem. **84**(2), 173–180 (2014))

47. Sadovnikov, S.I., Gusev, A.I.: Hydrochemical precipitation of nanocrystalline lead sulfide powders. Neorg. Mater. **51**(12), 1313–1318 (2015) (in Russian). (Engl. Transl.: Inorg. Mater. **51**(12), 1219–1224 (2015))

48. Noda, Y., Ohba, S., Sato, S., Saito, Y.: Charge distribution and atomic thermal parameters of lead chalcogenide crystals. Acta Crystallogr. B. **B39**(3), 312–317 (1983)

49. Noda, Y., Masumoto, K., Ohba, S., Saito, Y., Toriumi, K., Iwata, Y., Shibuya, K.: Temperature dependence of atomic thermal parameters of lead chalcogenide, PbS, PbSe, and PbTe. Acta Crystallogr. C. **C43**(8), 1443–1445 (1987)

50. Rempel, A.A., Kozhevnikova, N.S., Leenaers, A.J.G., van den Berghe, S.: Towards particle size regulation of chemically deposited lead sulfide (PbS). J. Cryst. Growth. **280**, 300–308 (2005)

51. Sadovnikov, S.I., Gusev, A.I.: A method of producing nanocrystalline lead sulfide, pp. 1–5. Patent No. 2591160 of Russian Federation (2016)

52. Jiang, Y., Wu, Y., Xie, B., Yuan, S.W., Liu, X.M.: Hydrothermal preparation of uniform cubic-shaped PbS nanocrystals. J. Cryst. Growth. **231**, 248–251 (2001)
53. Yang, Y.J., He, L.Y., Zhang, Q.F.: A cyclic voltametric synthesis of PbS nanoparticles. Electrochem. Commun. **7**(4), 361–364 (2005)
54. Sharon, M., Ramaiaha, K.S., Kumar, M., Neumann-Spallart, M., Levy-Clement, C.: Electrodeposition of lead sulphide in acidic medium. J. Electroanal. Chem. **436**, 49–52 (1997)
55. Yang, Y.J.: A novel electrochemical preparation of PbS nanoparticles. Mater. Sci. Eng. B. **131**(1–3), 200–202 (2006)
56. Pellegri, N., Trbojevich, R., de Sanctis, O.: Fabrication of PbS nanoparticles embedded in silica gel by reverse micelles and sol-gel routes. J. Sol-Gel Sci. Techn. **8**, 1023–1028 (1997)
57. Xu, L., Chen, X., Wang, L., Sui, Z.M., Zhao, J., Zhu, B.: Formation of lead sulfide nanoparticles via Langmuir-Blodgett technique. Colloids and Surface A: Physicochem. Eng. Aspects **257-258**, 457–460 (2005)
58. Yu, Y., Zhang, K., Sun, S.: One-pot aqueous synthesis of near infrared emitting PbS quantum dots. Appl. Surf. Sci. **258**, 7181–7187 (2012)
59. Gerion, D., Pinaud, F., Williams, S.C., Parak, W.J., Zanchet, D., Weiss, S., Alivisatos, A.P.: Synthesis and properties of biocompatible water-soluble silica-coated CdSe/ZnS semiconductor quantum dots. J. Phys. Chem. B. **105**(37), 8861–8871 (2001)
60. Gao, X., Cui, Y., Levenson, R.M., Chung, L.W.K., Nie, S.: In vivo cancer targeting and imaging with semiconductor quantum dots. Nature Biotechnology. **22**(8), 969–976 (2004)
61. Sathyamoorthy, R., Kungumadevi, L.: Facile synthesis of PbS nanorods induced by concentration difference. Advanc. Powd. Techn. **26**(2), 355–361 (2015)
62. Ding, B., Shi, M., Chen, F., Zhou, R., Deng, M., Wang, M., Chen, H.Z.: Shape-controlled synthesis of PbS submicro-/nanocrystals via hydrothermal method. J. Cryst. Growth. **311**(6), 1533–1538 (2009)
63. Emadi, H., Salavati-Niasari, M.: Hydrothermal synthesis and characterization of lead sulfide nanocubes through simple hydrothermal method in the presence of [bis(salicylate)lead(II)] as a new precursor. Superlatt. Microstr. **54**, 118–127 (2013)
64. Ni, Y., Liu, H., Wang, F., Liang, Y., Hong, J., Ma, X., Xu, Z.: Shape controllable preparation of PbS crystals by a simple aqueous phase route. Cryst. Growth Design. **4**(4), 759–764 (2004)
65. Huang, Q., Gao, L.: Simple route for synthesis of PbS dendritic nanostructured materials. Chem. Lett. **33**(10), 1338–1339 (2004)
66. Zhao, P.T., Chen, G., Hu, Y., He, X.L., Wu, K., Cheng, Y., Huang, K.X.J.: Preparation of dentritic PbS nanostructures by ultrasonic method. J. Cryst. Growth. **303**(2), 632–637 (2007)
67. Ma, Y., Qi, L., Ma, J., Cheng, H.: Hierarchical, star-shaped PbS crystals formed by a simple solution route. Cryst. Growth Design. **4**(2), 351–354 (2004)
68. Ding, Y.H., Liu, X.X., Guo, R.: Synthesis of hollow PbS nanospheres in pluronic F127/cyclohexane/H$_2$O microemulsions. Colloids Surf. A: Physicochem. Eng. Aspects. **296** (1–3), 8–18 (2007)
69. Leontidis, E., Orphanou, M., Kyprianidou-Leondidou, T., Krumeich, F., Caseri, W.: Composite nanotubes formed by self-assembly of PbS nanoparticles. Nano Letters. **3**(4), 569–572 (2003)
70. Li, G., Shi, G., Xu, H., Guang, S., Yin, R., Song, Y.: Nonlinear optical properties of the PbS nanorods synthesized via surfactant-assisted hydrolysis. Mater. Lett. **61**(8–9), 1809–1811 (2007)
71. Wang, W., Li, Q., Li, M., Lin, H., Hong, L.: Growth of PbS microtubes with quadrate cross sections. J. Cryst. Growth. **299**(1), 17–21 (2007)
72. Sun, J.-Q., Shen, X.-P., Guo, L.-J., Chen, K.-M., Liu, Q.: Microwave-assisted synthesis of flower-like PbS crystals. Physica E: Low-Dimens. Systems Nanostructures. **41**(8), 1527–1532 (2009)

73. Jiao, J., Liu, X., Gao, W., Wang, C., Feng, H., Zhao, X., Chen, L.: Synthesis of PbS nanoflowers by biomolecule-assisted method in the presence of supercritical carbon dioxide. Sol. State Sci. **11**(5), 976–981 (2009)
74. Shakouri-Arani, M., Salavati-Niasari, M.: A facile and reliable route to prepare of flower shaped lead sulfide nanostructures from a new sulfur source. J. Industr. Eng. Chem. **20**(5), 3141–3149 (2014)
75. Wu, M., Zhong, H., Jiao, Z., Li, Z., Sun, Y.: Synthesis of PbS nanocrystallites by electron beam irradiation. Colloids Surf. A: Physicochem. Eng. Aspects. **313–314**, 35–39 (2008)
76. Mozafari, M., Moztarzadeh, F., Seifalian, A.M., Tayebi, L.: Self-assembly of PbS hollow sphere quantum dots via gas-bubble technique for early cancer diagnosis. J. Luminesc. **133**, 188–193 (2013)
77. Schiener, A., Wlochowitz, T., Gerth, S., Unruh, T., Rempel, A., Amenitsch, H., Magerl, A.: Nucleation and growth of CdS nanoparticles observed by ultrafast SAXS. MRS Symp. Proc. **1528**, 1–6 (2013)
78. Schiener, A., Magerl, A., Krach, A., Seifert, S., Steinrück, H.-G., Zagorac, J., Zahn, D., Weihrich, R.: In-situ investigation of two-step nucleation and growth of CdS nanoparticles from solution. Nanoscale. **7**(26), 11328–11333 (2015)
79. Gusev, A.I.: Effects of the nanocrystalline state in solids. Uspekhi Fiz. Nauk. **168**(1), 55–83 (1998) (in Russian). (Engl. Transl.: Physics - Uspekhi. **41**(1), 49–76 (1998))
80. Gusev, A.I., Rempel, A.A.: Nanocrystalline Materials, p. 224. Nauka - Fizmatlit, Moscow (2000) (in Russian)
81. Gusev, A.I., Rempel, A.A.: Nanocrysnalline Materials, p. 351. Cambridge Intern. Science Publ, Cambridge (2004)
82. Gusev A.I.: Nanomaterials, Nanostructures, and Nanotechnologies. 3rd edn., p. 416. Nauka – Fizmatlit, Moscow (2009) (in Russian)
83. Okuno, T., Lipovskii, A.A., Ogawa, T., Amagai, I., Masumoto, Y.: Strong confinement of PbSe and PbS quantum dots. J. Luminesc. **87–89**, 491–493 (2000)
84. Wundke, K., Auxier, J., Schülzgen, A., Peyghambarian, N., Borrelli, N.F.: Room-temperature gain at 1.3 mm in PbS-doped glasses. App. Phys. Lett. **75**(20), 3060–3062 (1999)
85. Malyarevich, A.M., Gaponenko, M.S., Savitski, V.G., Yumashev, K.V., Rachkovskaya, G. E., Zakharevich, G.B.: Nonlinear optical properties of PbS quantum dots in boro-silicate glass. J. Non-Crystall. Solids. **353**, 1195–1200 (2007)
86. Kai, Xu, Heo, Jong: Precipitation of PbS quantum dots in glasses by thermal diffusion of Ag + ions from silver pastes. J. Non-Crystall. Solids. **387**, 76–78 (2014)
87. Del Monte, F., Xu, Y., Mackenzie, J.D.: Preparation and characterization of PbS quantum dots doped ormocers. J. Sol-Gel Sci. Techn. **17**, 37–45 (2000)
88. Pinero, M., de la Rosa-Fox, N., Erge-Montilla, R., Esquivias, L.: Small angle neutron scattering study of PbS quantum dots synthetic routes via sol-gel. J. Sol-Gel Sci. Techn. **26**, 527–531 (2003)
89. Krauss, T.D., Wise, F.W., Tanner, D.B.: Observation of coupled vibrational modes of a semiconductor nanocrystal. Phys. Rev. Lett. **76**(8), 1376–1379 (1996)
90. Haché, A.: LeBlanc Serge-Emile, LoCascio M., Martucci A. Optical switchings pectroscopy of PbS quantum dots with dual-wavelength pump-probe. Physica E. **17**, 104–106 (2003)
91. Ullrich, B., Wang, J.S.: Impact of laser excitation variations on the photoluminescence of PbS quantum dots on GaAs. J. Luminesc. **143**, 645–648 (2013)
92. Yu, Y., Zhang, K., Sun, S.: Effect of ligands on the photoluminescence properties of water-soluble PbS quantum dots. J. Molec. Str. **1031**, 194–200 (2013)
93. Pentia, E., Pintilie, L., Matei, I., Botila, T., Pintilie, I.: Combined chemical-physical methods for enhancing IR photoconductive properties of PbS thin films. Infrared Phys. Techn. **44**(3), 207–211 (2003)
94. Hodes, G.: Semiconductor and ceramic nanoparticle films deposited by chemical bath deposition. Phys. Chem. Chem. Phys. **9**(18), 2181–2196 (2007)

95. Sadovnikov, S.I., Gusev, A.I., Rempel, A.A.: New crystalline phase in thin lead sulfide films. Pisma v ZhETF. **89**(5), 279–284 (2009) (in Russian). (Engl. Transl.: JETP Lett. **89**(5), 238–243 (2009))

96. Sadovnikov, S.I., Rempel, A.A.: Crystal structure of nanostructured PbS films at temperatures of 293–423 K. Fiz. tverd. Tela. **51**(11), 2237–2245 (2009) (in Russian). (Engl. Transl.: Phys. Sol. State. **51**(11), 2375–2383 (2009))

97. Sadovnikov, S.I., Kozhevnikova, N.S., Gusev, A.I.: Optical properties of nanostructured lead sulfide films with a D03 cubic structure. Fiz. Tekhn. Poluprovodnikov. **45**(12), 1621–1632 (2011) (in Russian). (Engl. Transl.: Semiconductors. **45**(12), 1559–1570 (2011))

98. Sadovnikov, S.I., Kozhevnikova, N.S., Pushin, V.G., Rempel, A.A.: Microstructure of nanocrystalline PbS powders and films. Neorg. Mater. **48**(1), 26–33 (2012) (in Russian). (Engl. Transl.: Inorg. Mater. **48**(1), 21–27 (2012))

99. Sadovnikov, S.I., Kozhevnikova, N.S.: Microstructure and crystal structure of nanocrystalline powders and films. Fiz. tverd. Tela. **54**(8), 1459–1465 (2012) (in Russian). (Engl. Transl.: Phys. Sol. State. **54**(8), 1554–1561 (2012))

100. Sadovnikov, S.I., Gusev, A.I.: Structure and properties of PbS films. J. Alloys Comp. **573**, 65–75 (2013)

101. Sadovnikov, S.I., Rempel, A.A.: Method of producing thin films of lead sulfide, pp. 1–5. Patent No. 2553858 of Russian Federation (2015)

102. Fainer, N.I., Kosinova, M.L., Rumyantsev, YuM, Salman, E.G., Kuznetsov, F.A.: Growth of PbS and CdS thin films by low-pressure chemical vapour deposition using dithiocarbamates. Thin Solid Films. **280**(1–2), 16–19 (1996)

103. Chamberlin, R.R., Sharman, J.S.: Chemical spray deposition process for inorganic films. J. Electrochem. Soc. **113**(1), 86–89 (1966)

104. Thangaraju, B., Kaliannan, P.: Polycrystalline lead thin chalcogenide thin films grown by spray pyrolysis. Cryst. Res. Technol. **35**(1), 71–75 (2000)

105. Nicolau, Y.F.: Solution deposition of thin solid compound films by a successive ionic-layer adsorption and reaction process. Appl. Surf. Sci. **22**(23), 1061–1074 (1985)

106. Nicolau, Y.F.: Process and apparatus for the deposition on a substrate of a thin film of a compound containing at least one cationic constituent and at least one anionic constituent, pp. 1–3. US Patent No. 4675207 (1987)

107. Kanniainen, T., Lindroos, S., Ihanus, J., Leskela, M.: Growth of strongly orientated lead sulfide thin films by successive ionic layer adsorption and reaction (SILAR) technique. J. Mater. Chem. **6**(?), 161–164 (1996)

108. Puišo, J., Tamuleviius, S., Laukaitis, G., Lindroos, S., Leskclä, M., Snitka, V.: Growth of PbS thin films on silicon substrate by SILAR technique. Thin Solid Films. **403–404**, 457–461 (2002)

109. Puišo, J., Lindroos, S., Tamulevičius, S., Leskelä, M., Snitka, V.: Growth of ultra thin PbS films by SILAR technique. Thin Solid Films. **428**, 223–226 (2003)

110. Preetha, K.C., Murali, K.V., Ragina, A.J., Deepa, K., Remadevi, T.L.: Effect of cationic precursor pH on optical and transport properties of SILAR deposited nano crystalline PbS thin films. Curr. Appl. Phys. **12**(1), 53–59 (2012)

111. Yucel, E., Yucel, Y., Beleli, B.: Process optimization of deposition conditions of PbS thin films grown by a successive ionic layer adsorption and reaction (SILAR) method using response surface methodology. J. Cryst. Growth. **422**, 1–7 (2015)

112. Nair, P.K., Garcia, V.M., Hernandez, A.B., Nair, M.T.S.: Photoaccelerated chemical deposition of PbS thin films: novel applications in decorative coatings and imaging techniques. J. Phys. D: Appl. Phys. **24**(8), 1466–1472 (1991)

113. Zhukovskiy, M.A., Stroyuk, A.L., Shvalagin, V.V., Smirnova, N.P., Lytvyn, O.S., Eremenko, A.M.: Photocatalytic growth of CdS, PbS, and CuxS nanoparticles on the nanocrystalline TiO2 films. J. Photochem. Photobiol. A: Chem. **203**(2–3), 137–144 (2009)

114. Ananikov, V.P., Khemchyan, L.L., Ivanova, Y.V., Bukhtiyarov, V.I., Sorokin, A.M., Prosvirin, I.P., Vatsadze, S.Z., Medved'ko, A.V., Nuriev, V.N., Dilman, A.D., Levin, V.V., Koptyug, I.V., Kovtunov, K.V., Zhivonitko, V.V., Likholobov, V.A., Romanenko, A.V., Simonov, P.A., Nenajdenko, V.G., Shmatova, O.I., Muzalevskiy, V.M., Nechaev, M.S., Asachenko, A.F., Morozov, O.S., Dzhevakov, P.B., Osipov, S.N., Vorobyeva, D.V., Topchiy, M.A., Zotova, M.A., Ponomarenko, S.A., Borshchev, O.V., Luponosov, Y.N., Rempel, A.A., Valeeva, A.A., Stakheev, A.Y., Turova, O.V., Mashkovsky, I.S., Sysolyatin, S.V., Malykhin, V.V., Bukhtiyarova, G.A., Terent'ev, A.O., Krylov, I.B.: Development of new methods in modern selective organic synthesis: preparation of functionalized molecules with atomic precision. Russ. Chem. Rev. **83**(10), 885–985 (2014)
115. Chen, J.-H., Chao, C.-G., Ou, J.-C., Liu, T.-F.: Growth and characteristics of lead sulfide nanocrystals produced by the porous alumina membrane. Surf. Sci. **601**(22), 5142–5147 (2007)
116. Qadri, S.B., Yang, J., Ranta, B.R., Skelton, E.F., Hu, J.Z.: Pressure induced structural transitions in nanometer size particles of PbS. Appl. Phys. Lett. **69**(15), 2205–2207 (1996)
117. Knorr, K., Ehm, L., Hytha, M., Winkler, B., Depmeier, W.: The high-pressure & #x03B1;/β phase transition in lead sulphide (PbS). Eur. Phys. J. B. **31**(3), 297–303 (2003)
118. Zhang, J., Sun, L., Liao, S., Yan, C.: Size control and photoluminescence enhancement of CdS nanoparticles prepared via reverse micelle method. Solid State Commun. **124**(1–2), 45–48 (2002)
119. Metin, H., Esen, R.: Annealing studies on CBD grown CdS thin films. J. Cryst. Growth **258** (1–2), 141–148 (2003)
120. Wu, G.S., Yuan, X.Y., Xie, T., Xu, G.C., Zhang, L.D., Zhuang, Y.L.: A simple synthesis route to CdS nanomaterials with different morphologies by sonochemical reduction. Mat. Lett. **58**(5), 794–797 (2004)
121. Vorokh, A.S., Rempel, A.A.: Atomic structure of cadmium sulfide nanoparticles. Fiz. tverd. Tela. **49**(1), 143–148 (2007) (in Russian). (Engl. Transl.: Phys. Sol. State. **49**(1), 148–153 (2007))
122. Rempel, A.A., Magerl, A.: Non-periodicity in nanoparticles with close-packed structures. Acta Crystallogr. A. **A66**(4), 479–483 (2010)
123. Qadri, S.B., Singh, A., Yousuf, M.: Structural stability of PbS films as a function of temperature. Thin Solid Films. **431–432**, 506–510 (2003)
124. Fernandez-Lima, F.A., Gonzalez-Alfaro, Y., Larramendi, E.M., Fonseca Filho, H.D., Maia da Costa, M.E.H., Freire Jr., F.L., Prioli, R., de Avillez, R.R., da Silveira, E.F., Calzadilla, O., de Melo, O., Pedrero, E., Hernández, E.: Structural characterization of chemically deposited PbS thin films. Mater. Sci. Eng. B. **136**(2–3), 187–192 (2007)
125. Gotoh, Y., Onoda, M., Goto, M., Oosawa, Y.: Preparation and characterization of "PbVS3" a new composite layered compound. Chem. Lett. **18**(7), 1281–1282 (1989)
126. Wiegers, G.A., Meetsma, A., Haange, R.J., van Smaalen, S., de Boer, J.L., Meerschaut, A., Rabu, P., Rouxel, J.: The incommensurate misfit layer structure of (PbS)1.14NbS2 "PbNbS3" and (LaS)1.14NbS2 "LaNbS3": an x-ray diffraction study. Acta Crystallog. **B46** (3), 324–332 (1990)
127. Wullf, J., Meetsma, A., van Smaalen, S., Haange, R.J., de Boer, J.L., Wiegers, G.A.: Structure, electrical transport and magnetic properties of the misfit layer compound (PbS) 1.13TaS2. J. Solid State Chem. **84**(1), 118–129 (1990)
128. Wiegers, G.A.: Misfit layer compounds: Structures and physical properties. Progr. Solid State Chem. **24**(1–2), 1–139 (1996)
129. Sadovnikov, S.I., Rempel, A.A.: Nonstoichiometric distribution of sulfur atoms in lead sulfide structure. Dokl. Akad. Nauk. **428**(1), 48–52 (2009) (in Russian). (Engl. Transl.: Dokl. Phys. Chem. **428**(1), 167–171 (2009))
130. X'Pert Plus Version 1.0. Program for Crystallography and Rietveld analysis Philips Analytical B. V. © Koninklijke Philips Electronics N. V
131. Philips Analytical.: Philips Analytical X'Celerator. J. Appl. Crystallogr. **34**(4), 538 (2001)

132. Morton, R.W., Simon, D.E., Gislason, J.J., Taylor, S.: Managing background profiles using a new X'Celerstor detector. Adv. X-ray Anal. **46**, 80–85 (2003)
133. Rietveld, H.M.: A profile refinement method for nuclear and magnetic structures. J. Appl. Cryst. **2**(2), 65–71 (1969)
134. Sadovnikov, S.I., Rempel, A.A.: Correlation of sulfur atoms in nonmetal planes of lead sulfide films with the D03 structure. Fiz. tverd. Tela. **52**(12), 2299–2306 (2010) (in Russian). (Engl. Transl.: Phys. Sol. State. **52**(12), 2458-2466 (2010))
135. Gusev, A.I., Rempel, A.A., Magerl, A.J.: Disorder and Order in Strongly Nonstoichiometric Compounds. Transition Metal Carbides, Nitrides and Oxides, p. 608. Springer, Berlin (2001)
136. Sadovnikov, S.I, Rempel, A.A.: Simulation of pair and three-particle correlations in a binary solid solution with a hexagonal lattice. Fiz. tverd. Tela. **50**(6), 1085–1089 (2008). (in Russian). (Engl. Transl.: Phys. Sol. State. **50**(6), 1131–1136 (2008))
137. Moss, T.S.: Optical Properties of Semiconductors, p. 279. In: Hogarth, C.A. (ed.) Butterworths Sci. Publ. Ltd., London (1959)
138. Zemmel, J.N., Jensen, J.D., Schoolar, R.B.: Electrical and optical properties of epitaxial films of PbS, PbSe, PbTe and SnTe. Phys. Rev. **140**(1A), 330–342 (1965)
139. Ukhanov Yu, I.: Optical Properties of Semiconductors, p. 366. Nauka, Moscow (1977) (in Russian)
140. Gusev, A.I.: Nanocrystalline Materials: Production and Properties, p. 200. Ural Division of the RAS, Ekaterinburg (1998) (in Russian)
141. Sashchiuk, A., Lifshitz, E., Reisfeld, R., Saraidarov, T., Zelner, M., Willenz, A.J.: Optical and conductivity properties of PbS nanocrystals in amorphous zirconia sol-gel films. Sol-Gel Sci. Techn. **24**(1), 31–38 (2002)
142. Yu, B., Yin, G., Zhu, G., Gan, F.: Optical nonlinear properties of PbS nanoparticles studied by the Z-scan technique. Opt. Mater. **11**(1), 17–21 (1998)
143. Jana, S., Thapa, R., Maity, R., Chattopadhyay, K.K.: Optical and dielectric properties of PVA capped nanocrystalline PbS thin films synthesized by chemical bath deposition. Phys. E. **40**(10), 3121–3126 (2008)
144. Brus, L.E.: Electron-electron and electron-hole interactions in small semiconductor crystallites: The size dependence of the lowest excited electronic state. J. Chem. Phys. **80**(9), 4403–4409 (1984)
145. Wang, Y., Herron, N.: Nanometer-sized semiconductor clusters: materials synthesis, quantum size effects, and photophysical properties. J. Chem. **95**(2), 525–532 (1991)
146. Najdovski, M., Minceva-Sukarova, B., Drake, A., Grozdanov, I., Chunnilall, C.J.: Optical properties of thin solid films of lead sulfide. J. Mol. Struct. **349**(1), 85–88 (1995)
147. Parra, R.S., George, P.J., Sánchez, G.G., Jiménez González, A.E., Baños, L., Nair, P.K.: Optical and electrical properties of PbS+ In thin films subjected to thermal processing. J. Phys. Chem. Solids. **61**(5), 659–668 (2000)
148. Valenzuel-Jaureguia, J.J., Ramirez-Bon, R., Mendoza-Galvan, A., Sotelo-Lerma, M.: Optical properties of PbS thin films chemically deposited at different temperatures. Thin Solid Films **441**, 104–110 (2003)
149. Peterson, J.J., Krauss, T.D.: Fluorescence spectroscopy of single lead sulfide quantum dots. Nano Lett. **6**(3), 510–514 (2006)
150. Zhao, Y., Zou, J., Shi, W.: In situ synthesis and characterization of lead sulfide nanocrystallites in the modified hyperbranched polyester by gamma-ray irradiation. Mater. Sci. Eng. B. **121**(1–2), 20–24 (2005)
151. Sadovnikov, S.I., Kozhevnikova, N.S., Rempel, A.A.: The structure and optical properties of nanocrytalline lead sulfide films. Fiz. Tekhn. Poluprovodnikov. **44**(10), 1394–1400 (2010) (in Russian). (Engl. Transl.: Semiconductors. **44**(10), 1349–1356 (2010))
152. Tauc, J. (ed.): Amorphous and Liquid Semiconductors. Plenum, New York (1974)
153. Klingshirn, C.F.: Semiconductor Optics, p. 797. Springer, New York (2005)
154. Pankove, J.I.: Optical processes in semiconductors, 2nd edn, p. 428. Dover Publ, New York (1975)

155. Elliot, R.J.: Intensity of optical absorption by excitons. Phys. Rev. **108**(6), 1384–1389 (1957)
156. Mittleman, D.M., Schoenlein, R.W., Shiang, J.J., Colvin, V.L., Alivisatos, A.P., Shank, C. V.: Quantum size dependence of femtosecond electronic dephasing and vibrational dynamics in CdSe nanocrystals. Phys. Rev. B. **49**(20), 14435–14447 (1994)
157. Mozer, F., Urbach, F.: Optical absorption of pure silver halides. Phys. Rev. **102**(6), 1519–1523 (1956)
158. Kumara, D., Agarwal, G., Tripathi, B., Vyas, D., Kulshrestha, V.: Characterization of PbS nanoparticles synthesized by chemical bath deposition. J. Alloys Comp. **484**, 463–466 (2009)
159. Wang, Y., Suna, A., Mahier, W., Kasowski, R.: PbS in polymers. From molecules to bulk solids. J. Chem. Phys. **87**(12), 7315–7322 (1987)
160. Hellwege, K.-H., Madelung, O. (eds): Landolt-Börnstein: Zahlenwerte und Funktionen aus Naturwissenschaften und Technik – Neue Serie/Grouppe III: Kristall- und Festkorperphysik, Band 17f, pp. 155–162. Springer, Berlin (1983)
161. Mamiyev, Z.Q.: Balayeva N.O. Preparation and optical studies of PbS nanoparticles. Optic. Mat. **46**, 522–525 (2015)
162. Lifshitz, E., Sirota, M., Porteanu, H.: Continuous and time-resolved photoluminescence study of lead sulfide nanocrystals, ebmedded in polymer film. J. Cryst. Growth. **196**, 126–134 (1999)
163. Navaneethan, M., Sabarinathan, M., Harish, S., Archana, J., Nisha, K.D., Hayakawa, Y., Ponnusamy, S., Muthamizhchelvan, C.: Chemical synthesis and functional properties of multi-ligands passivated lead sulfide nanoparticles. Mat. Lett. **158**, 75–79 (2015)
164. Sharma, S.S:. Thermal expansion of crystals. Part VIII. Galena and pyrite. Proc. Indian Acad. Sci. Sect.A. **A34**(2), 72–76 (1951)
165. Novikova, S.I., Abrikosov, N.Kh.: Investrigation of thermal expansion of the lead chalcogenides. Fiz. Tved. Tela. **5**(7), 1913–1916 (1963) (in Russian) (Engl. Transl.: Sov. Phys. Solid State. **5**(7), 1558–1559 (1963)
166. Zhang, Yi, Ke, X., Chen, C., Yang, J., Kent, P.R.C.: Thermodynamic properties of PbTe, PbSe, and PbS: First-principles study. Phys. Rev. B. **B80**(2), 12 (2009) Paper 024304
167. Sadovnikov, S.I., Kozhevnikova, N.S., Rempel, A.A., Magerl, A.: Thermal expansion of a lead sulfide nanofilm. Thin Solid Films. **548**, 230–234 (2013)
168. Sadovnikov, S.I., Gusev, A.I.: Effect of particle size on the thermal expansion of nanostructured lead sulfide films. J. Alloys Comp. **610**, 196–202 (2014)
169. Sadovnikov, S.I., Gusev, A.I.: Thermal expansion of nanostructured PbS films and anharmonicity of atomic vibrations. Fiz. Tverd. Tela. **56**(11), 2274–2278 (2014) (in Russian). (Engl. Transl.: Phys. Sol. State. **56**(11), 2353–2358 (2014)
170. Ashcroft, N.W., Mermin, N.D.: Solid State Physics, pp. 492–494. Cornell University, New York (1976)
171. Petrov Yu, I. Physics of Small Particles, 360 pp. Nauka, Moscow (1982) (in Russian)
172. Bolt, R.H.: Frequency distribution of eigentones in a three-dimensional continuum. J. Acoust. Soc. Am. **10**(3), 228–234 (1939)
173. Maa, D.-Y.: Distribution of eigentones in a rectangular chamber at low frequency range. J. Acoust. Soc. Am. **10**(3), 235–238 (1939)
174. Montrol, E.W.: Size effect in low temperature heat capacities. J. Chem. Phys. **18**(2), 183–185 (1950)
175. Chudinov, A.A.: Dependence of velocity of ultrasound in monocrystals PbS on temperature in the range of 80–640 K. Fiz. tverd. Tela. **5**(5), 1458–1460 (1963). (in Russian). (Engl. Transl.: Sov. Phys. Sol. State. **5**(5), 1061–1062 (1963))
176. Padaki, V., Lakshimikumar, S., Subramanyam, S., Gopal, E:. Elastic constants of galena down to liquid helium temperatures. Pramana (J. Phys.) **17**(1), 25–32 (1981)
177. Li, W., Chen, J.-F., Wang, T.: Electronic and elastic properties of PbS under pressure. Physica B. **405**, 1279–1282 (2010)

178. Bhardwaj, P:. Investigation of structural phase transition of PbS. ISRN Cond. Matter Physics. **2012** (2012). Article ID 596397
179. Pei, Y.-L., Liu, Y.: Electrical and thermal transport properties of Pb-based chalcogenides: PbTe, PbSe, and PbS. J. Alloys Comp. **514**, 40–44 (2012)
180. Peresada, G.I., Ponyatovskii, E.G., Sokolovskaya, Zh.D.: Pressure dependence of the elastic constants of PbS. Phys. Status Sol. **35**(2), K177–K180 (1976)
181. Choudhury, N., Sarma, B.K.: Structural characterization of lead sulfide thin films by means of X-ray line profile analysis. Bull. Mater. Sci. **32**(1), 43–47 (2009)
182. Sadovnikov, S.I., Kozhevnikova, N.S., Rempel, A.A.: Thermal stability of lead sulfide nanocrystalline films. Fiz. Khim. Stekla. **35**(1), 74–82 (in Russian). (Engl. Transl.: Glass Phys. Chem. **35**(1), 60–66 (2009))
183. Sadovnikov, S.I., Kozhevnikova, N.S., Rempel, A.A:. Oxidation of nanocrystalline lead sulfide in air. Zh. neorg. Khimii. 56(12), 1951–1957 (2011) (in Russian). (Engl. Transl.: Russ. J. Inorg. Chem. **56**(12), 1864–1869 (2011))
184. Sadovnikov, S.I., Kozhevnikova, N.S., Rempel, A.A.: Stability and recrystallization of PbS nanoparticles. Neorg. Mater. **47**(8), 929–935 (2011) (in Russian). (Engl. Transl.: Inorg. Mater. **47**(8), 837–843 (2011))
185. Gertsman, V.Y., Birringer, R., Valiev, R.Z., Gleiter, H.: On the structure and strength of ultrafine-grained copper produced by severe plastic deformation. Scr. Met. Mat. **30**(2), 229–234 (1994)
186. Mikhlin Yu, L., Romanchenko, A.S., Shagaev, A.A.: Scanning probe microscopy studies of PbS surfaces oxidized in air and etched in aqueous acid solutions. Appl. Surf. Sci. **252**(16), 5645–5648 (2006)
187. Ihly, R., Tolentino, J., Liu, Y., Gibbs, M., Law, M.: The photothermal stability of PbS quantum dot solids. ACS Nano. **5**(10), 8175–8186 (2011)
188. García, V.M., Nair, M.T.S., Nair, P.K.: Optical properties of PbS·Cu$_x$S and Bi$_2$S$_3$·Cu$_x$S thib films with reference to solar control and solar absorber applications. Sol. Energy Mater. **23**(1), 47–59 (1991)
189. Nair, P.K., Gomezdaza, O., Nair, M.T.S.: Metal sulphide thin film photography with lead sulphide thin films. Adv. Mater. Opt. Electron. **1**(3), 139–145 (1992)
190. Loiko, P.A., Rachkovskaya, G.E., Zakharevich, G.B., Gurin, V.S., Gaponenko, M.S., Yumashev, K.V.: Optical properties of novel PbS and PbSe quantum-dot-doped alumino-alkali-silicate glasses. J. Non-Cryst. Solids **358**(15), 1840–1845 (2012)
191. Carrillo-Castillo, A., Salas-Villasenor, A., Mejia, I., Aguirre-Tostado, S., Gnade, B.E., Quevedo-Lopez, M.A.: P-type thin films transistors with solution-deposited lead sulfide films as semiconductor. Thin Solid Films. **520**, 3107–3110 (2012)
192. He, J., Luo, M., Hu, L., Zhou, Y., Jiang, S., Song, H., Ye, R., Chen, J., Gao, L., Tang, J.: Flexible lead sulfide colloidal quantum dot photodetector using pencil graphite electrodes on paper substrates. J. Alloys Comp. **596**, 73–78 (2014)
193. Slonopas, A., Alijabbari, N., Saltonstall, C., Globus, T., Norris, P.: Chemically deposited nanocrystalline lead sulfide thin films with tunable properties for use in photovoltaics. Electrochim. Acta. **151**, 140–149 (2015)
194. Sabet, M., Salavati-Niasari, M.: Deposition of lead sulfide nanostructure films on TiO2 surface via different chemical methods due to improving dye-sensitized solar cells efficiency. Electrocim. Acta. **169**, 168–179 (2015)
195. Jang, J., Song, J.H., Choi, Y., Baik, S.J., Jeong, S.: Photovoltaic light absorber with spatial energy band gradient using PbS quantum dot layers. Solar Energy Mater. Solar Cells. **141**, 270–274 (2015)
196. Zimin, S.P., Gorlachev, E.S.: Nanostructured Lead Chalcogenides, 230 pp. Yaroslavl's State University, Yaroslavl (2011) (in Russian)
197. Markov, V.F., Maskaeva, L.N.: Lead sulfide semiconductor sensing element for nitrogen oxide gas analyzers. Zh. Anal. Khimii. **56**(8), 846–850 (2001) (in Russian). (Engl. Transl.: J. Anal. Chem. **56**(8), 754–757 (2001))

198. Fu, T.: Research on gas-sensing properties of lead sulfide-based sensor for detection of NO2 and NH3 at room temperature. Sens. Actuators B. **140**(1), 116–121 (2009)
199. Bandyopadhyay, S.: Performance of nanocrystalline PbS gas sensor with improved cross-sensitivity. Particul. Sci. Technol. **30**(1), 43–54 (2012)
200. Karami, H., Ghasemi, M., Matini, S.: Synthesis, characterization and application of lead sulfide nanostructures as ammonia gas sensing agent. Int. J. Electrochem. Sci. **8**(10), 11661–11679 (2013)
201. Kaci, S., Keffous, A., Hakoum, S., Mansri, A.: Hydrogen sensitivity of the sensors based on nanostructured lead sulfide thin films deposited on a-SiC: H and p-Si(100) substrates. Vacuum **116**, 27–30 (2015)
202. Kullick, T., Quack, R., Röhrkasten, C., Pekeler, T., Scheper, T., Schügerl, K.: PbS-field-effect-transistor for heavy-metal concentration monotoring. Chem. Eng. Technol. **18**(4), 225–228 (1995)
203. Markov, V.F., Maskaeva, L.N., Zurabin, I.V., Zamaraeva, N.V.: Application of thin films of lead sulfide doped with halogens, to monitor the content of lead ions in aqueous media. Water: Chem. Ecolog. **6**, 80–85 (2012) (in Russian)
204. Xie, Y., Qiao, Z., Chen, M., Liu, X., Qian, Y.: Irradiation route to semiconductor/polymer nanocable fabrication. Adv. Mater. **11**(18), 1512–1515 (1999)
205. Garcia, O.P., de Albuquerque, M.C.C., da Silva Aquino, K.A., de Araujo, P.L.B., de Araujo, E.S.: Use of lead(II) sulfide nanoparticles as stabilizer for PMMA exposed to gamma irradiation. Mater. Res. **18**(2), 365–372 (2015)
206. Kirpichnikov, M.P., Kochetkov, S.N.: Chemistry and biomedicine: diversity and unity of goals. Russ. Chem. Rev. **84**(1), 1 (2015)
207. Povolotskaya, A.V., Povolotskiy, A.V., Manshina, A.A.: Hybrid nanostructures: synthesis, morphology and functional properties. Russ. Chem. Rev. **84**(6), 579–600 (2015)
208. Andreakou, P., Brossard, M., Bernechea, M., Konstantatos, G., Lagoudakis, P. Resonance energy transfer from PbS colloidal quantum dots to bulk silicon: the road to hybrid photovoltaics. In: Proceedings of SPIE "Physics, Simulation, and Photonic Engineering of Photovoltaic Devices", vol. 8256, pp. 82561L-1–82561L-6 (2012)
209. Narayanan, S., Sathy, B.N., Mony, U., Koyakutty, M., Nair, S.V., Menon, D.: Biocompatible magnetite/gold nanohybrid contrast agents via green chemistry for MRI and CT bioimaging. ACS Appl. Mater. Interfaces. **4**(1), 251–260 (2012)
210. Genuino, H., Huang, H., Njagi, E., Stafford, L., Suib, S.L.: A review of green synthesis of nanophase inorganic materials for green chemistry applications. In: Perosa, A., Selvav, M. (eds.) Handbook of Green Chemistry, vol. 8, pp. 217–244. Green Nanoscience. Wiley-VCH, Weinheim (2012)
211. Shen, A., Chen, L., Xie, W., Hu, J., Zeng, A., Richards, R., Hu, J.: Triplex Au–Ag–C core–shell nanoparticles as a novel Raman label. Adv. Funct. Mater. **20**(6), 969–975 (2010)
212. Lu, Y.J., Wei, K.C., Ma, C.C., Yang, S.Y., Chen, J.P.: Dual targeted delivery of doxorubicin to cancer cells using folate-conjugated magnetic multi-walled carbon nanotubes. Colloids Surf. B. **89**, 1–9 (2012)
213. Argyo, C., Weiss, V., Braeuchle, C., Bein, T. Multifunctional mesoporous silica nanoparticles as a universal platform for drug delivery. Chem. Mater. **26**(1), 435–451 (2014)
214. Fernandes, A.M., Abdalhai, M.H., Ji, J., Xi, B.-W., Xie, J., Sun, J., Noeline, R., Lee, B.H., Sun, X.: Development of highly sensitive electrochemical genosensor based on multiwalled carbon nanotubes-chitosan-bismuth and lead sulfide nanoparticles for the detection of pathogenic Aeromonas. Biosens. Bioelectr. **63**, 399–406 (2015)
215. Li, Q., Hu, X., Bai, Y., Alattar, M., Ma, D., Cao, Y., Hao, Y., Wang, L., Jiang, C.: The oxidative damage and inflammatory response induced by lead sulfide nanoparticles in rat lung. Food Chem. Toxicol. **60**, 213–217 (2013)
216. Cao, Y., Liu, H., Li, Q., Wang, Q., Zhang, W., Chen, Y., Wang, D., Cai, Y.: Effect of lead sulfide nanoparticles exposure on calcium homeostasis in rat hippocampus neurons. J. Inorg. Biochem. **126**, 70–75 (2013)

217. Huang, N., Zhao, Q., Xiao, J., Qi, L.: Controllable self-assembly of PbS nanostars into ordered structures: Close-packed arrays and patterned arrays. ACS Nano. **4**(8), 4707–4716 (2010)
218. Sadovnikov, S.I., Gusev, A.I., Rempel, A.A.: Nonstoichiometry of nanocrystalline monoclinic silver sulfide. Phys. Chem. Chem. Phys. **17**(19), 12466–12471 (2015)
219. Gusev, A.I., Sadovnikov, S.I., Chukin, A.V., Rempel, A.A.: Thermal expansion of nanocrystalline and coarse-crystalline silver sulfide Ag_2S. Fiz. Tverd. Tela. **58**(2), 246–251 (2016) (in Russian). (Engl. Transl.: Phys. Solid State. **58**(2), 251–257 (2016))
220. Sadovnikov, S.I. Gusev, A.I., Chukin, A.V., Rempel, A.A.: High-temperature X-ray diffraction and thermal expansion of nanocrystalline and coarse-crystalline acanthite α-Ag_2S and argentite β-Ag_2S. Phys. Chem. Chem. Phys. **18**(6), 4617–4626 (2016)
221. Chen, X.F., He, G., Liu, M., Zhang, J.W., Deng, B., Wang, P.H., Zhang, M., Lv, J.G., Sun, Z.Q.: Modulation of optical and electrical properties of sputtering-derived amorphous InGaZnO thin films by oxygen partial pressure. J. Alloys Comp. **615**, 636–642 (2014)

Chapter 3
Nanostructured Cadmium Sulfide CdS

Cadmium sulfide CdS is a single compound existing in the Pb–S system (Fig. 3.1). Cadmium sulfide contains ~ 50 at.% S and has the melting temperature 1678 K [1]. According to [2, 3], the melting temperature of CdS at a pressure 10.5 atm is 1748 ± 15 K. At atmospheric pressure, sublimation of CdS starts at 1253 K.

Low-temperature cubic (space group $F\bar{4}3m$) β-CdS and high-temperature hexagonal (space group $P6_3mc$) α-CdS polymorphs of bulk cadmium sulfide were reported in 1889 for the first time by Klobukow [4] and later were identified as the zinc-blende or sphalerite ZnS ($B3$ type) and wurtzite ZnS ($B4$ type) structures, respectively [5]. The coordinates of Cd and S atoms in the unit cells of cadmium sulfide with a cubic sphalerite type structure and hexagonal wurtzite type structure are given in Tables 3.1 and 3.2 in accordance with monograph [6]. Lattice constants of the unit cells of cubic β-CdS and hexagonal α-CdS are given according to data [7, 8], respectively. Transition from a cubic phase in hexagonal phase at heating above 653 K, and also from amorphous (disordered) structure in hexagonal phase was observed in work [9]. Under a pressure of 2.7 GPa, cadmium sulfide with the hexagonal (space group $P6_3mc$) $B4$ structure of wurtzite type acquires cubic (space group $Fm\bar{3}m$) $B1$ structure. At this transition, cadmium sulfide keeps semiconducting properties [10].

Both crystal structures β-CdS and α-CdS are built of alternating close-packed layers, each of them being comprised of sublayers of different-type ions (Fig. 3.2).

The sphalerite structure is a cubic close packing with layers arranged normal to four directions [111] of the face-centered cubic (fcc) lattice. Each anion is surrounded by six same anions in atomic planes and by twelve same-type anions in the three-dimensional space. Cations occupy 50% of tetrahedral voids, being equidistant from four anions. The wurtzite structure is characterized by hexagonal close packing (hcp) with layers arranged normal to [001] direction.

The polymorphs described above are typical of bulk CdS under normal conditions. Both structures are close-packed and only differ in the sequence of alternation of the packing layers on stacking. Namely, the structures can be obtained by

© Springer International Publishing AG 2018
S.I. Sadovnikov et al., *Nanostructured Lead, Cadmium, and Silver Sulfides*,
Springer Series in Materials Science 256, DOI 10.1007/978-3-319-56387-9_3

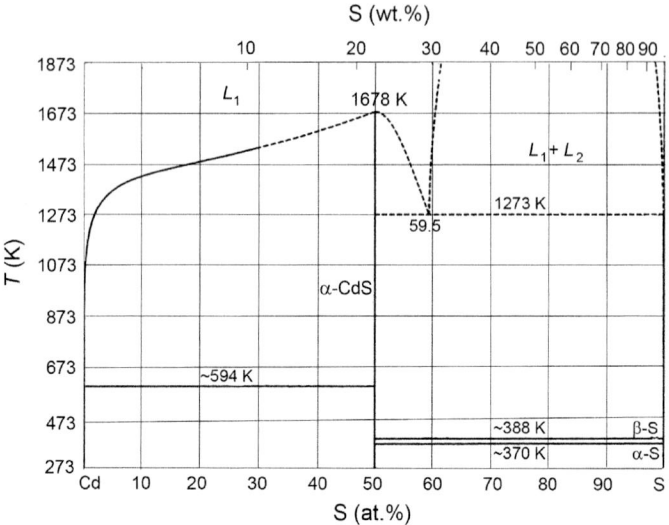

Fig. 3.1 Phase diagram of the Cd–S system [1]

Table 3.1 Low-temperature cubic (space group No. 216 – $F\bar{4}3m$ (T_d^2)) crystal structure of $B3$ type (zinc-blende or sphalerite ZnS type) of cadmium sulfide β-CdS at 298 K [6, 7]: $a = 0.582$ nm

Atom	Position and multiplicity	Atomic coordinates			Occupancy
		x/a	y/a	z/a	
Cd	4(a)	0	0	0	1
S	4(b)	0.25	0.25	0.25	1

Table 3.2 High-temperature hexagonal (space group No. 186 – $P6_3mc$ (C_{6v}^4)) crystal structure of $B4$ type (wurtzite ZnS type) of cadmium sulfide α-CdS at 298 K [6, 8]: $a = 0.414$ nm, $c = 0.672$ nm

Atom	Position and multiplicity	Atomic coordinates			Occupancy
		x/a	y/a	z/a	
Cd	2(b)	1/3	2/3	0	1
S	2(b)	1/3	2/3	0.385[*]	1

[*]or $3/\sqrt{24} \approx 0.615$

periodically repeating the *ABC* or *AB* sequence: ...*ABCABC*... for sphalerite (see Fig. 3.2a) or ...*ABABAB*... for wurtzite (see Fig. 3.2b). Periodicity in the alternation sequence of layers is a necessary condition for translational symmetry of the crystal structures.

Indeed, only two possibilities exist only for the stacking of the planes. If one marks the bottom plane as an *A* plane having the *x* and *y* coordinates of the atoms, then the step-by-step displacement of atoms of the second, third and following

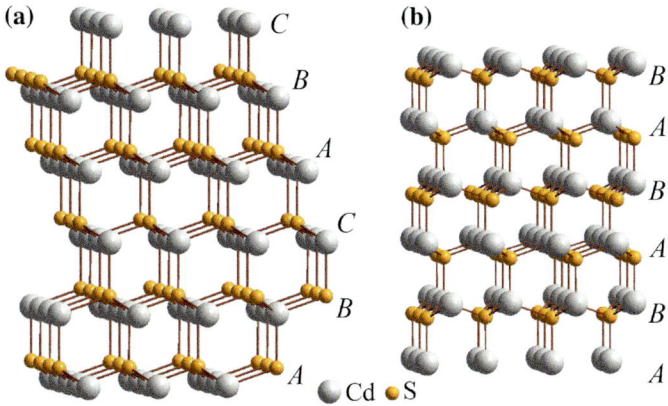

Fig. 3.2 Crystal structures of CdS: **a** cubic sphalerite and **b** hexagonal wurtzite

planes on the same translation vector (1/3, 2/3) leads to an *ABC* (…*ABCABC*…) stacking sequence and creates face-centered cubic structure (Fig. 3.2a). Displacement of atoms of the second and third planes on translation vectors (1/3, 2/3) and (2/3, 1/3), respectively, leads to an *AB* (…*ABAB*…) stacking sequence and creates a hexagonal close-packed structure (Fig. 3.2b). Both structures have identical atomic densities.

Noted that the experimental XRD patterns of CdS films 6–70 nm thick synthesized by different methods [11–18], as well as those of nanoparticles [19–21] and various complex forms of CdS [22–24] indicate that the structures of these systems differ from those of the polymorphs of bulk CdS (wurtzite or sphalerite).

In normal conditions, bulk coarse-crystalline CdS is a direct semiconductor. The band gap E_g of cadmium sulfide is rather strongly depending on temperature. Detailed analysis of temperature dependence $E_g(T)$ of cadmium sulfide was performed in study [25] with using data [26–29]. According to [25], $E_g(T)$ is monotonically decreasing from ~ 2.58 to ~ 2.49 eV when the temperature increases from 0 to 300 K. Conventional band gap of CdS at a temperature of 300 K is ~ 2.42 eV.

For bulk coarse-crystalline cadmium sulfide at 300 K, the electron and hole effective masses are equal $m_e = 0.19 m_0$ and $m_h = 0.80 m_0$ [30], $m_e = 0.21 m_0$ [31] or $m_e = 0.205 m_0$ and $m_h = 0.81 m_0$ [32], where $m_0 = 9.1 \times 10^{-31}$ kg is the free electron rest mass. Taking this into account, the reduced exciton mass $\mu_{ex} = m_e m_h / (m_e + m_h)$ for cadmium sulfide CdS is equal $\sim 0.16 m_0 = 1.46 \times 10^{-31}$ kg.

The characteristic size of the Wannier-Mott exciton (or the Bohr radius of the excitons) in the macroscopic semiconducting crystal is $R_{ex} \approx n^2 \hbar^2 \varepsilon / \mu_{ex} e^2 = (n^2 \varepsilon m_0 / \mu_{ex}) a_B$ where $a_B = \hbar^2 / m_0 e^2 = 0.0529$ nm is the Bohr radius. The dielectric constants ε_{11} and ε_{33} of cadmium sulfide measured by an ultrasound resonance at 300 K, are equal 8.67 and 10.2, respectively [33]. According to spectroscopic ellipsometry data [34], $\varepsilon_\perp = 8.28$ and $\varepsilon_\parallel = 8.73$ at 300 K. Taking into account generalized data [33–35], the value of dielectric constant ε for cadmium sulfide at a temperature of 300 K is equal ~ 9.0. For reduced exciton mass $\mu_{ex} = 0.16m_0$ and dielectric constant $\varepsilon_{CdS} = 9.0$, the radius R_{ex} of the first exciton state with $n = 1$ in coarse-crystalline cadmium sulfide CdS is ~ 3.0 nm; the diameter $D_{ex} \approx 6.0$ nm.

3.1 Chemical Deposition of Nanostructured Cadmium Sulfide CdS

Preparation of different forms of nanostructured lead sulfide PbS (colloidal solutions, nanopowders with different morphology of particles, quantum dots, thin films) by chemical deposition from aqueous solutions is considered in detail in Sect. 2.1. The same method is used widely for preparing nanostructured cadmium sulfide CdS [36].

The first classification of physical and chemical deposition methods was suggested by Chopra and Das [37], who showed that chemical bath deposition (CBD) can be considered as promising and self-contained method. In 2007, the first review concerning applications of CBD to preparation of nanocrystalline semiconductor films was reported [38]. This publication was followed by [39] of the compositions of the reaction mixtures for the synthesis of chalcogenides and oxides developed before 2011. Cadmium sulfide films are deposited using various sulfidizers as sources of sulfide ions (soluble sulfides and hydrosulfides, hydrogen sulfide gas or hydrosulphuric acid H_2S, diamide N_2H_4CS of thiocarbonic acid (thiourea) and their derivatives, sodium thiosulfate $Na_2S_2O_3$, thioacetamide $MeC(S)NH_2$, etc.).

In review [36], chemical bath deposition was considered as promising method for preparation of not only thin-film structures of cadmium sulfide, but also nanostructured semiconductor particles of different structure. Among the physical and chemical deposition methods, only chemical bath deposition yields various bulk and nanocrystalline hybrid forms of CdS not only as continuous or island-type films on surfaces of any size and configuration, but also as core-shell structures, agglomerated polycrystals of regular shape and isolated nanoparticles in colloidal solutions.

In general case, a reaction mixture for CBD represents a solution of one or more salts of the metal M^{n+}, a source of chalcogen X (X = S, Se, Te) and a complexing agent. According to [39], metal chalcogenide deposition involves four steps (see also Sect. 2.1.1): (1) formation of equilibrium homogeneous system complexing agent–water; (2) formation and dissociation of a complex metal compound $[ML_i]^{n+ik}$, where L denotes one or more ligands, n is the oxidation state of the metal M, i is the number of ligands, and k is the charge of the ligand; (3) hydrolysis of the source of chalcogen; and (4) formation of solid metal chalcogenide.

A feature of the synthesis in aqueous solutions is the possibility for side reactions resulting in almost water-insoluble oxygen-containing metal compounds to accompany the main reaction affording inorganic chalcogenide. The formation probability of side products is rather high because most metal ions are hydrolyzed in aqueous solutions. Hydrolysis significantly complicates the synthesis of pure compounds containing no basic salts and hydroxides. In this connection, a particular complexing agent for each metal ion should be used in stage 2 in order to suppress hydrolysis and to control the feed rate of metal ions to the reaction zone in stage of formation of solid-phase chalcogenide (stage 4).

Stage 3 is accompanied by a decrease in the concentration of metal cations due to the appearance of chalcogen ions X^{m-} in the solution and formation of solid chalcogenide M_mX_n. Stage 4 is an almost irreversible reaction.

Spontaneous formation of metal sulfide is possible at supersaturation

$$S_s = \mathrm{IP}/K_{sp} > 1, \tag{3.1}$$

where $\mathrm{IP} = [M^{2+}]_0 \cdot [S^{2-}]_0$ is the ionic product; $K_{sp} = [M^{2+}]_{eq} \cdot [S^{2-}]_{eq}$ is the solubility product; $[M^{2+}]_0$, $[S^{2-}]_0$, $[M^{2+}]_{eq}$, and $[S^{2-}]_{eq}$ are the initial and equilibrium concentrations of the M^{2+} and S^{2-} ions, respectively.

Thus, in order to discuss the possibility for metal sulfide formation to occur, one should determine the initial compositions of the reaction mixture in stages 2 and 3, i.e., the concentrations of the sulfur and metal ions.

It is the four mentioned stages taken altogether that are responsible for the mechanism of sulfide growth in solution and on substrate, for the chemical composition, structure, microstructure and type of the hybrid form (core-shell, agglomerated nanopowder, isolated particles in colloidal solutions) of solid sulfide formed in the reaction.

3.1.1   Formation of Aqueous Colloidal Solutions of CdS

Methods for synthesis of isolated nanocrystals based on Group II–VI element chalcogenides (CdSe, CdTe, CdS and ZnSe) in liquid media have long been known.

Most techniques for preparing isolated semiconducting nanocrystals or quantum dots use toxic precursors, which limits their utility in biological and medical applications. In study [40], chemical condensation method [41, 42] was used to obtain a stable aqueous colloidal solution of cadmium sulfide CdS nanoparticles. The problem arising from the nanoparticle agglomeration and sedimentation because of the high specific surface energy at interfaces was solved using stabilizers. As a stabilizer and complexing agent, authors of studies [40, 41] used disodium ethylenediaminetetraacetate Na_2H_2-EDTA $\equiv C_{10}H_{18}N_2Na_2O_{10} \equiv Na_2H_2Y$ (Trilon B), which ensured long-term stability of the solutions prepared.

Cadmium chloride $CdCl_2$ and sodium sulfide Na_2S served as the sources of Cd^{2+} and S^{2-} ions. To examine the effect of the cadmium/EDTA ratio on the stability of solutions, authors [40] carried out experiments aimed at preparing solutions at initial concentrations of cadmium and sulfur ions $C_{Cd^{2+}} = C_{S^{2-}} = 8$ mmol l^{-1} and EDTA ions from 1.6 to 16 mmol l^{-1}.

Since Na_2S, $CdCl_2$ и Na_2H_2Y are readily soluble in water, the first step of synthesis consisted in the obtaining solutions of strong electrolytes. As a result of the phenomenon of hydration, i.e., interaction of ions with surrounding water molecules, all the ions in electrolyte solutions become hydrated. Thus, a solution of sodium sulfide is composed of water molecules H_2O, hydrated sodium ions Na_{aq}^+, and hydrated sulfide ions S_{aq}^{2-}. Another solution, a solution of cadmium chloride, consists of water molecules H_2O, hydrated cadmium ions Cd_{aq}^{2+}, and hydrated chloride ions Cl_{aq}^-. The third solution, solution of disodium EDTA, is composed of water molecules H_2O, hydrated sodium ions Na_{aq}^+, and doubly charged ions $H_2Y_{aq}^{2-}$ of ethylenediaminetetraacetic acid.

At mixing these initial solutions, two reactions occur

$$Cd_{aq}^{2+} + H_2Y_{aq}^{2-} = CdY_{aq}^{2-} + 2H_{aq}^+, \tag{3.2}$$

$$CdY_{aq}^{2-} + S_{aq}^{2-} = CdS \downarrow + Y_{aq}^{4-}. \tag{3.3}$$

If all solutions had the same concentration and were taken in stoichiometric amounts, the total molecular equation of the formation of the disperse phase of CdS colloidal solution can be written as

$$CdCl_2 + Na_2S = CdS \downarrow + 2NaCl. \tag{3.4}$$

It is known that the basic property of disperse systems is their thermodynamic nonequilibrium associated with high surface energy of the developed interphase surface. Therefore, in these systems the process of coagulation occurs which is observed at the mixing of solutions of cadmium chloride and sodium sulfide in stoichiometric amounts in keeping with reaction (3.4). To prevent coagulation, i.e., to impart aggregative stability to the colloidal solution, as well as sedimentation stability, on the surface of solid particles should be created protective double ionic layers and the adsorption-solvate layers. These layers cause the electrostatic repulsion and prevent "adhesion" of the particles.

In studies [40, 41], the synthesis of aggregation- and sedimentation-stable colloidal solution was implemented by the reactions (3.2)–(3.4).

The size (hydrodynamic diameter D_{DLS}) and zeta potential ζ of disperse phase particles were determined directly in colloidal solutions using Dynamic Light Scattering (DLS) measurements on a Zetasizer Nano ZS instrument (Malvern Instruments Ltd.) at 298 K. The He–Ne laser with radiation wavelength 633 nm was used as a radiation source. The backscattered light detector was placed at an angle of 173° to the transmitted beam.

The particle size D_{DLS} determined by DLS measurements is the size of scattering centers composed of CdS nanoparticles, a stabilizing organic EDTA shell, and a solvation shell formed by molecules of the dispersion medium [43].

To assess the effect of the solution preparation sequence, two experiments were performed in study [40]. Experiment I included the premixing of an EDTA solution with a $CdCl_2$ solution. Next, an equimolar amount of the resultant mixture was added to a Na_2S solution. In experiment II, an EDTA-containing solution was premixed with a Na_2S solution, and then the mixture was added to an equimolar amount of a $CdCl_2$ solution. The sequence in which the starting solutions were mixed had a significant effect on a size of particles produced.

In the case of CdS formation by experiment I, DLS measurements showed that the solution remained stable at $C_{EDTA}/C_{Cd^{2+}}$ ions ratios from 0.4 to 1.2 (Fig. 3.3a).

Beyond this range (at lower or higher EDTA concentrations), the solutions contained agglomerates of CdS nanoparticles about 70 nm in size, the solutions were visually turbid, and there was a precipitate. The nanoparticles with a smallest D_{DLS} size 16 ± 5 nm were prepared in the $C_{EDTA}/C_{Cd^{2+}}$ range from 0.6 to 1 (Fig. 3.3a). At the same time, when an EDTA solution was mixed with a Na_2S solution and the mixture was then added to a cadmium chloride solution

Fig. 3.3 Dependences of **a** hydrodynamic D_{DLS} size and **b** zeta potential ζ of CdS nanoparticles on the $C_{EDTA}/C_{Cd^{2+}}$ ratio for different reactant mixing sequences [40]: *Filled circle* experiment I; *Open circle* experiment II

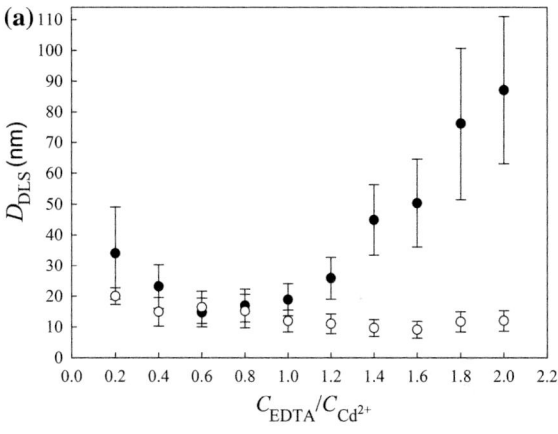

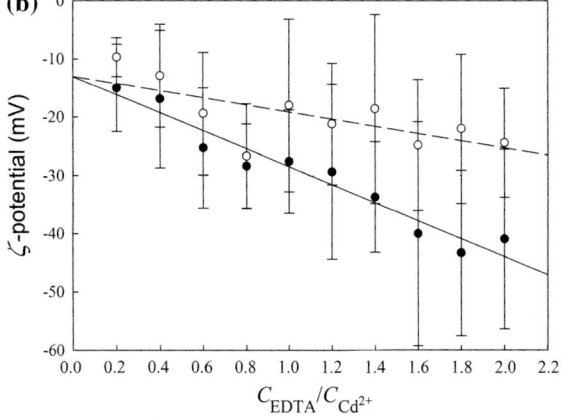

(experiment II), increasing the complexing (chelating) agent concentration caused no increase in nanoparticle size. The minimum size of the CdS nanoparticles was 10 ± 3 nm and remained constant at $C_{EDTA}/C_{Cd^{2+}}$ ions ratios from 1 to 2 (Fig. 3.3a).

The stability of a colloidal system can be assessed from the value of its zeta potential ζ. It is commonly accepted in the literature that a system is stable if its ζ-potential is less than -30 mV or greater than $+30$ mV. The measured ζ-potentials of CdS nanoparticles in aqueous solutions demonstrate that the disperse phase is stable and has a minimum particle size at ζ-potential values more than -30 mV. This can be understood in terms of the stabilization mechanism: the attachment of EDTA molecules to the surface of CdS nanoparticles.

The reason for the stability of the solution is that neither sodium, nor hydrogen, nor chlorine ions can be firmly attached to the surface of a crystal, whereas EDTA ions may form a strong complex with surface ions of a nanoparticle, imparting to it an excess negative charge.

Both aggregation and sedimentation stabilities of a colloidal solution are ensured by the formation of double ionic layer and an adsorption-solvation layer through the adsorption of EDTA ions on the surface of the disperse phase in a CdS solution, which in turn become surrounded by polar water molecules, thus increasing the solvation shell

The predominance of the stable complexes of cadmium ions with EDTA ions in the solutions prepared by experiment I tells on the measured zeta potential ζ, which is a stronger function of EDTA than in the case of the synthesis by experiment II (Fig. 3.3b). One possible reason for this is the large number of EDTA complexing agent ions located near the surface of the nanoparticles. The zeta potential ζ of the CdS nanoparticles in the absence of a complexing agent coincides is equal about -13 mV. This confirms an instability of system and its tendency toward agglomeration and sedimentation in the absence of EDTA.

Generally, measurements [40] have shown that the CdS colloidal solution remains stable at initial concentrations $C_{Cd^{2+}} = C_{S^{2-}} = 8$ mmol 1^{-1} and change of EDTA concentration in the range from 3.2 to 16 mmol 1^{-1}.

In study [44] for the first time a method for synthesis of colloidal chalcogenide nanoparticles in organic solvents in the presence of trioctylphosphine oxide (TOPO) was reported. This is the so-called high-temperature organometallic synthesis. Owing to small size of nanocrystals the synthesis is accompanied by intense agglomeration and subsequent coagulation and coalescence of nanoparticles. The problem is solved by coating nanoparticles with a TOPO layer but nanocrystals thus become soluble solely in organic solvents (chloroform, toluene, heptane, etc.). They can be used in biological media only after surface modification by attachment of hydrophilic groups. This underlies the development of methods for stabilization and solubilization of such particles.

For the solubilization of nanocrystals, the following three approaches are used: (1) utilization of bifunctional compounds capable of replacing TOPO molecules and containing hydrophilic groups that are exposed to solution after attachment to nanocrystals and thus provide solubility in water; (2) formation of a polymer shell surrounding the nanocrystal by penetration of hydrophobic fragments of solubilizing agents into the TOPO layer without removal of the TOPO molecules from the surface of nanocrystals; and (3) encapsulation of semiconductor nanocrystals into polymer microparticles (microspheres or microcapsules).

Replacement of TOPO by bifunctional ligands is based on the adsorption of ligands containing thiol or phosphine groups (mercapto acids, cysteine, etc.) or SH groups (chemically modified ethylene glycol, DNA, proteins or peptides) to form bonds with the surface metal or sulfur atoms [45, 46]. Solutions of these ligands are added to TOPO-coated nanocrystals in chloroform. The formation of bonds between sulfur atoms and the surface metal or sulfur ions leads to replacement of TOPO molecules by the solubilizing ligands. As a result, nanocrystals precipitate from the organic solvent. The precipitate is water soluble and the nanocrystal shell includes reactive groups for conjugation with biomolecules (antibodies, DNA, RNA, proteins, etc.) (Fig. 3.4).

A method is available that combines polyvalent bonding with subsequent formation of a polymer shell around the nanocrystals [47]. The surface of nanocrystals is coated with a silica/siloxane layer and then siloxane groups are polymerized to be covalently bound to methoxy compounds (for example, aminopropyltrimethoxysilane) (Fig. 3.5). Polysiloxane-coated nanocrystals are very stable.

The second approach (formation of 'insulating' polymer layer around nanocrystals) uses hydrophobic character of the TOPO molecules present on the surface of nanoparticles in order to bind hydrophobic fragments of bifunctional solubilizing polymers, e.g., multiblock copolymers, polyethylene glycol (PEG), etc. (see Fig. 2.16). Hydrophilic fragments of these molecules exposed to the surrounding aqueous solution form a polar shell [48] containing reactive groups for covalent attachment of biomolecules (antibodies, biotin, streptavidin, peptides, DNA).

Encapsulation of nanocrystals into submicrometre and micrometre polymer particles (third approach) not only makes it possible to utilize conventional methods of organic synthesis for covalent attachment to functional groups of biomolecules, but also enables isolation of nanocrystals from the environment and their stabilization in aqueous solutions. By filling each microsphere with a specified amount

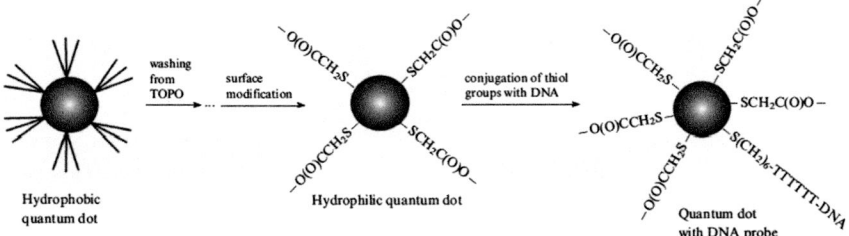

Fig. 3.4 Modification of quantum dots for conjugation with DNA [46]. The quantum dot surface is washed to remove TOPO and then is covered with sulfide groups to bind biomolecules

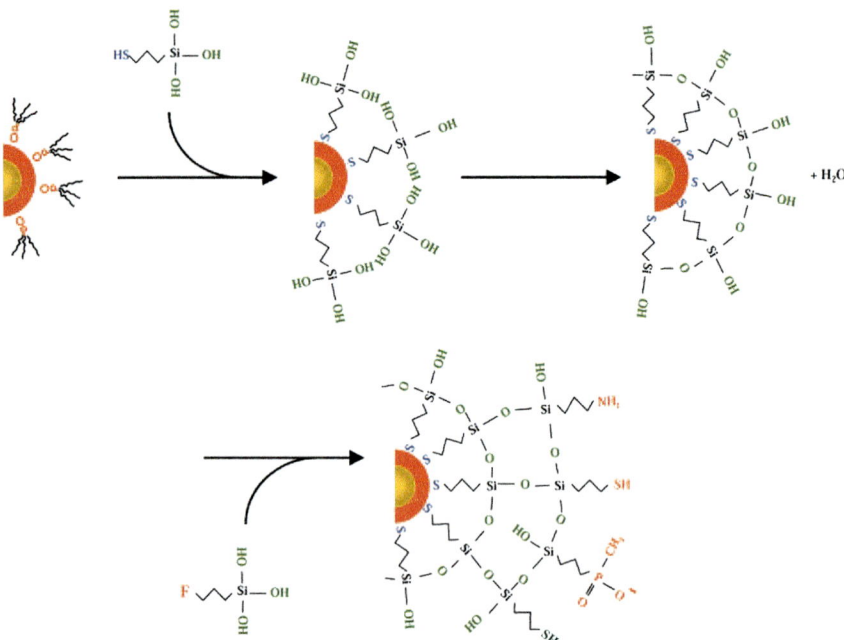

Fig. 3.5 Scheme of the typical synthesis of silica-coated nanoparticles [47]. The TOPO-capped CdS/ZnS core/shell particles are dissolved in pure mercaptopropyltris(methyloxy)silane (MPS). The MPS replaces the TOPO molecules on the surface. The methoxysilane groups (Si-OCH$_3$) hydrolyze into silanol groups (Si-OH), and form a primary polymerization layer. Heat strengthens the silanol-silanol bridges by converting them into siloxane bonds. Then, fresh silane precursors are incorporated into the shell and may tailor the nanocrystal surface functionality

of nanocrystals with particular photoluminescence color one can practically implement multiple spectral encoding. To embed semiconductor nanocrystals into polymer matrices, it was proposed to use molecular encapsulation (formation of micelles with encapsulated nanocrystals). The technique is based on the growth of nanocrystals immediately in the polymer matrix. Cadmium sulfide nanoparticles can be stabilized by chelating groups (Fig. 3.6) [49] or by amino groups of block copolymers [50], as well as by Starburst (PAMAM) dendrimers based on amine macromolecules [51] with the amino groups acting as stabilizers (Fig. 3.7).

The major drawbacks of the solubilization of nanocrystals include complexity and long duration of the process, high loss of materials and a low yield of solubilized nanocrystals. Semiconductor nanoparticles can be synthesized by CBD involving preparation of aqueous colloidal solutions of CdS with high aggregative and sedimentation stability using the *one-pot* approach. Unlike organometallic synthesis, chemical bath deposition is carried out at room temperature with aqueous CdCl$_2$ solution as the source of cadmium instead of Cd(Me)$_2$, which is toxic, pyrophoric and unstable at room temperature. Coagulation of CdS nanoparticles is prevented by introducing stabilizing agents (ligands for Cd^{2+} ions) including

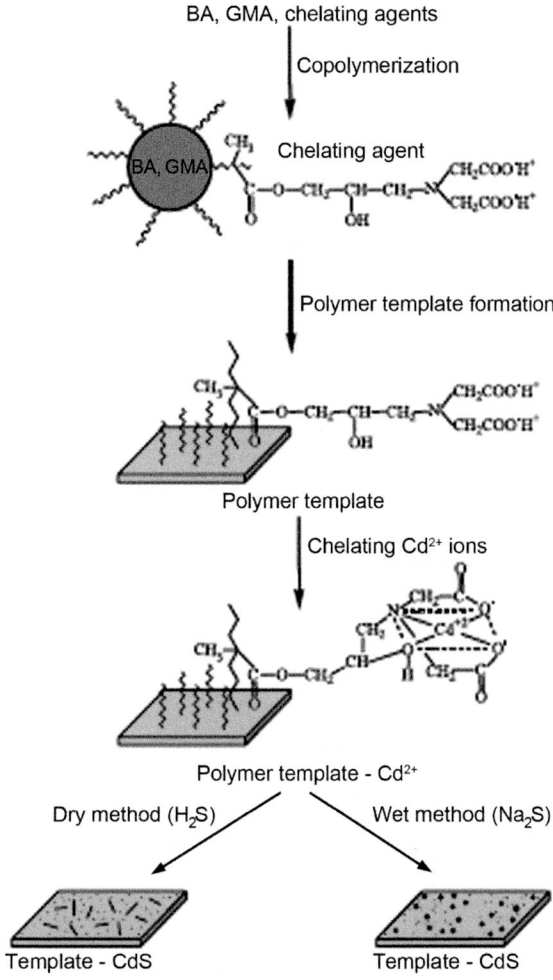

Fig. 3.6 Scheme of preparation of CdS nanoparticles stabilized by chelating groups on polymer tempates (BA and GMA are N-butilacrylate and glycidyl methacrylate, respectively) [49]

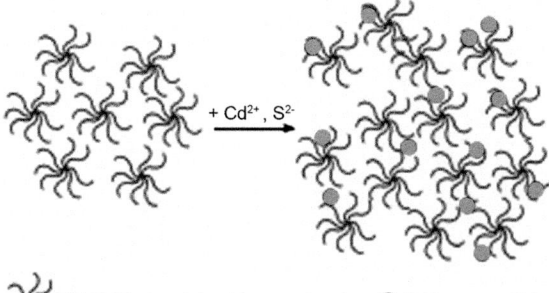

Fig. 3.7 Stabilization of CdS nanoparticles in aqueous solution by molecules of a Starburst (PAMAM) dendrimer based on polyaminonamine (generation 4.0) [51]

phosphates, thiols, amino acid derivatives and hydrophilic polymers into the solution and using H₂S, thiourea and Na₂S as the sources of sulfide ions [41, 43, 52–58]. Among stabilizers, a particular position is occupied by two biomolecules, cysteine and glutathione (GSH), which simultaneously impart hydrophilic and bio-compatible properties to CdS nanoparticles [52, 54, 58]. During the synthesis, glutathione plays the role of a polydentate ligand [54, 58] with respect to Cd^{2+} ions. Recently, it was proposed to use glutathione as sulfidizing agent [58]. Chemical bath deposition of water-soluble CdS nanoparticles in alkaline glutathione solutions at different pH values is schematically shown in Fig. 3.8.

The first step (see Fig. 3.8a) involves the formation of a Cd^{II}–GSH coordination polymer using coordination interaction between Cd^{2+} and GSH ions. Here, Cd^{2+} binds to thiol groups and can form a peptide bond between cyteine amino group and glutamine carboxyl group. In the next step, the addition of NaOH to the Cd^{II}–GSH colloidal solution is followed by cleavage of the peptide bond, isolation of sulfur and formation of the CdS–GSH quantum dot.

Different types of coordination bond can be formed in the first step depending on the pH value (see Fig. 3.8b). As the pH value increases, the size of spherical particles of the Cd^{II}–GSH polymer decreases while that of CdS nanocrystals increases [58].

The reasons for water solubility of hydrophobic quantum dots prepared by chemical bath deposition are as follows. Stabilizing agents on the surface of quantum dots orient the surrounding polar water molecules and create an additional

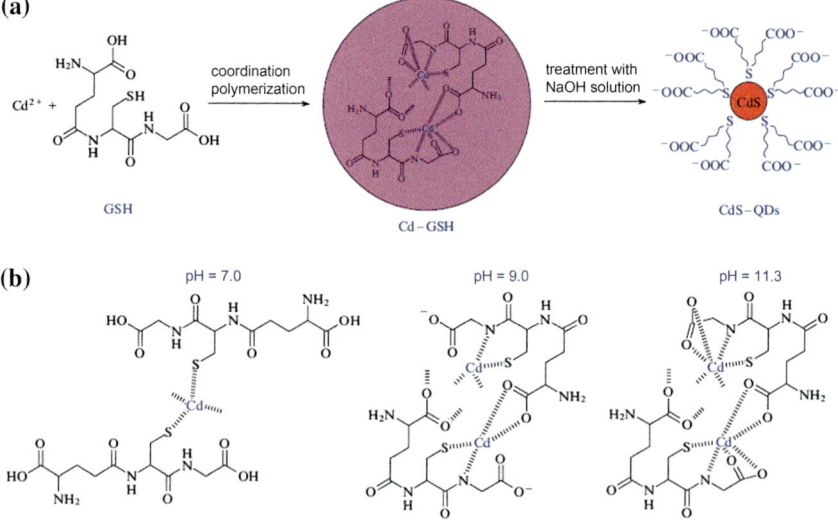

Fig. 3.8 Scheme of formation of water-soluble CdS–GSH (glutathione) nanoparticles [58]: **a** alkaline hydrolysis of Cd^{II}–GSH coordination polymer at pH = 11.3; **b** types of coordination bond between Cd^{2+} ions and GSH molecules in Cd^{II}–GSH particles at different initial pH values of the GSH solutions

hydrophilic solvation (hydration) shell. Thus one deals with the adsorption-solvation stability factor of the colloidal solution.

However, the formation of strong complexes may also cause more intense decomposition (dissolution) of the dispersed phase formed by nanoparticles. For instance, EDTA as chelating agent [41, 43] can retard the formation and growth of CdS nanoparticles during synthesis of CdS colloidal solutions. EDTA present in the solution is responsible for the occurrence of two competing reactions, namely, complexation between Cd^{2+} ions and EDTA and dissolution of CdS nanoparticles. As a result, the colloidal solution is characterized by a dynamic equilibrium between dissolution and formation of the dispersed CdS phase.

It is convenient to describe colloidal solutions using the term "micelle" to show that the case in point is the structure of hydrophobic particles of insoluble solid CdS present in the aqueous solution and to distinguish them from powder particles formed after removal of the dispersed medium.

Authors of study [43] used the dynamic light scattering (DLS) method for the micelles investigating directly in the solution. The studies of the size distribution function of the scattering centers showed that most micelles in the solution have a size of about 15 nm (Fig. 3.9). The measured ζ-potential of nanoparticles varied in the range from -75 to -21 mV, thus indicating a rather high stability of the solution and a low extent of coagulation.

High contrast between the electron density of the CdS nanoparticles and the electron density of the organic shell, as well as between the electron densities of the dispersed phase and the dispersion medium allowed to use the small-angle X-ray scattering (SAXS) for the determination of the diameter of the CdS-core of micelle, which was found to be equal to ~ 3 nm. Using small-angle neutron scattering (SANS) and high contrast between the scattering cross-section of the organic shell and the scattering cross-section of the dispersion medium (heavy water was used as the solvent), authors [43] determined the thickness of the organic shell (~ 1 nm) and confirmed the diameter of the CdS nanoparticle determined by SAXS.

Fig. 3.9 The size distributions of (*1*) CdS cores and (2) CdS micelles [43]

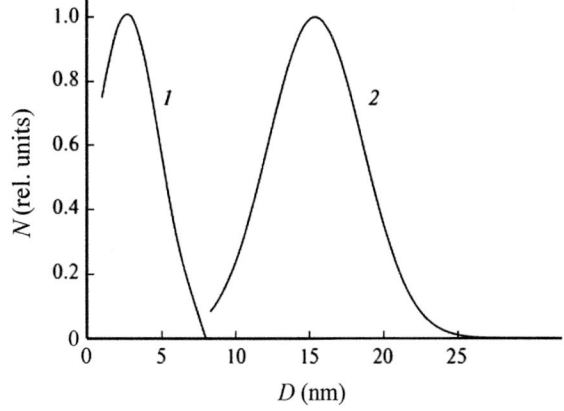

Taking into account measured negative value of the ζ-potential, it was proposed in study [43] to denote the micelle of CdS colloidal solution prepared in the presence of EDTA as

$$\{[CdS]_m \cdot 2nCd(EDTA^{2-}) \cdot (2n - x)Na_{aq}^+ \cdot (2n - x)H_{aq}^+\}^{2x-} \cdot xNa_{aq}^+ \cdot xH_{aq}^+, \quad (3.5)$$

where $m = 1$.

The $Cd(EDTA)^{2-}$ chelates adsorbing on the surface of CdS nanoparticles and imparting a negative charge to the micelle cores interfere with the approach of S^{2-} anions to the cores and preclude aggregation of nanoparticles and growth of the CdS nanocrystal. Cadmium sulfide nanoparticles with anions adsorbed on them become aggregatively stable. From the micelle formula given above it follows that the micelle contains the CdS aggregate consisting of CdS nanoparticles, an organic shell, as well as the adsorption and diffuse layers of hydrated counterions (solvation shell) (Fig. 3.10).

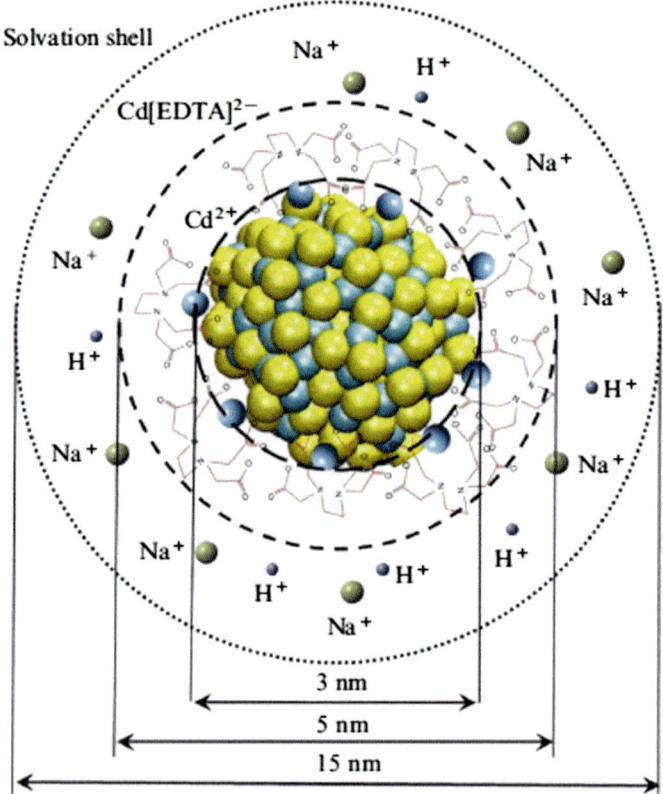

Fig. 3.10 Double-shell structure of the CdS micelle [43]: the shell consisting of $Cd(EDTA)^{2-}$ anions and the solvation shell. The scaled correlation of diameters is given according to the results of measurements of diameters of the core and two shells

3.1.2 Formation of CdS Nanoparticles

Chemical bath deposition of nanopowders belongs to "bottom-up" methods of fabrication of nanostructured materials [59, 60]. Here, nanoparticles are formed from atoms and molecules, i.e., by growth of seeds (nuclea) to nanoparticles. Cadmium sulfide nanoparticles can be synthesized by aqueous solution deposition using reactions of soluble metal compounds with sulfidizing agents. Various groups of sulfidizers only differ in that the precipitating agent (S^{2-} ion) is either added to the solution containing Cd^{2+} or $[CdL_i]^{2+ik}$ ions being deposited or is gradually formed in the course of hydrolysis proceeding in the bulk solution. Solid particles that are formed in the reactions with hydrolyzable sulfidizers not only form the film on the substrate surface or on reactor walls, but also coagulate with the formation of powder-like deposit in solution. According to [61] in this case, typical kinetic curves have an S-shape characteristic of autocatalytic processes in which the role of catalyst is played by the surface of the solid phase CdS.

The rate of sulfide formation in the system is proportional to the surface area of the solid phase nucleating in bulk reaction mixture and to the concentrations of the main reactants, that is, the complex salt $[CdL_i]^{2+ik}$ and thiourea N_2H_4CS. Therefore, the reaction rate can be described by the model second-order equation written using the concentrations of metal and N_2H_4CS in the solution [61, 62]:

$$\omega = \frac{d(c_{M,0} - c_M)}{d\tau} = K_{exp}(c_{M,0} - c_M)^{2/3} c_M c_{thio},$$ (3.6)

where ω is the rate of metal sulfide formation in the solution; K_{exp} is the experimental rate constant for the reaction; $c_{M,0}$ and c_M are the concentrations of metal ions in the solution at the initial time ($\tau = 0$) and at an arbitrary instant of time ($\tau > 0$), respectively; $c_{thio,0}$ and c_{thio} are the concentrations of thiourea N_2H_4CS at the same instant of time $\tau = 0$ and $\tau > 0$, respectively. The amount of sulfide formed during the time interval τ since the beginning of the reaction is equal to $(c_{M,0} - c_M) = (c_{thio,0} - c_{thio}) = x$. According to this, value $(c_{M,0} - c_M)^{2/3} = x^{2/3}$ is the quantity proportional to the surface area of the metal sulfide particles seeded in the homogeneous solution.

If only one sulfide crystalline phase is formed, CdS in N_2H_4CS solutions forms much more slowly than in the solution of Na_2S containing the "available" ions S^{2-} and HS^-. For instance, the formation of CdS using Na_2S at room temperature is observed a few seconds after the beginning of the reaction. For N_2H_4CS solutions, first CdS nanoparticles are observed a few minutes after the mixing of reactants. The use of thiourea N_2H_4CS leads to low initial supersaturation, small amount of crystal nuclei, while subsequent hydrolysis of N_2H_4CS causes growth of the available particles rather than nucleation of CdS. This suggests the existence of a limiting stage of hydrolysis of N_2H_4CS. The growth rate of CdS particles is described by the second-order equation with respect to the main components involved in the reaction (see (3.6)). Also, the growth rate of CdS particles is proportional to the initial concentrations of reactants, i.e., to the chemical affinity A of the deposition reaction. Thus, in N_2H_4CS solutions, nuclei of the solid phase grow much faster than nucleation occurs and the particle size increases with increasing chemical affinity A. This leads to formation of larger particles compared to those produced by deposition using Na_2S [60]. If initially the starting solution contains S^{2-} ions produced by dissociation of Na_2S in the preparation step, the formation of sulfide nanoparticles obeys the classical theory of precipitation and the Weymarn-Haber precipitation rule, according to which the nanoparticle size decreases with increasing chemical affinity A [63].

The insights into the influence of the chemical affinity A on the nanoparticle size are valid if only one crystalline phase, e.g., CdS is formed in the solution. However, the first solid phase formed in alkaline solutions is often $Cd(OH)_2$ [64]. Since it is formed in bulk solution, it is clear that not only the film, but also powder-like precipitate will contain cadmium hydroxide $Cd(OH)_2$. The $Cd(OH)_2$ phase can influence the morphology of CdS single crystals and polycrystalline CdS particles.

In Fig. 3.11, the concentration region between the surfaces I and II corresponds to formation of a single phase, namely, CdS. Chemical bath deposition using N_2H_4CS produced only wurtzite CdS powders in this concentration region; no CdS films were formed. The average size of the CdS particles formed in this concentration region was determined using XRD data and averaged over all (hkl) directions. It appeared to be larger than 9 nm. The concentration region of H^+ ions and

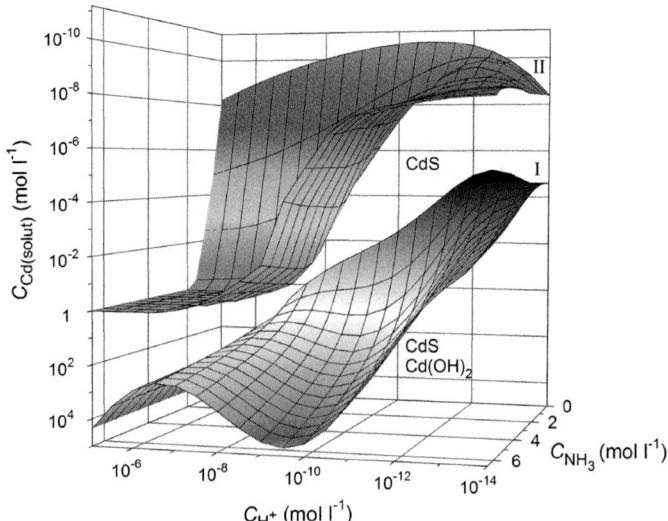

Fig. 3.11 Dependence of the equilibrium total concentration $C_{Cd(solut)}$ of Cd^{2+} ions in aqueous . solution on the concentrations of H^+ hydrogen ions and complexing agent (ammonia) [64]. The concentration C_{NH_3} is given in absolute units, and the concentrations C_{H^+} and $C_{Cd(solut)}$ are plotted on a logarithmic scale. CdS particles are prepared by chemical bath deposition using thiourea N_2H_4CS (diamide of thiocarbonic acid N_2H_4CS)

complexing agent (ammonia NH_3) below surface I corresponds to co-deposition of CdS and $Cd(OH)_2$ with the formation of CdS powders with particle size smaller than 9 nm and CdS films.

Thus, aqueous solution deposition using N_2H_4CS under conditions where no Cd $(OH)_2$ is formed produces wurtzite CdS powders with particles larger than 9 nm in size. If the formation of $Cd(OH)_2$ is possible, both CdS powders and thin films can form. Here, CdS particles have a size <9 nm. Cadmium sulfide particles synthesized using Na_2S as the source of sulfide ions are also smaller than 9 nm.

Cadmium sulfide particles produced by chemical bath deposition can form polycrystals of regular shape (see, e.g., hexagonal prisms with an edge length of 2 μm shown in Fig. 3.12a) [21]. According to [21], the presence of so large regular agglomerates was explained by self-organization of CdS nanoparticles with a size about 6 nm. Groups of regular CdS agglomerates are shown in Fig. 3.12b, c, d, e.

The alternative explanation of the regular form of CdS agglomerates was supposed in work [36].

Fig. 3.12 Scanning electron microscopy images of agglomerated CdS particles [21]. **a** The agglomerate has a regular shape of a hexagonal prism with an edge length of ∼2 μm and consists of coagulates 200–300 nm in size. According to transmission electron microscopy and XRD data, coagulates comprise CdS nanocrystals with a size about 6 nm; **b, c, d,** and **e** groups of CdS agglomerates

According to [36], Cd(OH)$_2$ particle deposited under the same conditions as CdS particles [21], except introduction of sulfidizer into the solution, had a hexagonal prismatic shape though this statement has no experimental verification. According to [36], this hexagonal prismatic shape is inherited by agglomerates of CdS nanoparticles formed in the course of sulfidizing of Cd(OH)$_2$ in aqueous solution using SC(NH$_2$)$_2$. The formation of CdS particles of regular spherical but not hexagonal shape in the presence of Cd(OH)$_2$ was observed earlier [65–67].

The mechanism of formation of regularly shaped spherical monodisperse polycrystalline CdS particles with a size about 40 nm in aqueous Cd(OH)$_2$ sol using thioacetamide as source of sulfide ions and ammonia and/or ethylenediaminetetraacetic acid (EDTA) as complexing agents was for the first time studied by Sugimoto et al. [66]. It was shown in studies [65–67] that the addition of MeC(S)NH$_2$ to the Cd(OH)$_2$ sol is followed by instantaneous formation of CdS nuclei that grow without coagulation, i.e., the number of CdS nuclei remains the same. Cadmium sulfide nuclei are formed in the reaction between S^{2-} and Cd^{2+} ions produced upon hydrolysis of MeC(S)NH$_2$ and dissolution of Cd(OH)$_2$, respectively. Cadmium hydroxide is dissolved through complexation.

According to [66], nucleation and growth of nuclei can be described by the overall reaction

$$MeC(S)NH_2 + Cd(OH)_2 \rightarrow CdS + MeCN + 2H_2O. \qquad (3.7)$$

Studies on the kinetics of Cd(OH)$_2$ transformation to CdS in aqueous thiourea solutions using NH$_3$ as complexing agent showed that the relative content of Cd

Fig. 3.13 XRD patterns of the core-shell Cd(OH)$_2$/CdS heterostructure at various precipitation times t: (*1*) 6, (*2*) 20, (*3*) 120, and (*4*) 2880 min [64]. The reflections of CdS phase and of the hexagonal Cd(OH)$_2$ hydroxide are marked with *solid* and *dashed lines*, respectively. The figure is plotted using logarithmic XRD intensity and time scales. All the XRD patterns are recorded in CuK$\alpha_{1,2}$ radiation

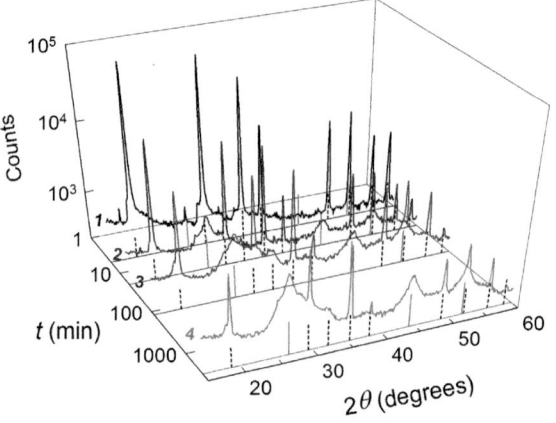

$(OH)_2$ in the deposited powder decreases while that of CdS increases with deposition time [64].

From the data shown in Fig. 3.13 it follows that almost no cadmium sulfide is formed within 1 min since the start of deposition. Then, CdS is synthesized in the reaction between free cadmium ions and sulfide ions. The formation of insoluble sulfide shell on $Cd(OH)_2$ particles precludes further dissolution of cadmium hydroxide, the concentration of cadmium ions decreases and the sulfidizing process slows down. The CdS to $Cd(OH)_2$ ratio in the $Cd(OH)_2/CdS$ heterostructure depends on the deposition time, all things considered, and can be fixed at any time by terminating the reaction. However, the form of $Cd(OH)_2$ particles on which layer CdS grows was not determined and studied by authors [64]. Therefore the explanation [36] of hexagonal shape of agglomerated CdS particles by inheritance of the hexagonal prismatic shape of $Cd(OH)_2$ particles has no experimental verification.

3.1.3 Deposition of CdS Film on Substrates

It is commonly accepted [11, 36, 38, 68–70] that chemical bath deposition of films may follow the ion exchange, cluster (hydroxide or colloidal) and ion-cluster mechanisms.

The ion exchange mechanism is schematically shown in Fig. 3.14. The formation of CdS film in alkaline solution N_2H_4CS is described by the reactions

$$Cd^{2+} + iL^k \leftrightarrow (CdL_i)^{2+ik}, \tag{3.8a}$$

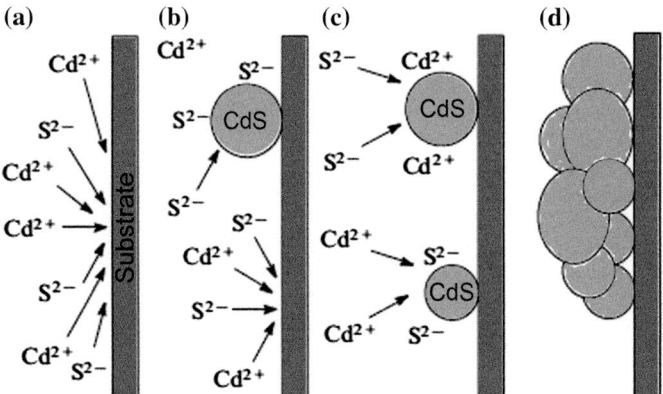

Fig. 3.14 Schematic diagram showing the probable steps involved in the ion-by-ion mechanism [69]: **a** diffusion of Cd and S ions to the substrate; **b** nucleation of the Cd^{2+} and S^{2-} ions facilitated by the substrate to form CdS nuclei; **c** growth of the CdS nucleii by adsorption of Cd^{2+} and S^{2-} ions from solution and nucleation of new CdS crystals; **d** continued growth of CdS crystals, which adhere to each other through van der Waals forces (possibly also chemical interactions)

$$N_2H_4CS + 2HO^- \leftrightarrow S^{2-} + CN_2H_2 + 2H_2O, \tag{3.8b}$$

$$Cd^{2+} + S^{2-} \rightarrow CdS. \tag{3.8c}$$

Reaction (3.8c) is possible when the ionic product IP exceeds the solubility product K_{sp} for cadmium sulfide CdS (see (3.1)). Once this condition is met, CdS particles are formed directly on the substrate (Fig. 3.14) and adhesion of solid CdS to the substrate is, generally, due to van der Waals forces.

The cluster (colloid) mechanism assumes nucleation of a solid phase, formation of solid-phase clusters immediately in the homogeneous solution and subsequent growth, coagulation and adsorption of clusters on the substrate surface [71]. The chemical composition of the clusters is governed by the synthesis conditions. Namely, chemical bath deposition using MeC(S)NH$_2$ and thiosulfate ion S$_2$O$_3^{2-}$ as sources of sulfide ions at pH < 5 leads to formation of sulfide clusters and metal hydroxide clusters in alkaline solutions. The formation of metal hydroxide is hard to avoid under experimental conditions despite the presence of complexing agents [72]. Therefore, this mechanism is sometimes called hydroxide mechanism (Fig. 3.15) [69]. In this case, complexation (3.8a) and hydrolysis of N$_2$H$_4$CS (3.8b) are accompanied by the reaction

$$Cd^{2+} + 2OH^- \rightarrow Cd(OH)_2, \tag{3.9}$$

and the formation of metal sulfide is not described by (3.8c). Following reaction takes place:

$$Cd(OH)_2 + S^{2-} \rightarrow CdS + 2OH^-. \tag{3.10}$$

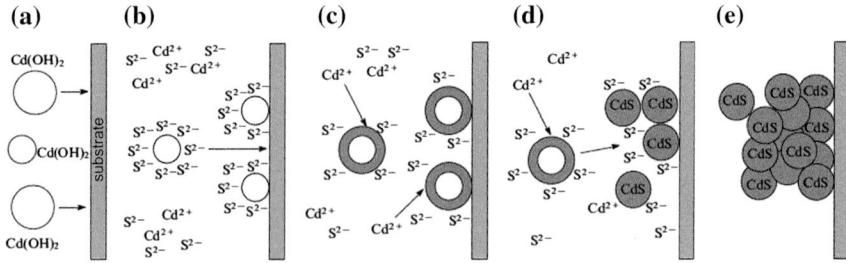

Fig. 3.15 Schema of cluster (hydroxide) mechanism of CdS film formation on substrate [69]: **a** diffusion of Cd(OH)$_2$ hydroxide colloidal particles to the substrate; **b** adsorption of particles; **c** interaction of Cd(OH)$_2$ particles with S^{2-} ions (the S^{2-} ions are formed either upon homogeneous hydrolysis of the source of sulfide ions or in heterogeneous reaction where Cd(OH)$_2$ particles can act as catalyst of hydrolysis of the source of sulfide ions); **d** exchange of hydroxide ions for sulfide ions proceeds on the substrate surface and in bulk solution; **e** the primary particles of CdS adhere to each other and form an aggregated CdS film

Transformation of metal hydroxide to metal sulfide is only possible if the solubility product K_{sp} for metal hydroxide is larger than K_{sp} for metal sulfide because the solubility product of CdS (pK_{sp} = 26.10) much exceeds that of freshly deposited Cd(OH)$_2$ (pK_{sp} = 13.66). Earlier it was established that the formation of CdS films using N$_2$H$_4$CS as source of sulfide ions proceeds via the metal hydroxide route. On the one hand, the hydroxide layer has undesired effect on the functionality of semiconductor devices, in particular, Cu-In-Ga-S-Se solar cells where CdS thin films are used as the "window layer" [73]. On the other hand, with no hydroxide layer the CdS film loses its adhesive properties.

The ionic and colloid mechanisms assume homogeneous nucleation. However, studies on the kinetics of deposition under spontaneous solid phase formation showed that formation of metal sulfides is a heterogeneous autocatalytic process [61, 62, 74]. This is an indication of more complicated, mixed mechanisms of film formation. It should be noted that the fundamental difference between the mixed mechanism and the simple cluster (hydroxide, colloidal) mechanism consists in that metal sulfide is formed not only due to consumption of free S^{2-} ions (see reaction (3.8b)). The mixed mechanism assumes the possibility for the source of sulfide ions, in particular, N$_2$H$_4$CS, to adsorb on the surface of solid Cd(OH)$_2$ particles or on (CdOH)$_2$(NH$_3$)$_2$ particles [75], thus forming metastable surface complex compounds which then decompose with the formation of CdS [76]. Based on the available experimental data, the hydroxide mechanism of CdS film deposition was represented in study [76] as

$$Cd(NH_3)_4^{2+} + 2OH^- + Si\text{-site} \leftrightarrow Cd(OH)_{2,Ads} + 4NH_3, \quad (3.11a)$$

$$Cd(OH)_{2,Ads} + N_2H_4CS \rightarrow \left[Cd(OH)_2N_2H_4CS\right]_{Ads}, \quad (3.11b)$$

$$\left[Cd(OH)_2N_2H_4CS\right]_{Ads} \rightarrow CdS + H_2CN_2 + 2H_2O + Si\text{-site}, \quad (3.11c)$$

or in the following way in study [77]

$$Cd^{2+} + Si\text{-site} + 2OH^- \rightarrow Cd(OH)_{2,Ads}, \quad (3.12a)$$

$$Cd(OH)_{2,Ads} + N_2H_4CS \rightarrow C^* \rightarrow CdS + site, \quad (3.12b)$$

where Ads are compounds adsorbed on the substrate; C* is the intermediate compound; Si-site is the surface of a silicon or glass substrate onto which the CdS film is deposited.

It should be noted that reaction scheme (3.12a, b) describing the mixed mechanism of film formation includes two undefined phases Cd(OH)$_{2,Ads}$ and C* [77]. According to [78], gradual adsorption of atomic layers of the Cd(OH)$_2$ film on the surface of the oxygen-containing substrate results in the formation of a common layer of composition $Cd(OH)_2/SiO_2$. Since direct interaction of SiO_2 with the cadmium salt in aqueous alkaline solutions mainly results in cadmium silicate Cd $(SiO_2)_n$ with a SiO_2:Cd mole ratio of about 1:1, then the $Cd(OH)_{2,Ads}$ phase in (3.12a, b) corresponds to $Cd(OH)_2/Cd(SiO_2)_n/SiO_2$ provided that deposition occurs onto a silicon or glass substrate. Subsequent sulfidizing of the surface of this cadmium-containing layer on the substrate produces a precursor layer (denoted by C*) representing a $CdS/Cd(OH)_2/Cd(SiO_2)_n/SiO_2$ intermediate layer. Initial adhesion of the bottom atomic layers of $Cd(OH)_2$ to the oxygen-containing substrate film and partial sulfidizing of the top layers of $Cd(OH)_2$ ensure growth of the CdS film and strong adhesion to the substrate. X-ray grazing incidence diffraction patterns of the early stage of the growth of CdS films confirm the discussed mechanism of CdS film growth (Fig. 3.16).

The films were deposited at 325 K with varying deposition times. After 2 min from reaction starting no cadmium compound is found on the substrate, while a cadmium hydroxide film is clearly seen to have deposited on the substrate after 3 min. It takes around 5 min for $Cd(OH)_2$ sulfidizing and for CdS film formation. Note that a complete conversion of the hydroxide to sulfide causes the adhesive force to vanish and the CdS film can be removed from the substrate with ease.

According to [36, 78], chemical deposition of CdS thin films from aqueous alkaline solutions onto silicon or glass substrates necessarily involves the formation of compound $Cd(OH)_2$ as the intermediate layer provided that the substrate was not pretreated [68, 79]. Thus, the mixed mechanism of film formation which is described by (3.12a, b) seems to be the most plausible. In revised form, this mixed mechanism was presented by authors of studies [30, 72] in following form:

$$Cd^{2+} + 2OH^- + Si\text{-site} \rightarrow Cd(OH)_2 \xrightarrow{\quad SiO_2, \text{adhesion} \quad} Cd(OH)_2/Cd(SiO_2)_n/SiO_2,$$

$$(3.13a)$$

Fig. 3.16 XRD patterns of CdS films deposited at 325 K for 2, 3, and 5 min [79]. All the XRD patterns are recorded by X-ray grazing incidence diffraction method. The diffraction reflections of the hexagonal $Cd(OH)_2$ and nanostructured CdS are marked with *short* and *long lines*, respectively

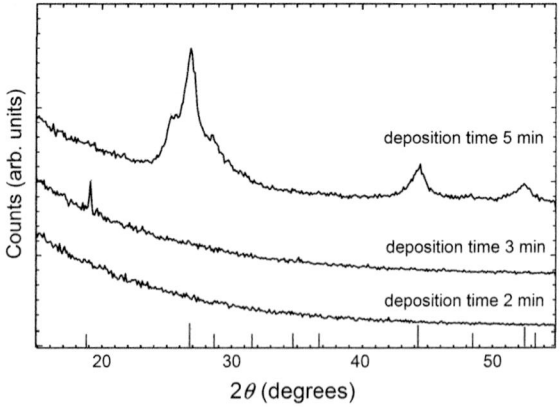

$$S^{2-} + Cd(OH)_2/Cd(SiO_2)_n/SiO_2 \rightarrow Cd(OH)_2/Cd(SiO_2)_n/SiO_2 \xrightarrow{OH^-,\text{sulfide formation}} CdS - SiO_2/Si,$$

(3.13b)

$$Cd^{2+} + S^{2-} + Cd(OH)_2/Cd(SiO_2)_n/SiO_2 \rightarrow CdS + Cd(SiO_2)_n/SiO_2 + Si\text{-site}.$$

(3.13c)

Thus, combined analysis of the grazing incidence diffraction and reflectometry data which was performed in study [78] suggests that during deposition of the CdS film under thermodynamic stability conditions of cadmium hydroxide on the first minutes of deposition a thin $Cd(OH)_2$ film with a thickness of 1–3 nm forms on the silicon substrate. Due to gradual atomic layer adsorption of the hydroxide film by the oxygen-containing substrate surface, the formation of a single $Cd(OH)_2/SiO_2$ layer occurs, which corresponds to the $Cd(OH)_{2,Ads}$ phase in the hydroxide scheme of film formation. The partial sulfidizing of the upper layers of $Cd(OH)_2$ provide the growth and strong adhesion of CdS film to the substrate.

3.2 Synthesis of CdS Nanoparticles in Glass Matrix

The optical behavior and properties of quantum-size semiconductors allow considering them as a promising sensitive probe for the effect of quantum confinement of the electron and hole wave functions by a deep potential well on the excitonic and band gap energy structures. In the last decade, several groups have conducted experimental and theoretical studies of the optical behavior of confined Wannier excitons in CdS [80–87], CdSe [88, 89], and CdS_xSe_{1-x} [89–93], in the form of nanocrystalline particles in such amorphous matrixes as glasses.

As shown by Potter and Simmons [80], it is reasonable to investigate quantum confinement effects using composites containing nanoparticles only of one species: either CdS or CdSe. The reason for this is that it is impossible to control changes in composition and stoichiometry during the growth of forming nanoparticles in the case of composites containing CdS_xSe_{1-x} solid solutions, because the size effects and the nonstoichiometry contribution are then difficult to separate.

Systematic study of CdS nanoparticles in a silicate glass matrix, and investigation of the structure and properties of the synthesized composite material has been carried out by Kuznetsova et al. [84–87]. As a matrix, the glass containing 64, 13, 11, 9, and 3 wt% of SiO_2, ZnO, K_2O, Na_2O, and B_2O_3, respectively, has been used. Coarse-crystalline CdS powder at an amount of 0.9 wt%. was added into the glass charge [84, 86, 87]. Unlike works [84, 86, 87], in the study [85] ZnS and CdO were used as the initial components to introduce the coloring agent and to form CdS. The glasses were melted in a platinum crucible at a temperature of 1710 K in air for 3 h. The synthesized molten glasses were poured into molds and annealed within 30 min at temperature 820 K. Then glasses were cooled to a room temperature within 20 h for removal of internal stresses. Under similar conditions, CdS-free glass of the same composition was prepared as a reference.

To precipitate the semiconductor phase, transparent colorless samples of the as-prepared CdS-containing glass were annealed isothermally at a temperature of 870 K in air for 6–96 h [84–86]. Figure 3.17 presents the change of appearance of the CdS-containing glasses after isothermal annealing at 870 K during different time from 0 to 96 h. For comparison, the sample of glass synthesized without cadmium sulfide is shown also.

After cooling, both the CdS-free and CdS-containing glasses were colorless and transparent. The colorlessness and transparency of CdS-free glass means that the starting materials for glass preparation contained no coloring impurities. The quenched CdS-containing glass samples are colorless because no CdS particles were formed and the cadmium and sulfur ions are uniformly distributed over the glass matrix on a molecular scale. In quenched glass, the number of cadmium-sulfur chemical bonds is extremely small.

The formation of CdS nanoparticles is initiated by isothermal annealing, whose conditions determined the particle size. Heat treatment leads to ion redistribution over the glass matrix and an appearance of Cd–S chemical bonds. As a result, the glass took on a yellow color (see Fig. 3.17), typical of CdS. After heat treatment under similar conditions, the reference CdS-free glass remained transparent and colorless. This confirms that the yellow color of the cadmium-containing glass is due to the formation of cadmium sulfide particles.

According to [84, 86], the chemical analysis confirmed the sulfur presence in the molten glass samples. In spite of the fact that the found sulfur quantity in the glasses is a factor of ~ 5 smaller than the introduced one, it is sufficient for binding cadmium and producing CdS nanoparticles.

According to [84], under the assumption of random distribution of the CdS nanoparticles over the glass matrix and their volume concentration 0.2 vol.% in the glass, the average distance between the nanoparticles is determined to be ~ 20 nm.

Fig. 3.17 Appearance of CdS-free glass and change of appearance of the CdS-containing glasses as a result of isothermal annealing at 870 K within 0, 24, 48 and 96 h

This suggests that the nanoparticles are isolated from each other by the dielectric matrix, which prevented nanoparticle agglomeration and coalescence (Fig. 3.18).

Figure 3.19 shows the XRD pattern of a CdS-containing glass after isothermal heat treatment favoring nanoparticle formation. Duration of glass annealing at a temperature of 870 K is 48 h. Analysis of XRD data confirmed that the glasses are in an amorphous state: Their diffraction patterns contained no peaks, in particular, no peaks from CdS. This is due to the low content of the sulfide phase (under 1%) and small size of the nanoparticles. The small particle size leads to large broadening of diffraction reflections, so that they cannot be de detected in standard XRD measurements.

In studies [84, 86], the values of the energy gap E_g were found from the optical absorption spectra of CdS-containing glasses. For different annealing conditions, the energy gap E_g is equal 2.55, 3.91, and 3.66 eV. The average size of the CdS nanoparticles was estimated from the values of the energy gap. In the calculations,

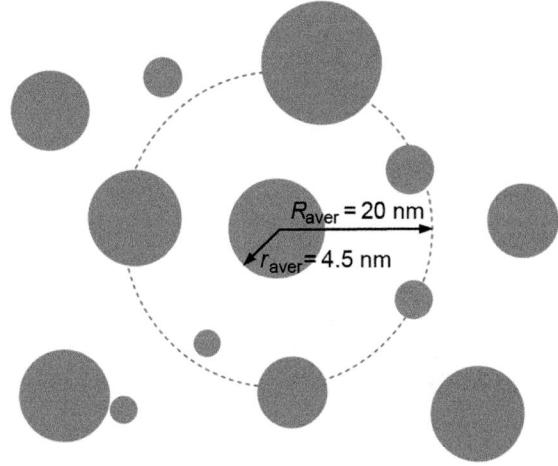

Fig. 3.18 Disordered arrangement of CdS nanoparticles with an average radius $r_{aver} = 4.5$ nm in a silicate glass matrix with allowance for the interparticle distance $R_{aver} = 20$ nm [84]. CdS nanoparticles with minimal, middle and maximum sizes are shown

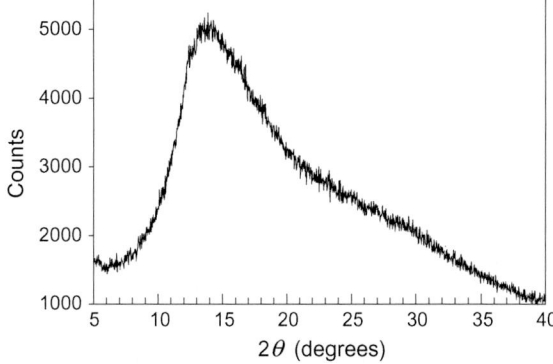

Fig. 3.19 XRD pattern of CdS-containing glass sample after isothermal heat treatment favoring CdS nanoparticle precipitation [84]. Duration of glass annealing is equal to 48 h

theoretical model [94] was used. According to [94], the effective width of the energy gap E_g of the nanoparticle increases with a decrease of its radius R:

$$E_g = E_b + \frac{\pi^2 \hbar^2}{2\mu_{ex}R^2},$$ (3.14)

where $E_b \equiv E_{g0}$ is the optical energy band gap of the bulk coarse-crystalline semiconductor ($E_b = 2.42$ eV for the bulk coarse-crystalline CdS), μ_{ex} is the mass of the exciton, and $\hbar$ is the Planck constant. Consideration of the dispersion of nanoparticle size leads to a more accurate expression [94]:

$$E_g = E_b + 0.71 \frac{\pi^2 \hbar^2}{2\mu_{ex}R^2}.$$ (3.15)

According to calculation without considering size dispersion [84], the diameter of the CdS nanoparticles after 6 h of heat treatment at a temperature of 870 K is about 9 nm. According to calculations [86], the average diameter of the CdS nanoparticles after slow cooling of synthesized glasses for removal of internal stresses and after additional 24 h of annealing at a temperature of 870 K are equal to 2.5 and 2.7 nm, respectively.

To study the formation of CdS nanoparticles, the structure of CdS-glass composite was simulated by authors [86]. The computer model for obtaining images of the internal structure of CdS-containing glasses is based on the following conditions: the CdS-glass composite is strongly diluted, the CdS nanoparticles are spherical, and their distribution in a limited three-dimensional space of glass matrix is disordered with the exclusion of overlapping. The Borland Delphi 7 program was used [95]. The developed computer simulation program allowed an automatic output of the graphs of the quantitative and spatial particle size distribution, as well as that of the function of the pair distribution of the distances between the adjacent nanoparticles.

According to the results of computer simulation [86], the CdS nanoparticles with an average diameter about 5 nm and a volume concentration about 0.2% fill homogeneously a glass matrix volume (Fig. 3.20). An optimal thickness of dielectric interlayer between CdS nanoparticles is equal about 15 nm.

Based on the results of studies [84–86], one can suggest the following scheme of the formation of a composite material from silicate glass doped with CdS. A solution oversaturated with CdS is formed upon the cooling of the melted glass. At first the centers of crystallization are formed, and, possibly, there is some growth of the nuclei of CdS phase. The diameter of the nuclei is about 2.5 nm. During the heat treatment of CdS-containing glasses at temperature of 820–920 K, particle growth is observed. Such heat treatment enables one to obtain nanoparticles of the average diameter of 2.7 nm and a narrow size distribution.

At low nanoparticle density and large distances between them, the emission properties of composite system are mainly determined by individual quantum dots,

Fig. 3.20 Model of the
internal structure of strongly
diluted CdS-glass composite
[86]. The size of cube edge is
equal to 300 nm, the circles'
diameters are proportional to
an average diameter of CdS
nanoparticles which equal to
$\sim 5 \pm 1$ nm

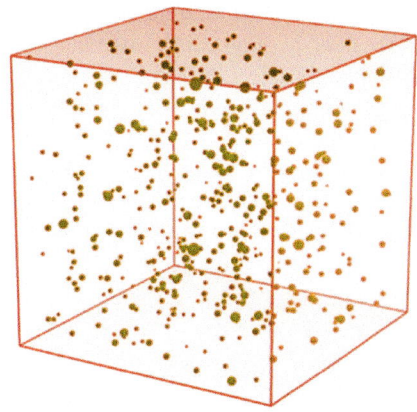

i.e., individual nanoparticles not bound to each other. As the average distance
between CdS nanoparticles is about 15 nm and the concentration of the nanoparticles in contact with each other is negligibly small [86], one can assume that the
emission properties of strongly diluted CdS-glass composites is formed mainly by
individual CdS quantum dots.

In the most studies devoted to the synthesis and investigation of sulfide
nanoparticles in the glass matrix, it is supposed that all the nanoparticles are
spherical. The features of the structure, actual form, and the size of CdS
nanoparticles in glass matrix has been studied for the first time by small angle X-ray
and neutron scattering (SAXS and SANS) in work [87].

It is well known that small angle scattering (SAS) of X-rays, electrons, and
neutrons constitutes a group of highly informative diffraction methods used to study
the internal structure of compounds and materials with heterogeneities of their
densities and sizes in the range from 1 to ~ 200 nm [96, 97]. These methods are
used effectively in the physics of condensed matter, material science, and chemistry
for studying of highly dispersive systems and nanomaterials. Small angle X-ray
scattering (SAXS) and small angle neutron scattering (SANS) are one of the
methods belonging to this group and are suitable for analyzing different
nanostructures.

In study [87], all SAS measurements performed on the samples of
CdS-containing glasses with a thickness about 150 μm. Samples were annealed at a
temperature of 870 K in air for 12–48 h. SAXS measurements were carried out
with the use of a BRUCKER NANOSTAR X-Ray Diffraction System with
monochromatic CuKα radiation ($\lambda = 1.54$ Å) in the range of diffraction scattering
vector $q = (4\pi\sin\theta)/\lambda$ from 0.005 to 0.3 Å^{-1}. SANS measurements were performed
on VSANS (very small angle scattering) diffractometer which is installed at KWS-3
beamline at the Research Neutron Source Heinz Maier-Leibnitz (FRM II) of
Technical University of Munich (Garching, Germany). VSANS diffractometer is
equipped by the focusing mirror. SANS measurements were carried with neutron
wavelength 12 Å in the q range between 0.0003 and 0.035 Å^{-1}. Measurement

results were fitted with the program IRENA (IgorPro) [98] which allows analyzing SAS data by immediate modeling of the scattering from several heterogeneity populations selecting forms.

Formation of CdS nanoparticles in the glass was confirmed by the optical absorption spectra (Fig. 3.21).

As a result of annealing and the formation the CdS nanoparticles in the glass, color of the samples has changed from colorless to yellow. The absorption spectra have showed two absorption edges at 330 and 530 nm that correspond to the amorphous glass matrix and CdS nanoparticles, respectively. The absence of other peaks in the absorption spectra attributed to other phase components contribution can be observed. This is a strong indication that only CdS is crystallized in glass. A reference glass sample without doping by Cd and S treated with the same annealing conditions remained transparent and colorless and did not show a yellow color. A slight absorption of CdS-containing glasses within a transmission region above 530 nm in comparison with the reference glass (Fig. 3.21) is dealt with the presence of the CdS crystalline phase. A non-ideal phase boundary between the crystalline phase and amorphous matrix leads to additional scattering.

SAS spectra of CdS-containing glasses are shown in Fig. 3.22. SAS spectra are normalized on the thickness of samples. All samples were measured under equal conditions.

To describe the SAS data, authors of study [87] considered a system composed out of the glass matrix and CdS nanoparticles.

The scattering signal from large-scale fluctuations of glass was modeled using two particle populations represented by a sphere with radius of about 1700 nm or 17 000 Å (population 1) and elongated spheroid with a short and a long radius of about 25 and 1700 nm or 250 and 17,000 Å (population 2), respectively. The sum of these contributions is shown in Fig. 3.22 as line 1. These sizes may not be related directly to the real physical extensions of the glass heterogeneities, but they were used to provide a description of the glass background.

Fig. 3.21 Optical absorption of a reference glass (*1*) and CdS nanoparticles containing glass annealed for 12 (*2*), 24 (*3*) and 48 (*4*) hours at 870 K [87]. An *arrow* specifies the fundamental absorption edge of bulk coarse-crystalline CdS. Change in the optical absorption near 310 nm is caused by a change of the light source in the spectrophotometer

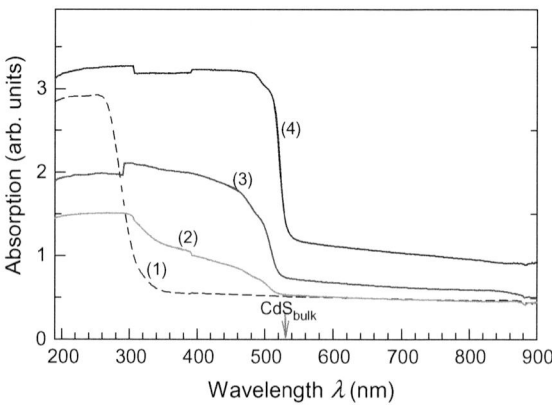

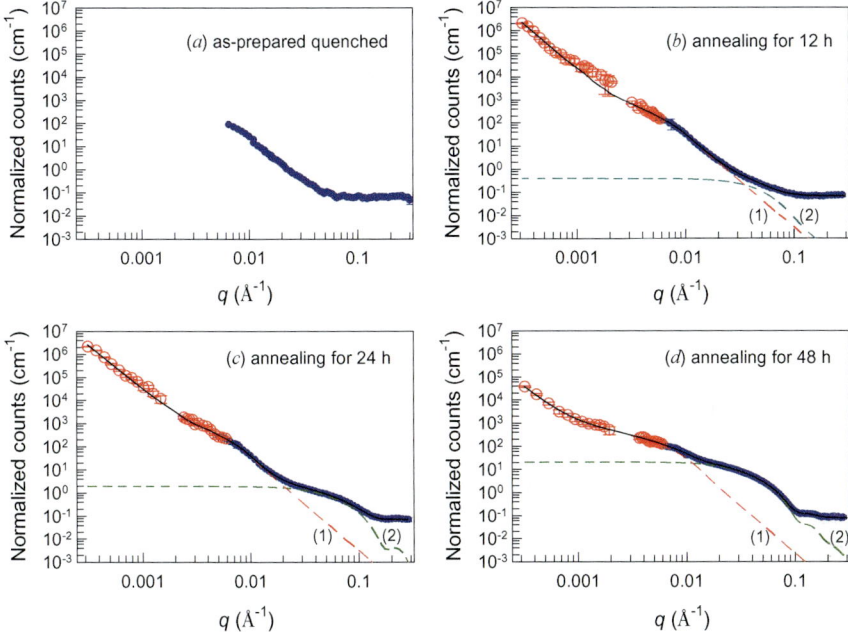

Fig. 3.22 Experimental SANS (*open circle*) and SAXS (*filled circle*) intensities for samples annealed at 870 K for different periods of time [87]. **a** as-prepared quenched CdS-containing glass; **b**, **c**, and **d** CdS-containing glasses annealed for 12, 24, and 48 h at 870 K. *Dashed red (1)* and *dark green (2) lines* represent model curves composed out of glass background and CdS nanoparticles (*see text for details*)

All samples show pronounced scattering at q below 0.02 Å^{-1} (see Fig. 3.22) which is associated with large scale inhomogeneity due to density and concentration fluctuations. This is common in multicomponent glasses [99]. The modeling for this q region allows supposing the existence of long thin fibers, which size gradually decreases during the annealing. Such glass melt defect referred to as a gnarls that has commonly a fiber shape and can gradually dissolute during heat treatment.

The scattering intensity from the as-quenched Cd and S doped glass only shows the scattering from the glass matrix without CdS nanoparticles (Fig. 3.22a). This implies that the quenching is suppresses efficiently the formation of nanoparticles.

The salient change in the SAS patterns is an increasing intensity with annealing time for q values above 0.01 Å^{-1} (see Fig. 3.22b, c, d), which reflects directly the formation and growth of CdS nanoparticles during the annealing. The scattering signal from small-scale heterogeneities of CdS nanoparticles was modeled using particle population 3. According to [87], for the annealing time of 12 h the scattering can be described as scattering by small-scale heterogeneities in the form of a sphere with a radius of 34 ± 4 Å. Thus, the analysis of the CdS nanoparticle size suggests that the annealing at 870 K during the first 12 h period causes the nucleation of spherical CdS nanoparticles with a radius of about 34 Å. However, for longer annealing times a description with a spheroid is needed. Prolonged heat treatment of 24 h and beyond leads to the change of the nanoparticle shape from spherical to spheroidal with an aspect ratio of 1:3 and reaching after 24 h a short and a long axis of 24 and 72 Å, respectively. After annealing for 48 h spheroidal CdS nanoparticles have a short and a long axis of 40 and 120 Å, respectively [87]. The contributions to the scattering caused by CdS nanoparticles are shown in Fig. 3.22b, c, d as lines 2.

The volume increase of CdS nanoparticle between 12 and 24 h and between 24 and 48 h is about a factor of 8 and 2, respectively. According to [87], the increase of annealing duration from 12 to 24 h leads along with the increase of particle number to particle shape changing from sphere to spheroid which is likely more energy preferable for a hexagonal structure of CdS nanoparticles [100].

On the whole, the combination of SANS and SAXS has shown the formation of CdS nanoparticles in glass matrix. The size and shape of CdS nanoparticles changed with annealing time from spherical into spheroidal with the short and long radii up to 40 and 120 Å, respectively. Thus, long annealing of CdS-containing glass initiates the formation of CdS nanoparticles with anisotropic shape of spheroid, exhibiting the particular character of the oriented crystal growth.

3.3 Features of Crystal Structure of CdS Nanoparticles

Transition of semiconducting sulfides into nanocrystalline state is accompanied not only by transformation of their properties due to the size effects, but also by structural changes and nonstoichiometry enhancement. So, cadmium sulfide CdS in the form of single crystals and coarse-grained powders has crystal structures of hexagonal wurtzite (*B4* type) [14] or cubic sphalerite (*B3* type) [15, 22]. However, the crystal structure of CdS in thin films, nano- and ultradispersed powders does not coincide with the structure of these modifications. The authors of works [101–103] showed that CdS nanoparticles have a peculiar disordered hexagonal structure with random alternation of close-packed atomic planes.

3.3.1 Non-periodicity in Sulfide Nanoparticles with Close-Packed Structures

Nonstoichiometry in nanoparticles has some peculiarities connected with non-periodic distribution of close-packed atomic layers. Therefore, before discussing non-periodicity in nanoparticles let us briefly remind which degree of space filling in crystals is the largest. It is known that in case of equal tangent balls, close-packed planes with hexagonal symmetry allow the greatest degree of space filling to be achieved. According to the Kepleri hypothesis [104], hexagonal close packing (hcp) and face-centered cubic (fcc) packing have the maximum density of the packing of identical balls.

The degree of space filling is equal to the ratio of the volume of n atoms entering the unit cell to the unit cell volume V_c. Each atom with radius r has volume $V_{at} = 4\pi r^3/3$. If the unit cell contains n atoms with radius r, the degree of filing is $p = nV_{at}/V_c$.

At first, let us consider the monoatomic hexagonal layer (Fig. 3.23a). In this layer, a hexagonal (space group No 191 – $P6/mmm$) unic cell with lattice constants $a = 2r$ and $c = 2r$ can be distinguished (Fig. 3.23a), which has volume $V_c = a^2 c \times \sin(2\pi/3) = 4\sqrt{3}r^3$ and includes one atom with coordinates (0 0 0) in the hexagonal axes. The degree of filling of one hexagonal layer is equal to $p = V_{at}/V_c = \pi\sqrt{3}/9 \approx 0.6046$. The degree of filling of hexagonal layer can be determined also as the ratio of the volume of one atom to the volume of the hexagonal prism, whose base edge and height are $\sqrt{3}r/3$ and $2r$, respectively (see Fig. 3.23a, at the top). The volume of this hexahedral prism equals $4\sqrt{3}r^3$ and coincides with the unit cell volume.

In order to achieve the highest packing density in three dimensions, close-packed hexagonal planes can be superimposed on each other by two different ways called the hexagonal closest packing AB (Fig. 3.23b) and cubic closest packing ABC (Fig. 3.23c). The letters A, B and C denote the sequence of layer alternation. In both packings, each sphere (atom) has 12 neighbours.

In the hexagonal closest packing (Fig. 3.23b), a hexagonal (space group No 187 – $P\bar{6}m2$) unit cell can be distinguished, which incudes two atoms with coordinates $(0\ 0\ 0)_{hex}$ in the A layer and $(1/3\ 2/3\ 1/2)_{hex}$ in the B layer. Three atoms from the A layer and an atom from the B layer located above them form a tetrahedron with

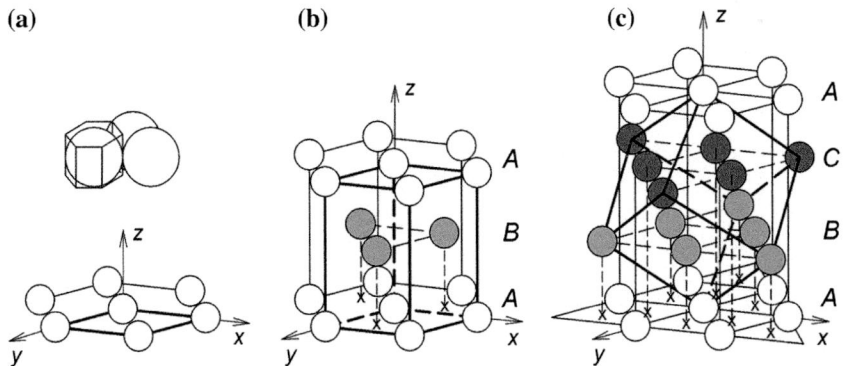

Fig. 3.23 The closest spherical packings and space filling p: **a** hexagonal close-packed plane with $p = \pi\sqrt{3}/9 \approx 0.6046$; **b** hexagonal closest packing $ABAB...$ with $p = \pi\sqrt{2}/6 \approx 0.74048$; **c** cubic closest packing $ABCABC...$ with $p = \pi\sqrt{2}/6 \approx 0.74048$. All atoms have a radius r, unit cells are limited by *thick lines*. Projections of atoms of layers B and C on $(x\ y\ 0)$ plane are shown as (*multiplication sign*)

lateral length $2r$. The height of this tetrahedron is $2\sqrt{6}r/3$ and equals $c/2$, therefore the lattice spacing is $c = 4\sqrt{6}r/3$. The unit cell spacing a is equal to $2r$. The volume of the hexagonal (space group $P\bar{6}m2$) unit cell is equal to $V_{\text{c-hcp}} = a^2 c \times \sin(2\pi/3) = 8\sqrt{2}r^3$. Considering this and the number of atoms in the unit cell, the degree of filling of the hexagonal closest packing is equal to $p = 2V_{\text{at}}/V_{\text{c-hcp}} = \pi\sqrt{2}/6 \approx 0.74048$.

In the cubic closest packing (Fig. 3.23c), the second hexagonal plane is plane-parallelly displaced relative to the first hexagonal plane as in the hexagonal closest packing, i.e. by vector $(1/3\ 2/3)_{\text{hex}}$, and the third hexagonal plane is plane-parallelly displaced relative to the second plane by the same vector $(1/3\ 2/3)_{\text{hex}}$. As a result of the same displacement of the fourth plane, the position of atoms in it coincides with the position of atoms in the first plane. The atoms in planes A, B and C have hexagonal coordinates $(0\ 0\ 0)_{\text{hex}}$, $(1/3\ 2/3\ 1/3)_{\text{hex}}$ and $(2/3\ 1/3\ 2/3)_{\text{hex}}$. The hexagonal axis z coincides with the direction $[111]_{\text{cub}}$ of the cubic (space group No 225 – $Fm\bar{3}m$) unit cell, which includes four atoms. As seen from Fig. 3.23c, the face diagonal of the cubic cell has length $4r$, therefore the lattice constant a (edge length) of the cubic cell is equal to $2\sqrt{2}r$. The volume $V_{\text{c-ccp}}$ of this cubic (space group $Fm\bar{3}m$) unit cell is equal to $16\sqrt{2}r^3$, and the degree of filling of the cubic closest packing $p = 4V_{\text{at}}/V_{\text{c-ccp}} = \pi\sqrt{2}/6 \approx 0.74048$ coincides with that of the hexagonal closest packing.

The difference between the two closest packings is in the fact that in the hexagonal structure there is only one $[001]$ direction, perpendicularly to which hexagonal close-packed layers are located, whereas the cubic structure has four such directions: $[111]$, $[\bar{1}11]$, $[1\bar{1}1]$ and $[11\bar{1}]$.

Most monoatomic substances crystallize in one of the two close-packed structures: in the hexagonal structure with closest packing AB (Fig. 3.23b) or in the cubic structure with the closest packing ABC (Fig. 3.23c). Many binary substances form more complicated, but also periodic close-packed structures composed of two fcc or two hcp packings – one for each element. These packings are displaced relative to each other so that the atoms of one type are either in octahedral or tetrahedral environment of atoms of another type. The most widespread binary close-packed structures are the cubic $B1$ structure with octahedral environment, the cubic $B3$ structure (zinc blende (sphalerite)) with tetrahedral environment and hexagonal $B4$ structure (wurtzite) with tetrahedral environment.

The attempts [105] to prove the Kepleri hypothesis revealed that the structures with a higher density than that in the fcc and hcp structures are possible for the case of a finite number of spherical atoms [106]. Such denser structures can be energetically favorable for nanoparticles. In particular, nanoparticles of metals can exist in the form of clusters with icosahedral structure [107]. In icosahedral groupings, every kth atomic shell (layer) contains $(10\ k^2 + 2)$ atoms, and the total number of atoms of the icosahedral cluster is $n = (2N + 1) + 10 \sum_{k=1}^{N} k^2$, where k is the order number of the atomic layer (shell), and N is a number of atomic shells [108]. Thus, the number of atoms in иicosahedral cluster is equal 13, 55, 147, 309, 561 and so on depending on number of atomic layers. It should be mentioned that the icosahedral particles are characterised by 6-angle profile on electron microscopic images.

According to estimations [109, 110], the energy of icosahedral clusters is lower than the energy of fcc clusters. Therefore, fcc clusters spontaneously transfer to the icosahedral form.

Rempel and Magerl [100] have analyzed the structure of nanocrystalline sulfides, which normally have $B3$ and $B4$ structures. In work [100] they showed that in sulfide nanoparticles with a small amount of close-packed atomic planes the principle of regular sequence of packings is not observed and, consequently, the crystal periodicity is lacking. This means that in nanocrystals the disordered structure is more important than the periodic arrangement of packing layers. In the opinion [100], ordering of regular sequences of packings, i.e. formation of well-defined crystal structure, is similar to the formation of superstructures on the basis of disordered basis network, i.e. in particular it is similar to ordering in nonstoichiometric carbides, nitrides and oxides. Usually different sequences of close-packed planes are called stacking faults or polytypes.

Over the years, great efforts have been made to find the answer to the question why close-packed substances may have several polytypes. Silicon carbide SiC is a binary semiconducting compound with the largest number of polytypes (more than one hundred). The most widespread polytypes among them are 4-, 6- and 15-layer hexagonal and rhombohedral polytypes $4H$, $6H$ and $15R$ [111, 112]. Cadmium sulfide CdS has at least two polytypes with $B3$ and $B4$ structures. The authors of study [100] supposed that substances forming polytypes have a hypothetical disordered structure with a higher symmetry than the symmetries of each polytype of this substance. Such disordered structure can be stable at high temperatures close to the melting temperature. The possibility to observe such disordered structures at ambient temperature can be achieved by decreasing their crystal size to nanolevel. The Mackay icosahedra [113] which consist of 20 close-packed trigonal pyramids sharing a single vertex can be considered as an example of non-crystallographic high-symmetry packing. The point symmetry $\bar{5}\,\bar{3}m$ of this icosahedral structure with 120 symmetry operations is much higher than it is for structures with cubic ($m\bar{3}m$) or hexagonal ($6/mmm$) point symmetry, which have 48 and 24 symmetry operations, respectively [114]. The degree of space filling p for Mackay icosahedron tends to 0.68818 when the number of atomic layers in the icosahedron increases. It is smaller than the degree of filling $p = \pi\sqrt{2}/6 \approx 0.74048$ for closest-packed hcp and fcc packings. At the same time, the value of p for Mackay icosahedron is larger than $p = 0.6046$ for AA packing, in which hexagonal close-packed atomic planes are superimposed on each other without displacement. Mackay has assumed that small crystals having a non-periodic, i.e. a non-translational structure, can easily transform into a crystallographic structure when the size of the crystal increases [113].

The short-range order is the same for hexagonal and cubic close-packed stacking sequences with first and second coordination numbers of 12 and 6. If the interactions between the atoms in a substance are central or pair wise and short ranged like van der Waals, covalent, etc. then the enthalpy will be independent of the sequencing. However, the entropy term for disordered layer sequences is higher than for a truly periodic stacking and will thus reduce the free energy of the solid. It can be one of the reasons for a non-periodic alternation of the closest packing in nanoparticles.

Let H and S are the enthalpy and the entropy of the particle without taking into accounts an anisotropy term and long-range order interactions. The enthalpy H_{dis} of the particle with a disordered sequence of planes is equal to TS_{dis} where S_{dis} is the configurational entropy of the disordered particle. If the phononic contributions to the free energy can be neglected, then the free energy of the disordered particle is equal to $G_{dis} = H - T(S + S_{dis})$ and the free energy of the ordered particle is equal to $G_{ord} = H - H_{ord} - TS$. The configurational entropy S_{dis} of the disordered particle is proportional to the number of non-equivalent sequences Z in which the planes can be assembled and this can easily be calculated for a given number n of stacking planes. In our case $Z = 2^{n-2}$ and $S_{dis} = k_B \ln Z$. Contribution to the enthalpy H which corresponds to this entropy is equal $k_B T \ln Z$, where T is a temperature for which disordered distribution of close-packed planes is fixed experimentally.

Cadmium sulfide CdS was considered by the authors [100] to illustrate the concepts of nonperiodic structure of nanoparticles. Depending on the synthesis method, coarse-grained cadmium sulfide has cubic $B3$ structure (sphalerite) or hexagonal $B4$ structure (wurtzite). However, the crystal structure of nanostructured cadmium suldife CdS differs from the structure of coarse-grained CdS.

Experimental XRD patterns from CdS thin films [12, 15] and nanopowders [21, 22, 24] obtained using $CuK\alpha_{1,2}$ radiation have the following features: The main intensity maximum is observed in the angular range $2\theta \approx 26.5-27°$, and two peaks with much lower intensity are located at angles near $44°$ and $52°$. Shoulders are observed in some experimental XRD patterns on both sides of the main maximum. There are three main possible interpretations of these diffraction pattern features that are typical for the nanostructured cadmium sulfide. First, the nanostructured CdS film or nanopowder is a mixture of crystallites or domains of the cubic and hexagonal phases; in this case, the XRD pattern is described by the superposition of reflections from two phases with their partial content taken into account [15, 21]. Second, strong orientation along the c axis, which is directed along the normal to the substrate surface, occurs in the nanostructured CdS film; in this case, the phase is identified as hexagonal [12]. Third, the CdS structure is not the crystal structure of wurtzite or sphalerite, but is a peculiar type of polytype structure [11, 17].

The main deficiency of the first two interpretations of CdS diffraction patterns is the absence of a number of reflections from both the wurtzite structure (for example, (102) and (103)) and the sphalerite structure (for example, (200)). In spite of significant disadvantages, two mentioned interpretations of the CdS structure are predominant in the published papers. The concept of polytypism proposed in the third version suggests the alternation of domains of the hexagonal and cubic phases or the periodic repetition of domains with a random sequence of planes. Such an approach is applicable to the bulk coarse-grained structures, where the number of packing planes is equal to several hundreds of thousands or more. This interpretation cannot be assumed to be satisfactory for description of the nanoparticle structure, where the number of packing planes is small.

All three approaches to the understanding of the structure of nanostructured CdS are based on the closed packing principle which is common for $B3$ and $B4$ crystal structure. Indeed, all the above-mentioned structures contain alternating close-packed planes consisting of either Cd atoms or S atoms. The superposition of planes occurs in such a way that all cadmium atoms are in the tetrahedral environment of the sulfur atoms, and all sulfur atoms are in the tetrahedral environment of the cadmium atoms. The distinction between the $B3$, $B4$, and polytype structures is only in the alternation sequence of close-packed planes. In particular, the closest packing $ABC...$ exists in cubic $B3$-type structure, the closest packing $AB...$ is realized in hexagonal $B4$-type structure, and the planes alternate randomly in the disordered CdS structure (Fig. 3.24).

To calculate the positions of the diffraction reflections on the XRD pattern of such a disordered structure, the authors of studies [100–103, 115, 116] have considered its "average" crystal lattice. This "average" lattice has hexagonal symmetry and is described by space group $P6$ and by a unit cell with the parameters $a = a_0/\sqrt{3} = 0.238$ nm and $c = 0.337$ nm, where a_0 is the distance between identical atoms of close-packed plane, and c is the distance between the nearest close-packed planes of cadmium or sulfur in the ordered CdS crystal. The unit cell of disordered CdS contains two atoms, a cadmium atom with the coordinates (000) and a sulfur atom with the coordinates $(0\ 0\ 1/\sqrt{2}4)$ (Table 3.3). The occupancy of all atomic positions is equal to 1/3.

Figure 3.25 shows the positions of the peaks of the line spectrum from the disordered structure with space group $P6$ in comparison with the experimental spectrum (Fig. 3.25a) of nanostructured cadmium sulfide prepared in works [101,

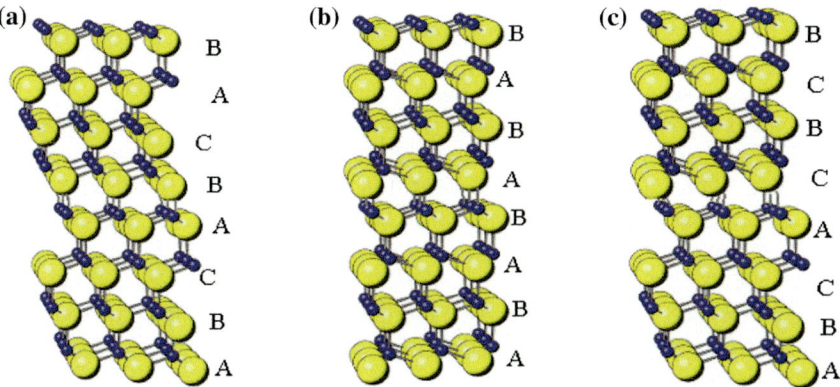

Fig. 3.24 The close-packed stacking sequence of atomic planes in ordered and disordered phases of CdS cadmium sulfide [100]: **a** cubic zinc-blende $B4$-type structure, **b** hexagonal wurtzite $B3$-type structure, and **c** disordered structure

Table 3.3 Hexagonal (space group No. 168—$P6$ (C_6^1)) disordered close-packed structure of CdS nanoparticles with a size of 3–9 nm and with "average" random lattice at 298 K [117, 122]: $a = 0.236$ nm, $c = 0.334$ nm

Atom	Position and multiplicity	Atomic coordinates			Occupancy
		x/a	y/a	z/c	
Cd	1(a)	0	0	0	1/3
S	1(a)	0	0	$1/\sqrt{24} \approx 0.204$	1/3

Fig. 3.25 Comparison of the experimental XRD pattern of the CdS nanopowder **a** and the calculated line spectra [101, 102]: **b** the disordered structure with an "average" hexagonal (space group $P6$) lattice; **c** the line spectrum for hexagonal (space group $P6_3mc$) B4-type wurtzite structure; **d** the line spectrum for cubic (space group $F\bar{4}3m$) B3-type sphalerite structure

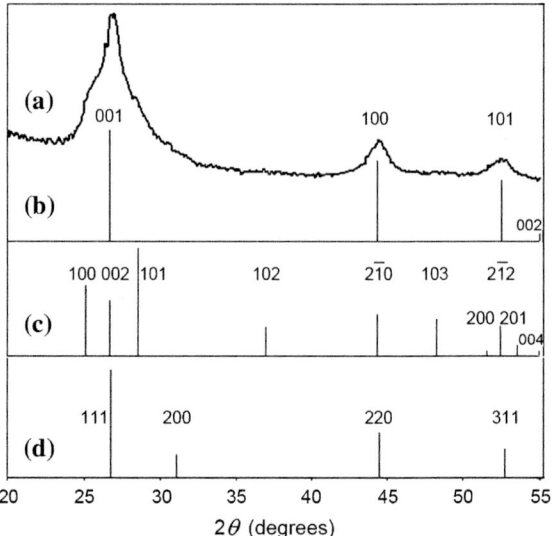

102, 115, 116] by chemical deposition from aqueous solutions. The synthesis method is described in [21].

The experimental spectrum has features characteristic of the spectra from thin films and nanopowders of CdS. The large widths of the peaks in the experimental CdS XRD pattern (Fig. 3.25a) are due to the small size of the coherent scattering regions (CSR) in the nanostructured CdS. The parameters of the unit cell of CdS with an "average" hexagonal lattice which were taken for modeled calculation are $a = 0.236$ nm and $c = 0.334$ nm. As seen in Fig. 3.25b, the position and relative intensity of the diffraction reflections of XRD pattern calculated for an "average" lattice are in good agreement with the experimental data. Calculated line spectra of cadmium sulfide with wurtzite (Fig. 3.25c) and sphalerite (Fig. 3.25d) crystal structures differ considerably from the experimental XRD pattern.

For a full-profile analysis of the experimental XRD patterns of CdS nanopowders, and to analyze the features of XRD patterns, which are associated with small size of CSRs, a method [117, 118] for calculating XRD patterns has been used in works [100–103, 115, 116]. The calculation algorithm is based on the Debye formula [119]:

$$I(q) = \sum_{v} N_v f_v^2(q) + 2 \sum_{j=1}^{N} \sum_{k>j}^{N} f_j(q) f_k(q) \frac{\sin(2\pi q R_{jk})}{2\pi q R_{jk}}, \qquad (3.16)$$

where v specifies the kind of atom (Cd or S), N_v is the number of atoms of the vth kind satisfying the equality $N_{Cd} + N_S = N$; $R_{jk} = |r_j - r_k|$ where r_j and r_k are the radius vectors of the jth and kth atoms in the particle, respectively; $q = |q|$ is the length of the wave vector q; $f_j(q)$ and $f_k(q)$ are the atomic scattering factors for the jth and kth atoms. The structural periodicity, the unit cell, and the basis, which are specified in standard methods of calculation, are not important if the Debye formula is used. Therefore, the diffraction pattern for the compound with an arbitrary atomic structure (for example, a disordered one) can be calculated. The typical features of the nanoparticle XRD pattern, in particular, the broadening of peaks or "shoulders," are a direct consequence of the calculation using the Debye formula.

The calculation using the Debye formula was carried out for spherical particles with the disordered alteration of pairs of the Cd and S close-packed planes. At first, the imposition of planes was carried out, and then the ball-shaped particle was cut. In the calculation, 50 spherical particles with a different alteration order of planes were used; this order was chosen using a random-number generator. The spectra were calculated for each of the 50 spherical particles, and the averaging was performed over all spectra.

The relative difference for linear particle sizes in three different directions is determined by the aspect ratio $A = d/h$, where d and h are the horizontal and vertical sizes of the object, respectively. The analysis of the experimental XRD patterns recorded in works [11, 21, 24], has shown, that the aspect ratio for nanoparticles of CdS powders lays in limits from 1 to 3.

Ab initio calculations of XRD powder patterns using the Debye equation (3.16) and a comparison with the experimental spectra shows that the atomic structure of

Fig. 3.26 Three-dimensional model of the CdS nanoparticle consisting of 3038 atoms with a disordered alteration *ACABCBC* of 7 close-packed planes (only the cadmium sublattice is shown) [102]. The height of the nanoparticle is equal to 2.3 nm, and the aspect ratio is $A = 2.5$

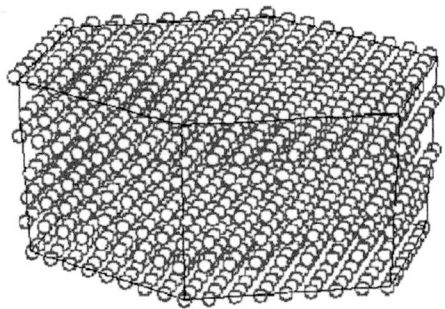

Fig. 3.27 Comparison of the **a** experimental and **b** simulated XRD patterns of CdS nanopowder [100]. The simulation was performed using the Debye equation for a powder of nanoparticles with a height about 3 nm and aspect ratio $A = 1.6$; each nanoparticle contained 3380 atoms arranged in ten CdS layers

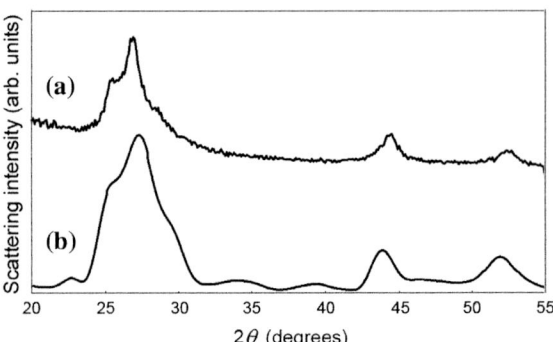

CdS nanoparticles smaller than 5 nm is disordered. The model CdS nanoparticles included from 7 to 12 close-packed planes formed only by Cd atoms and only by S atoms. A height of the model nanoparticles was from 2 to 3 nm (Fig. 3.26). The best coincidence with experiment has been reached when the CdS nanoparticles were modelled as hexagonal prisms with height about 3 nm and aspect ratio $A = 1.6$. Such nanoparticles contained 3380 atoms in ten close-packed Cd-S layers with a disordered relative arrangement (Fig. 3.27).

Both synthesis of CdS nanopowder and X-ray data collection were performed at a temperature of 295 K. Since the total number n of stacking planes is equal to 10, there is a total of $Z = 2^{n-2} = 256$ different versions of disordered sequences of close-packed planes. In consequence, the entropy term gives 140 meV per pair of Cd and S atoms [100]. This value should be comparable to the energy difference between the disordered and the ordered phase and it should also be close to the energy difference between the phases with $B3$ and $B4$ structures. However, these values are not known for CdS.

3.3.2   Visualization of "Average" Long-Range Order in the Non-periodic Structure of CdS Nanoparticle

According to [100–103], a disordered close-packed structure is peculiar to CdS nanoparticles.

Such structures formally have no long-range order and translation symmetry. For this reason, Junkermeier et al. [120], revealing the presence of the short-range order and the absence of the long-range order in CdS nanoparticles, concluded that their structure is amorphous. Ab initio molecular dynamics calculations were used to simulate spherical CdS nanoparticles with the wurtzite and sphalerite structures and to relax and to anneal them at 300 K [120]. As an example, modeled $Cd_{132}S_{132}$ nanoparticle with cubic sphalerite-type structure is shown in Fig. 3.28a. Using the pair distribution function of atoms in the calculated particles, it was found that the relaxed 2-nm nanoparticles lost their crystal structure and became amorphous

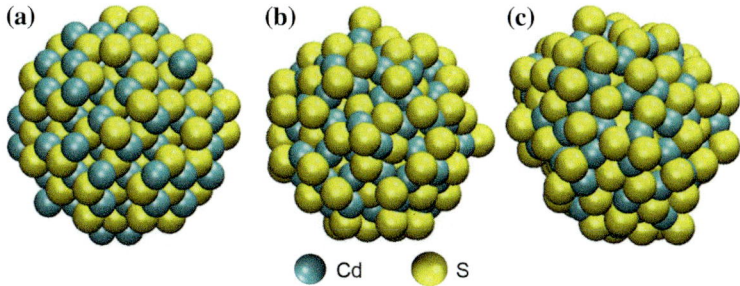

Fig. 3.28 Structure of the $Cd_{132}S_{132}$ nanoparticle of radius 1.164 nm [120]: **a** cubic sphalerite-type (zinc-blende) structure, **b** the same nanoparticle after relaxation, **c** the same nanoparticle after virtual "annealing" at 300 K

(Fig. 3.28b, c). According to [120], such structure with essentially short-range order is characteristic of particles smaller than 6 nm in size.

However, XRD patterns of nanopowders and thin films of cadmium sulfide usually contain three characteristic reflections in a 2θ angular range of $15°$–$60°$, which indicate the presence of the specific long-range order in the structure of CdS nanoparticles. The calculation of X-ray scattering from the nanopowder of model nanoparticles [100–102, 116] with the Debye formula (3.16) has shown that the diffraction reflections provide the "average" position of atoms in the disordered close-packed structure of a CdS nanoparticle, which is neither crystal nor amorphous. Nevertheless, information obtained by diffraction methods is insufficient for identifying the non-crystalline structure of nanostructured substances for two reasons. First, scattering occurs from a set of nanoparticles of the powder or domains of the film; as a result, the spatial position of atoms in the non-crystalline structure is averaged over all of the particles of the powder. Second, in contrast to crystal structures, the unambiguous correspondence between the representations of a non-crystalline structure in the direct and reciprocal spaces is absent.

For this reason, XRD patterns from the disordered close-packed structure of CdS nanoparticles [100–102] and from the icosahedral structure of metal clusters [107] are almost the same. These two close-packed structures can be distinguished only by visualization in the direct space of the non-crystalline structure of a single nanoparticle. The direct method of visualization is high resolution transmission electron microscopy (HRTEM).

A method for identification of non-crystalline structure of CdS nanoparticles in the direct space by means of the interpretation of the HRTEM image was proposed in study [121]. The nanocrystalline CdS powder was obtained by chemical deposition from an aqueous alkaline solution by the procedure described in [21]. The atomic structure of CdS nanoparticles in the direct space was studied on a JEM3010 (JEOL) transmission electron microscope at a voltage of 400 keV (the wavelength is 2 pm). The aperture (diaphragm) was 225 nm and the instrument constant was

2.05 μm nm. The HRTEM images were analyzed with the "Gatan Digital Micrograph 3.9.0 code for GMS 4.0" program for processing the images.

Although nanostructured CdS is characterized by a non-periodic sequence of the packing layers (for example, *ABCBACA*), the short-range order in CdS nanoparticles is identical with the short-range order in CdS with fcc structure of *B*3-type (sphalerite) and hcp structure of *B*4-type (wurtzite).

In spite of the disorder and the absence of translation symmetry, the non-crystalline close-packed structure of nanostructured cadmium sulfide has the "average" long-range order [100–102, 116], which is manifested in the form of pronounced maxima of the intensity in the diffraction experiment. In scattering from stochastically alternating layers of the packing, information on the arrangement of layers disappears due to the interference of waves and only information that atoms are located in the packing layers that can be located in any of three spatial positions, *A*, *B*, and *C*, remains. This is equivalent to the representation that a layer is in all three positions with the same probability 1/3, remains. Owing to this approach, a lattice with hexagonal symmetry (space group No. 168—*P*6 (C_6^1)) and the unit cell parameters $a = 0.236$ nm and $c = 0.334$ nm, where a is the distance between the projections of all of the atoms on the packing plane and c is the distance between the nearest close-packed layers of the atoms of the same kind (Cd or S), can be associated with the non-crystalline structure. The Cd and S atoms in the unit cell of the "average" lattice occupy single positions $1(a)$ with the coordinates $(0\ 0\ z)$, where $z = 0$ and $1/\sqrt{24}$, respectively; the occupancy of these positions is 1/3 (Table 3.3). The $1(a)$ positions have the point symmetry group 6 (C_6), which includes six symmetry elements H_1–H_6 of the total hexagonal group 6/*mmm* [122–124]. Since the "average" lattice is primarily used to describe diffraction and does not describe the real positions of atoms in the particle, the determination of the physical positions of the atoms in the non-crystalline structure requires the consideration of the particle in the direct space.

Visualization of non-crystalline structure of nanostructured CdS in direct space was carried out by means of the interpretation of the HRTEM image of the CdS nanoparticle (Fig. 3.29) with the clearly seen columns of atoms. In order to exactly determine the positions of the atomic columns, the intensity maxima were determined by contrasting the image; circles were imposed on these maxima taking into account the interatomic and interplanar distances. The measurement of all of the possible characteristic interatomic and interplanar distances indicates that the image presents the close-packed layers "in the section". In other words, the electron beam passes in parallel to the layers of the packing of the nanoparticle and coincides with the translation in the [100] direction for the "average" lattice of the disordered structure with space group *P*6. The atomic columns on the HRTEM image (see Fig. 3.29) are columns of cadmium atoms, which are less penetrable for the electron beam. The columns of cadmium and sulfur atoms can be distinguished only if the distance between them is larger than the sum of the covalent radii of Cd and S [125, 126]; for the coordination number 4, this sum is 0.245-0.252 nm. In the projection shown in Fig. 3.29, the distance between the column of Cd atoms and the nearest

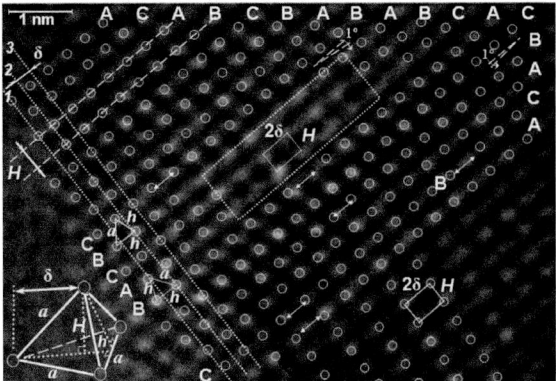

Fig. 3.29 HRTEM image of the CdS nanoparticle in the projection perpendicular to the packing layers [121]. The *circles* mark the intensity maxima. A tetrahedron whose parameters a, h, H, and δ correspond to the characteristic distances between the atomic columns is shown in the lower *left corner*. The packing layers are arranged at an angle of 45° to the *horizontal line*. The layers intersecting with lines *1*, *2*, and *3* are marked as A, B, and C, respectively. Displacements of the close-packed layers are marked by *arrows*. Also the divergences of the layers by an angle of 1° are shown. The *dotted regions* in the center and lower right corner of the image constitute a lattice with a $H \times 2\delta$ rectangle unit cell describing the sites of the "average" lattice (space group $P6$) of the disordered CdS structure

column of S atoms should be equal to ~ 0.145 nm. The distance between Cd atoms in nanostructured CdS is usually 0.409 nm. The interatomic distance determined from the HRTEM image is 0.41±0.01 nm and, consequently, corresponds to the distance between the columns of cadmium atoms.

The short-range order in cadmium sulfide with close-packed structures of $B3$- and $B4$-type is determined by the tetrahedral environment. A sulfur atom is located at the center of mass of a tetrahedron whose vertices are formed by cadmium atoms. This tetrahedron is shown in the lower left corner in Fig. 3.29. It is clearly seen that all of the characteristic distances between the atomic lines and columns on the HRTEM image correspond to the distances in such a tetrahedron. Three closest atomic columns on the HRTEM image constitute the projection of the tetrahedron on the plane dissecting it through one edge and the heights of two faces; the lines of this section are shown by the dotted lines on the tetrahedron in Fig. 3.29. Correspondingly, the distances between three columns are a, h, and h, where $h = a\sqrt{3}/2 = 0.36$ nm is the height of the face of the tetrahedron. The atomic columns at the distance a belong to different layers of the packing, the distance between neighboring columns in the same layer is h, and two tetrahedron edges with the lengths a and h correspond to the distances between the nearest columns of different layers. Owing to a change in the character of the alternation of the layers, the symmetry of the projection of the tetrahedron changes (see Fig. 3.29); this phenomenon is similar to the twinning phenomenon in coarse-crystalline samples.

Since three of the four vertices of the tetrahedron belong to the same layer of the packing and the fourth vertex belongs to another layer, the height of the tetrahedron $H = a\sqrt{2/3} = 0.34$ nm is the distance between the packing layers. The packing layer, which corresponds to the equidistant series of circles, was translated to the distance H between the layers and was superposed with the intensity maxima in this series in order to identify these layers. As a result, most of the intensity maxima were described in the framework of the close-packed structure [121].

The character of the alternation of the packing layers was determined in study [121] as follows. The shift of the close-packed layers with respect to each other corresponds to the distance between the projection of the vertex of the tetrahedron on the plane and any of the three vertices in the same plane, which is $\delta = h$ $2/3 = a/\sqrt{3} = 0.24$ nm. Parallel lines 1, 2, and 3, separated by distance δ, are parallel to the [100] zone axis of the lattice with space group $P6$ and, correspondingly, are perpendicular to the packing layers (see Fig. 3.29). The intersection of the atomic column belonging to a particular packing layer by line 1, 2, or 3 makes it possible to identify layer by index A, B, or C, respectively. Thus, the layers with the same index are intersected by the same line and are arranged strictly over each other. Owing to the indexing in the observed CdS nanoparticle, the sequence of the packing layers is determined and presented along the upper edge of Fig. 3.29. The total sequence of the packing layers presented on the HRTEM image (see Fig. 3.29) has the form $ACABCBABABCACBACA$. Since this sequence is non-periodic, the structure is certainly non-crystalline close-packed.

The detailed analysis of the HRTEM image makes it possible to determine microstrains even in a non-crystalline structure. The translation of atomic columns is violated in some packing layers and, then, is recovered in the same layer or with the shift to another layer. In particular, it is seen in Fig. 3.29 that the central fragment of the series $ACABCBA$–$BABCA$–$CBACA$ is entirely transformed into the sequence $CBCAB$, i.e., is completely displaced to the distance δ with respect to the initial position (layer A changes to B, B to C, and C to A). In addition, almost all of this fragment is deviated from the remaining sequence by $1°$ (see Fig. 3.29).

Thus, the section without microstrains is relatively small and is determined by the beginning of the series $ACABCBA$–$BABCACBACA$. It can be stated that only this fragment is identical to the region of coherent scattering in the diffraction experiment in agreement with the characteristic size of the particles of 5 nm, which is determined from the experimental broadening of the XRD reflections [100–102, 116], and with the calculations by the Debye formula. On the whole, the determination of the positions of the atomic columns, packing layers, and their alternation make it possible to identify the short-range order in the non-crystalline structure of CdS nanoparticle in the direct space and to reveal the presence of microstrains.

Let us consider the dotted regions in the center and lower right corner of the image shown in Fig. 3.29, which cannot be identified as the closest packing. The intensity maxima in these sections are also located on the lines corresponding to the packing layers with the same interplanar distance H. However, the distances

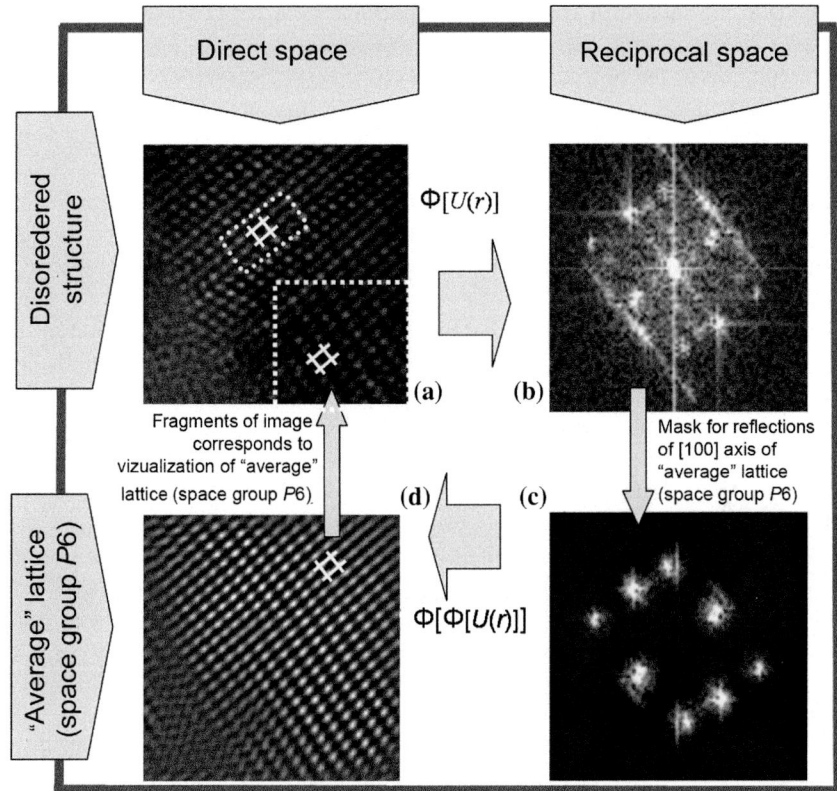

Fig. 3.30 Scheme of the processing of the HRTEM image shown in Fig. 3.29 [121]. **a** initial HRTEM image with the rectangle unit cell which is shown by the *dotted lines*; **b** Fourier transformation of the initial image limited by the *dotted lines* to the reciprocal space; **c** imposition of a mask on the Fourier transform; **d** inverse Fourier transform in the direct space. Imposition of a mask and inverse Fourier transformation have allowed to explain the observed rectangle unit cells as result of visualization of the "average" lattice (space group *P*6) of the disordered CdS structure where an identical rectangle unit cell appears

between the maxima in the same series do not coincide with the translation of the columns in the CdS packing layer, which is $a = 0.41$ nm. The distribution of the intensity maxima in both indicated sections is on the whole the same and is determined by the periodic lattice with the rectangular unit cell with the sides H and 2δ.

Figure 3.30 shows the scheme of the interpretation of the discussed sections of the HRTEM image. The Fourier transform $\Phi[U(r)]$, which is equivalent to the electron diffraction from the separated region of the studied CdS nanoparticle, is calculated for the initial HRTEM image (see Fig. 3.30a). The analysis of the Fourier transform of the electron diffraction (Fig. 3.30b) indicates that the intensity maxima in the reciprocal space are the reflections from the [100] axis of the "average" (space group $P6$) lattice of the disordered structure.

In addition to the reflections of the lattice with space group $P6$, there are two wide diffuse bands, which do not pass through the structure sites of the reciprocal lattice, are parallel to the [100] direction, and lie at distances corresponding to the sphalerite $B3$-type structure. A mask imposed on the Fourier transform (Fig. 3.30c) covers the diffuse bands and zero*th* reflection and makes it possible to separate the reflections of the [100] axis of the "average" (space group $P6$) lattice. The periodic diffuse effects that do not pass through the structure sites of the reciprocal lattice can be attributed to the short-range order and to the atomic-displacement waves [124, 127, 128], which appear in this case owing to the non-periodic alternation of the packing layers. Thus, the diffuse bands in the reciprocal space present information on a particular sequence of the packing layers. For this reason, their positions at distances corresponding to the sphalerite $B3$-type structure are explained by the presence in nanoparticles of sections of the type $ABCAB$, which are small sequences of the packing layers similar to the cubic structure. In other words, the diffuse effects are attributed to the presence of specific tetrahedral clusters.

The inverse Fourier transform $\Phi[\Phi[U(r)]]$ of the mask of the Fourier transform provides the direct-space image of the periodic lattice with a $H \times 2\delta$ rectangle unit cell (Fig. 3.30d), which is identical to the lattice in the dotted regions in Fig. 3.29. Note that there is a one-to-one correspondence between the quantities δ and H and the parameters a and c of the unit cell of the "average" (space group $P6$) lattice. Thus, despite the obvious difference of the structure of the fragments indicated by the dotted lines in Fig. 3.30a from the disordered close-packed structure, these fragments constitute the visualization of the "average" (space group $P6$) lattice of the non-crystalline cadmium sulfide.

The mapping of the "average" (space group $P6$) crystal lattice on the HRTEM image is the direct-space visualization of the "average" long-range order in the non-crystalline structure of the CdS nanoparticle. In this projection, under the superposition of packing layers, the distance between the atomic columns is $\delta = 0.24$ nm, which is smaller than the linear length 0.6 nm of two neighboring Cd atoms. It is obvious that the resolution of the microscope does not provide the visualization of two or three packing layers superimposed in the plane of the passing beam. Since the distance between the columns is small, the electron beam undergoes scattering and leaves the interference result in the form of the "average" lattice on the HRTEM image.

This effect is a manifestation of the wave nature of electrons. However, in contrast to moiré appearing due to the superposition of the two-dimensional projections of the lattices of crystal grains, the "average" lattice effect in Fig. 3.29 appears due to the superposition of the one-dimensional projections of the packing layers A, B, and C. Thus, the moiré effect makes it possible to visualize the "average" long-range order in non-crystalline structures, in particular, under the superposition of one-dimensionally ordered objects, e.g., packing layers.

The interpretation of the image obtained by means of high resolution transmission electron microscopy makes it possible to identify the non-crystalline structure of the CdS nanoparticle as disordered close-packed. This structure is not amorphous, because it has not only the short-range order, but also the "average" long-range order. The long-range order is manifested in the direct space of the HRTEM image as the distribution of the sites of the "average" lattice (space group $P6$) of the disordered close-packed structure of the nanoparticle. This effect can be described as moiré for the one-dimensional case.

The vast majority of experimental data point that, irrespective of the method of synthesis and morphology of nanostructured cadmium sulfide, there is a correlation between size and structure and size of CdS nanoparticles (Fig. 3.31). For instance, nanocrystalline CdS powders with an average nanoparticle size of about 5 nm have a disordered close-packed structure [19, 20, 129, 130]. Dendritic samples [22]

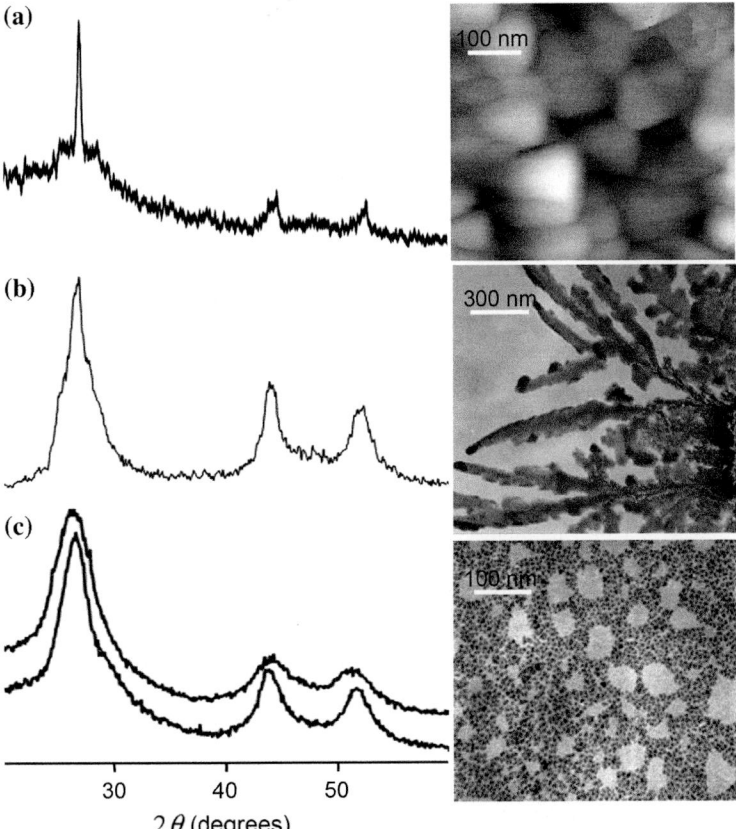

Fig. 3.31 Different forms of nanostructured CdS cadmium sulfide with different morphology. **a** XRD pattern of CdS thin film on glass substrate and atomic-force microscopy image of this film [133]; **b** XRD pattern of dendritic-like CdS structure and its TEM image [22]; **c** XRD patterns of CdS nanoparticles capped by n-trioctylphosphine oxide (TOPO) [20]. Diffraction reflections which are characteristic of a disordered close-packed structure with an "average" hexagonal (space group $P6$) of CdS nanoparticles are present on all XRD patterns

consist of nanoparticles with the same random close-packed structure characterized by a Gaussian-like size distribution in the range from 2 to 20 nm with a peak at 10 nm. However, large nanoparticles mainly have the wurtzite structure. In particular, spherical nanoparticles and nanofilaments 20-60 nm in size have well-defined hexagonal wurtzite structure [131]. Nanofilaments 100 nm in diameter have both hexagonal and, probably, cubic structure (nonequilibrium under normal conditions) [132].

The dependence of the structure on the particle size is also observed for CdS thin films. Based on the XRD patterns [11, 17, 133, 134], the structure of a film with nanoscale domains can be identified as random close-packed. As the domain size in the film increases to 60 nm and more [135–137], the domain structure can be identified as wurtzite.

In the whole, it follows from performed analysis of cadmium sulfide structure that CdS nanoparticles of 3–9 nm in size have a disordered close-packed structure which is characterized by an "average" long-range order [101, 102, 116, 121].

3.4   Optical Properties of Nanostructured CdS

As the size of a semiconductor crystal decreases to nanoscale level in one, two or three dimensions, the electronic structure and properties of the material change essentially [108, 138]. Discrete structure of the electronic energy levels in semiconductor nanoparticles allows them to be considered as "artificial atoms". Early in the 1980s, soviet researchers Ekimov, Onushchenko, Efros et al. predicted and then discovered the blue shift of the absorption edge upon size reduction of CdS nanoparticles [31, 94]. This also implies the possibility of controlling other properties, in particular, luminescence. However, changes in the electronic, optical and other properties with decreasing particle size are due to the influence of two factors including the size confinement effect and an increase in the surface-area-to-volume ratio. Let us consider the effect of these factors on the properties of nanoparticles in order to demonstrate the possibility of creation of quantum dots based on CdS nanoparticles.

Quantum dots are quasi-zero-dimensional ("0" D) structures. In such structures, the charge carrier motion is limited spatially in three directions and these structures are characterized by completely discrete energy spectra. Indeed, the limiting case of quantum-size effect in semiconductors is realized in structures with spatial confinement of the charge carriers in all three dimensions. The quantum dots afford the possibility of the most radical modification of the electron spectrum in comparison with the case of a bulk superconductor. The electron spectrum of an ideal quantum dot is a set of discrete levels, separated by regions of forbidden states, and corresponds to the electron spectrum of a single atom. However, the real quantum dot (or "superatom") may consist of hundreds of thousands of atoms.

The blue shift of the absorption edge upon size reduction of CdS nanoparticles was observed for the first time in study [31]. For CdS nanoparticles from 1.5 to

40 nm in size dispersed in a dielectric glass matrix, the blue shift increased with decreasing particle size and reached a value of 0.8 eV for nanoparticles 1.5 nm in size [31]. The band gap increases due to quantization of states of an exciton in the potential well with infinitely high walls (the potential well boundaries are specified by the surface of the spherical semiconductor particle). Quantization due to spatial confinement manifests itself for particles whose size is comparable or smaller than the diameter of an exciton. For coarse-crystalline cadmium sulfide, the diameter D_{ex} of an exciton is ~ 5.8-6.0 nm. According to [94], the absorption edge as a function of the size (radius R) of nanoparticles is described by (3.15).

A more detailed expression for the dependence of the exciton energy E on the radius $R = D/2$ of monodisperse semiconductor nanoparticles was proposed in study [30] and has the form (2.39). Having experimentally determined the band gap, one can estimate the average nanoparticle radius R as

$$R = \pi\hbar/(2\mu_{ex}\Delta E_g)^{1/2}, \tag{3.17}$$

where $\Delta E_g = E_g - E_b$. The reduced exciton mass $\mu_{ex} = m_e m_h/(m_e + m_h)$ for cadmium sulfide CdS is equal $\sim 0.16 m_0$ ($m_0 = 9.1 \times 10^{-31}$ kg is the free electron rest mass). Taking exciton mass into account, the formula (3.17) for CdS nanoparticles can be transformed to a following form

$$D = 3.06/(\Delta E_g)^{1/2}, \tag{3.18}$$

where D is CdS nanoparticle diameter in nanometers (nm), and ΔE_g is the experimental value of blue shift measured in electron-volt (eV).

The blue shift of the absorption edge with a decrease of CdS nanoparticle size has been observed in study [139]. CdS nanoparticles were synthesized via deposition from an aqueous solution containing cadmium chloride, thiourea and ammonia. Thiophenol as a capping agent was added to the solution. According to [139], CdS nanoparticles with a size ~ 9-10 and ~ 5 nm have a hexagonal wurtzite structure, and CdS nanoparticles with a size ~ 2 nm and smaller have a cubic

Fig. 3.32 Optical absorption for CdS nanoparticles with different size [139]. All absorption edges are blue-shifted with respect to the band gap of bulk hexagonal CdS. The vertical arrow notes absorption edge of bulk (coarse-crystalline) hexagonal cadmium sulfide

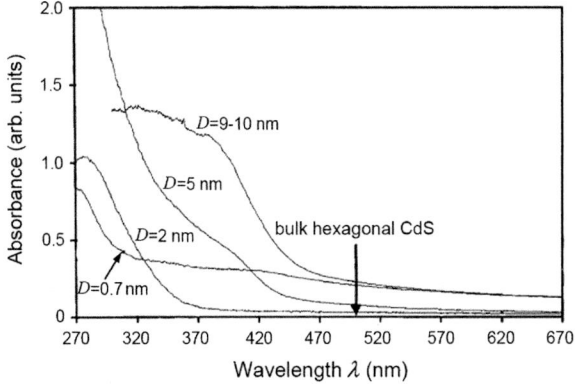

sphalerite structure. Authors [139] stated that a reduction in the CdS nanoparticles size is accompanied by a transformation from the hexagonal wurtzite-type structure to the cubic zinc-blende-type (sphalerite-type) structure. The critical transformation size is around 4–5 nm. With taking into account data [100–103] (see also Sect. 3.3.1), it is more probably that CdS nanoparticles with a disordered random close-packed structure were studied in work [139].

The optical absorption data for CdS nanoparticles [139] are shown in Fig. 3.32. The absorption edge corresponding to the band gap in bulk hexagonal CdS is indicated by vertical arrow.

It is evident from Fig. 3.22 that all four samples exhibit absorption edges which are blue-shifted with respect to the bulk CdS band gap, arising from quantum confinement effects in the nanoparticles. According to [139], the band gaps E_g corresponding to the absorption edges for CdS nanoparticles with a size ~ 9–10, ~ 5, ~ 2, and ~ 0.7 nm are equal to 2.59, 2.76, 3.57, and 3.85 eV, respectively.

Linearized muffin-tin orbital (LMTO) calculations [140, 141] of the density of electronic states in CdS nanoclusters showed that the band gap in CdS nanoparticles increases owing to the shift of the conduction band minimum (CBM) towards higher energies and shift of the valence band maximum (VBM) towards lower energies (Fig. 3.33).

Experimental confirmation of the shift of the valence band maximum and conduction band minimum in thiol-stabilized CdS nanoparticles with decreasing particle size (see Fig. 3.33) was obtained in a soft X-ray spectroscopy study [142].

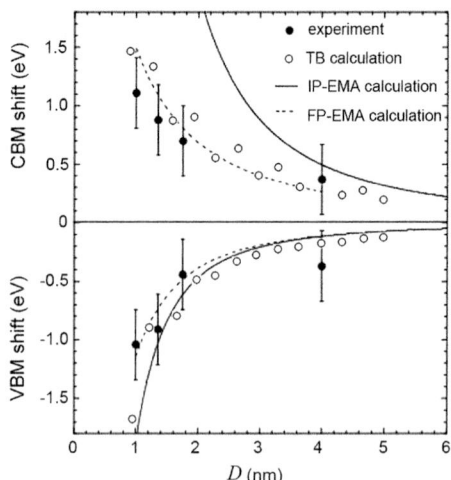

Fig. 3.33 Contributions of the occupied and unoccupied states to the total band gap in comparison with calculations in effective mass approximation (EMA) [142]: *Filled circle* experimental data [142]; *Solid line* the IP-EMA calculation using the model of an infinitely high confining potential [30, 94]; *Dash line* the FP-EMA calculation [143] using the finite height of the potential walls; *Open circle* the tight-binding (TB) calculation [144]. VBM and CBM are valence band maximum and conduction band minimum, respectively

It was shown that two effects contribute to the increase in the band gap. These are narrowing of the valence band from 4.7 to 3.5 eV and separation in the conduction band of contributions from the size confinement effect of excitons and from quantization of the density of states in the conduction band. The best correlation with experimental data was obtained in FP-EMA calculations with taking into account the finite height of the potential walls [143].

Figure 3.34 shows the dependences of the band gap energy E of CdS quantum dot with the core size of 4.7 nm from a thickness of capped shells from HgS and CdS; calculations [143] have been performed in different approaches. From Fig. 3.34 it is easily seen that the band gap energy decreases rapidly with increasing CdS outer layer thickness. The results of calculation [144], performed in the tight-binding approximation, are close enough to the data [142, 143].

According to [140], the best fit of the dependence of the band gap change, ΔE_g, on the CdS nanoparticle size (diameter) D is given by the following expression

$$\Delta E_g = 2.83 \exp(-D/8.22) + 1.96 \exp(-D/18.07), \tag{3.19}$$

where diameter D is measured in Å and ΔE_g is measured in eV. Similar expression for CdS quantum dots is proposed in study [141]:

$$\Delta E_g = 36.9/D^{1.2} + 269.7/D^{2.2}. \tag{3.20}$$

Both expressions describe the decrease in the shift of the absorption edge with increasing nanocrystal size.

Fig. 3.34 Calculated band gap E for CdS/HgS/CdS quantum-dot quantum well with 4.7-nm CdS cores as a function of HgS and CdS shell thicknesses [143]: **a** calculation with infinite potentials and with neglecting the Coulomb interaction; **b** calculation with finite potentials and with neglecting the Coulomb interaction; **c** calculation, including Coulomb interaction and finite potentials at the nanoparticle boundaries

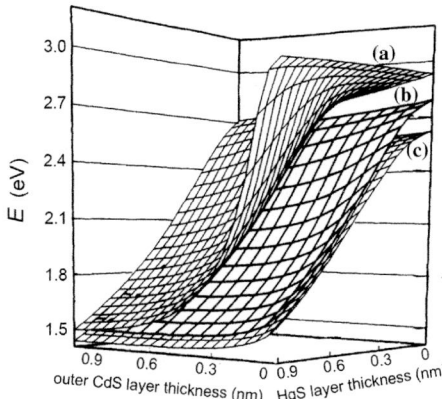

Fig. 3.35 Dependence of the band gap change, ΔE_g, on the size (diameter) D of CdS nanoparticle [36]: *Filled circle* experimental data [142]; (*1*) calculation according to (2.40), proposed in studies [30, 94]; (*2*) calculation according to (3.20), proposed in study [141]; (*3*) calculation according to (3.19), proposed by Sapra and Sarma[140]

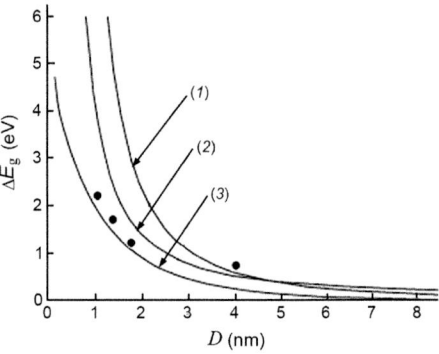

Figure 3.35 shows the dependencies of the band gap change, ΔE_g, on the CdS nanoparticle size D which are constructed and calculated in study [36]. Dependencies (*1*) and (*2*) have been calculated using the formulas (2.40) and (3.20), proposed in studies [30, 94] and [141], respectively. Obviously, both dependencies give overestimated ΔE_g value, especially in the region of $D < 2$ nm (see Fig. 3.35). The realistic estimations of the ΔE_g values in the region of small nanoparticle sizes and the best agreement with the experimental data [142] have been obtained by calculation based on the formula (3.19), proposed in study [140].

The blue shift of the optical absorption edge with decreasing nanoparticle size implies a corresponding blue shift of the luminescence band. Such effect is of importance for practical applications of quantum dots as luminophores or biomarkers. Calculations predict that a band gap E_g for CdS nanoparticles with a size of ~ 2 nm reaches 3.5 eV; correspondingly, luminescence should occur in the UV region at wavelength $\lambda = hc/E_g \approx 354$ nm. However, in most studies the blue shift of luminescence is not observed, or does not reach the characteristic values, or one deals with a red shift relative to the peak at ~ 510 nm characteristic of bulk coarse-crystalline cadmium sulfide.

Probably, disagreement between the absorption and luminescence bands is due to structural disorder. However, a large shift of luminescence maximum to 725 nm was observed in the case of crystalline CdS nanoparticles 3–6 nm in size [145] synthesized in reverse micelles. Observed red shift cannot be attributed to low degree of crystallinity of nanoparticles. On the contrary, such a large red shift indicates that perfect crystal structure of CdS nanoparticles leads to numerous defect surface states. Indeed, amorphization of the surface of particles synthesized in reverse micelles was also reported in study [146].

Theoretically, the red shift of luminescence maximum relative to the absorption edge was described for solid solutions of cadmium chalcogenides in which quantum dots are formed by the vacancy clusters [147]. According to calculations, the

influence of different types of electron-phonon coupling causes additional oppo-
sitely directed Stokes shifts of the absorption and luminescence bands.

Disordering in the sequence of close-packed layers is accompanied by a decrease
in the number of dangling bonds and the red shift should decrease too. Indeed,
luminescence maximum of nanoparticles whose structure is characterized by
twinning is observed at 550 nm [148]. Note that such structure with twinning can
be identified as random close-packed. The band gap in the thin CdS films annealed
at 720 K decreased from 3.50 to 3.46 eV, while the nanoparticle size increased
from 2.60 to 2.64 nm [149]. Annealing was followed by a red shift of the luminescence maximum from 520 to 544 nm, which was associated with an increase in
the number of surface defect states in the sulfur sublattice.

Theoretical calculation [150] of the behaviour of the surface sulfur atoms at the
cleavage of a CdS crystal revealed small atomic displacements relative to the upper
layer of cadmium atoms. These displacements of sulfur atoms on the surface of
crystals lead to formation of clearly seen surface states. Calculation [140] revealed
strong influence of uncompensated dangling bonds on the energy band structure.
Numerous unoccupied orbitals may give rise to electronic states in the band gap. In
the case of photoluminescence, these states can act a electron traps and,

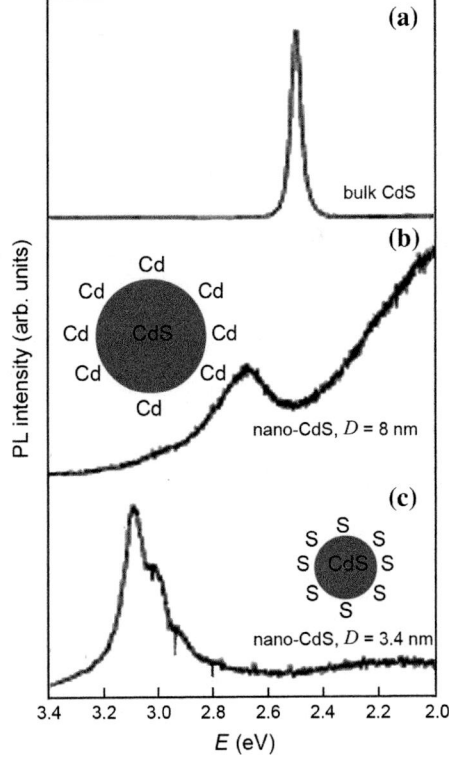

Fig. 3.36 Blue shift of
photoluminescence of CdS
nanoparticles due to
suppression of defect states in
the band gap by passivation of
the surface of CdS
nanoparticles [151]. **a** bulk
CdS (single crystal); **b** CdS
nanoparticle after etching the
surface by hydrochloric acid
for removal the surface sulfur
layer; **c** surface of CdS
nanoparticle is passivated
with a sulfur shell

consequently, induce a red shift. After elimination of unoccupied surface orbitals in the model nanoparticles, the electron states within the band gap disappeared.

The influence of passivation of the surface layer on the luminescence of CdS nanoparticles was confirmed experimentally [151]. The surface sulfur layer was removed by etching the surface of particles with hydrochloric acid and the Cd:S ratio became as high as 3:1. In the presence of excess cadmium the proportion of defect surface states increased, and along with increase in band gap E_g owing to reduction of the CdS nanoparticles size a red shift of the luminescence peak was observed. As a result, the total increasing band gap in comparison bulk CdS was small (Fig. 3.36b). If the surface of nanoparticles was passivated with a sulfur shell 3 nm thick, red shift has been suppressed and appreciable blue shift of the luminescence peak was observed (Fig. 3.36c).

Thus, the red shift or, more exactly, the absence of the expected blue shift, is due to the formation of surface trap states in the band gap and to electron-phonon coupling. To induce a blue shift, one should suppress the formation of surface trap states in the band gap of CdS quantum dots.

The most widely used and efficient method for elimination of surface states is to passivate the nanoparticle surface with an organic or inorganic shell surrounding the CdS core. At the same time, coating a nanoparticle with an organic shell may give rise to luminescence maxima in the red and infrared regions. For instance, luminescence of CdS nanoparticles surrounded by oleic acid molecules is characterized by three broad peaks in the region from 515 to 1030 nm. A maximum at 704 nm (1.76 eV) is produced by the surface defect states characteristic of crystalline nanoparticles. Two maxima at 576 nm (2.15 eV) and 905 nm (1.72 eV) are a result of the influence of the bonds between the surface atoms of the nanoparticle with the organic shell [152].

The blue shift of the luminescence peak from 515 to 420 nm in the synthesis of 1.5-nm CdS nanoparticles in a matrix of Langmuir-Blodgett film is accompanied by broadening of the peak owing to the influence of defect states at the quantum dot/matrix interface [153]. The intensity of luminescence of CdS nanoparticles with a size of 5 nm which characterized by the peak at 420 nm (2.95 eV) appreciably increases upon passivation of their surface in various solutions [129]. For the nanoparticles with a size of 8 nm, synthesized in reverse micelles, the blue shift of the luminescence peak from 550 to 410 nm and peak narrowing originate from removal of water from micelles [24].

The photoluminescence intensity maximum in the region of 420 nm (2.95 eV) can be interpreted in different manner. On the one hand, one deals with the blue shift relative to the luminescence peak at 510 nm (2.42 eV) of bulk CdS. On the other hand, such a behaviour of the luminescence of nanoparticles with a size smaller than 4 nm can be treated as red shift compared to the absorption edge shift and to an increase in the band gap by more than 0.5 eV.

Weak blue shift of photoluminescence is usually explained by nonradiative recombination of excitons from the remaining surface trap states. The exciton energy is 3-5 eV depending on the quantum dot size. However, recombination of an exciton leads to an expense of energy in nonradiative processes and on emission from the surface trap states.

Thus, the synthesis of quantum dots with specified luminescent properties should be carried out taking into account both the influence of the surface trap states and nonradiative Auger processes. Efros, who has predicted blue shift in CdS nanoparticles in a pioneering study [94], wrote in 2009: "In the early sixties, the one-dimensional carrier confinement achieved in semiconductor quantum wells and superlattices brought about a revolution in solid-state device technology... However, the application of nanostructures to real-world devices has been strongly curtailed by the enhancement of dissipative Auger processes that undergird all aspects of carrier relaxation and recombination. In particular, Auger processes have been attributed to the decrease of the photoluminescence quantum efficiency,... photoluminescence degradation and photoluminescence blinking in nanocrystal quantum dots" [154].

Auger processes in nanoparticles were ignored for long because the probability for such processes to occur in bulk crystals is negligible. Permanent irradiation of nanoparticles is accompanied by random charge-discharge processes. Absorption of a photon or creation of an electron-hole pair followed by emission of a photon do not influence electrical neutrality of nanoparticles. But if a nanoparticle bears an excess charge, nonradiative Auger recombination occurs. The rate of Auger recombination is a few orders of magnitude higher than that of radiative exciton recombination; correspondingly, Auger recombination is much more probable. Also, Auger processes involving surface trap states can occur [155].

High efficiency of Auger processes is due to abrupt termination of the electron potential at the particle boundaries [154]. Electron and hole tend to localize at the quantum dot boundaries. In the case of a charged nanoparticle, it leads to appearance of a surface trap states [156].

Thus, the properties of CdS nanoparticles depend strongly on their size and on the proportion of surface bonds.

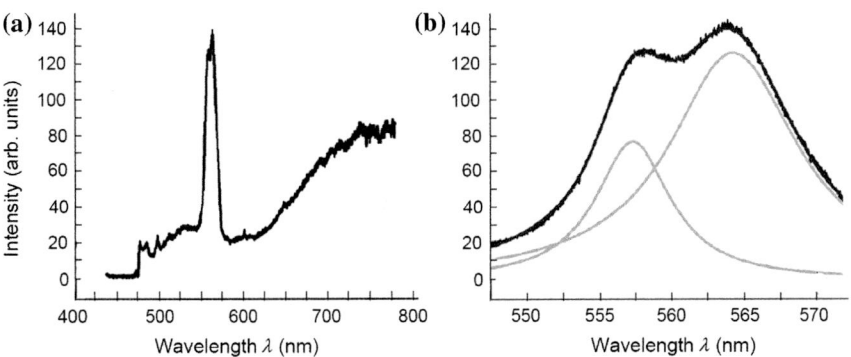

Fig. 3.37 Fluorescence spectrum of CdS nanoparticles [157]: **a** general view of spectrum; **b** fitting of two close peaks by the sum of two Lorentzians

Fluorescence of CdS quantum dots has been studied in work [157]. Fluorescence is a form of luminescence. It is the emission of light by a substance that has absorbed light or other electromagnetic radiation. Cadmium sulfide quantum dots with a size from 2 to 6 nm were prepared by chemical condensation [41] from aqueous colloidal solutions. The fluorescence spectrum of CdS quantum dots was measured by scanning confocal laser microscopy. The wavelength of the excitation light was 472.8 nm. The spectrum included a pretty narrow fluorescence peak with the maximum at about 557 nm and a broad long-wave peak (Fig. 3.37a). The detailed investigation of the spectra in the vicinity of the fluorescence maximum by means of a diffraction grating with a spectral resolution of 0.027 nm resolved two close peaks (Fig. 3.37b). Fitting of the spectrum to two Lorentzians yields the positions of the peaks to be 557 and 563 nm. The distance between the lines in the luminescence spectrum corresponds to 0.02 eV.

In studies [100–103], it was demonstrated that CdS nanoparticles similar to those studied in work [157] have a disordered structure. Their lattice consists of alternating atomic layers typical for hexagonal α-phase or cubic β-phase of cadmium sulfide. If only first three coordination spheres are taken into account, this disordered structure supports two possible surroundings of a cadmium atom. Presumably, the appearance of two peaks in the fluorescence spectrum separated by 0.02 eV is associated with the disordered nature of the structure of CdS quantum dots and, consequently, with two different surroundings of a cadmium atom, which are inherent to hexagonal wurtzite and cubic sphalerite (zinc-blende) structures.

References

1. Sharma, R.C., Chang, Y.A.: Cd-S (Cadmium-Sulfur). In: Massalski, T.B., Okamoto, H., Subramanian, P.R., Kacprzak, L. (eds.) Binary Alloy Phase Diagrams, vol. 2, 3rd edn, pp. 1020–1021. ASM International Publications, Materials Park, Ohio, (1992)
2. Woodbury, H.H.: Measurement of the Cd-CdS liquidus. J. Phys. Chem. Solids **24**(7), 881–884 (1963)
3. Abrikosov, N.Kh., Bankina, V.F., Poretskaya, L.V., Skudnova, E.V., Chizhevskaya, S.N.: Semiconducting Chalcogenides and Their Alloys, 220 pp. Nauka, Moscow (1975) (in Russian)
4. Klobukow, N.V.: Beitraege zur Kenntniss der auf nassem Wege entstehenden Modificationen des Cadmiumsulfides. Ztschr. praktische Chemie. Bd.39. No 2. S.413–421 (1889)
5. Ulrich, F., Zachariasen, W.: Über die Kristallstruktur des α- und β-CdS sowie des Wurtzits. Ztschr. Kristallogr. Mineralog. Bd.A62. S.270–273 (1925)
6. Wyckoff, R.W.G.: Crystal Structures, 2nd edn. Interscience Publications, New York (1963)
7. Traill, R.J., Boyle, R.W.: Hawleyite, isometric cadmium sulphide, a new mineral. Am. Mineralogist. **40**(7–8), 555–559 (1955)
8. Razik, N.A.: Use of a standard reference material for precise lattice parameter determination of materials of hexagonal crystal structure. J. Mater. Sci. Lett. **6**, 1443–1444 (1987)
9. Andrushko, A.F.: Defects of packing in crystals of powders of cadmium sulfide. Fiz. Tverf. Tela. **4**(3), 582–586 (1962) (in Russian)

10. Zelaya-Angel, O., de Castillo-Alvarado, L.F., Avendaño-López, J., Escamilla-Esquivel, A., Contreras-Puente, G., Lozada-Morales, R.: Raman studies in CdS thin films in the evolution from cubic to hexagonal phase. Solid State Commun. **104**(3), 161–166 (1997)
11. Gibson, P.N., Oezsan, M.E., Lincot, D., Cowache, P., Summa, D.: Modelling of the structure of CdS thin films. Thin Solid Films **361–362**, 34–40 (2000)
12. Su, B., Wei, M., Choy, K.L.: Microstructure of nanocrystalline CdS powders and thin films by Electrostatic Assisted Aerosol Jet Decomposition/Deposition method. Mater. Lett. **47**(1), 83–88 (2001)
13. Senthil, K., Mangalaraj, D., Narayandass, S.K., Kesavamoorthy, R., Reddy, G.L.N., Sundaravel, B.: Investigations on nitrogen ion implantation effects in vacuum evaporated CdS thin films using Raman scattering and X-ray diffraction studies. Phys. B **304**, 175–180 (2001)
14. Zhang, J., Sun, L., Liao, S., Yan, C.: Size control and photoluminescence enhancement of CdS nanoparticles prepared via reverse micelle method. Solid State Commun. **124**(1–2), 45–48 (2002)
15. Metin, H., Esen, R.: Annealing studies on CBD grown CdS thin films. J. Cryst. Growth **258**(1–2), 141–148 (2003)
16. Conde, O., Rolo, A.G., Gomes, M.J.M., Riccoleau, C., Barber, D.J.: HRTEM and GIXRD studies of CdS nanocrystals embedded in Al_2O_3 films produced by magnetron RF-sputtering. J. Cryst. Growth **247**, 371–380 (2003)
17. El Maliki, H., Berne'de, J.C., Marsillac, S., Pinel, J., Castel, X., Pouzet, J.: Study of the influence of annealing on the properties of CBD-CdS thin films. Appl. Surf. Sci. **205**(1), 65–79 (2003)
18. Rusu, M., Rumberg, A., Schuler, S., Nishiwaki, S., Wurz, R., Babu, S.M., Dziedzina, M., Kelch, C., Siebentritt, S., Klenk, R., Schedel-Niedrig, Th, Lux-Steiner, MCh.: Optimisation of the CBD CdS deposition parameters for $ZnO/CdS/CuGaSe_2/Mo$ solar cells. J. Phys. Chem. Solids **64**, 1849–1853 (2004)
19. Wang, W., Liu, Z., Zheng, C., Xu, C., Liu, Y., Wang, G.: Synthesis of CdS nanoparticles by a novel and simple one-step, solid-state reaction in the presence of a nonionic surfactant. Mater. Lett. **57**(18), 2755–2760 (2003)
20. Pan, D., Jiang, S., An, L., Jiang, B.: Controllable synthesis of highly luminescent and monodisperse CdS nanocrystals by a two-phase approach under mild conditions. Adv. Mater. **16**(12), 982–985 (2004)
21. Rempel, A.A., Kozhevnikova, N.S., Van den Berghe, S., Van Renterghem, W.: Self-organization of cadmium sulfide nanoparticles on the macroscopic scale. Phys. Stat. Sol. (b) **242**(7), R61–R63 (2005)
22. Wu, G.S., Yuan, X.Y., Xie, T., Xu, G.C., Zhang, L.D., Zhuang, Y.L.: A simple synthesis route to CdS nanomaterials with different morphologies by sonochemical reduction. Mater. Lett. **58**(5), 794–797 (2004)
23. Chen, C., Zhu, C., Hao, L., Hu, Y., Chen, Z.: Preparation and characterization of the CdS/PSA core-shell "egg". Inorg. Chem. Commun. **7**, 322–326 (2004)
24. Li, C., Yang, X., Yang, B., Yan, Y., Qian, Y.: Growth of microtubular complexes as precursors to synthesize nanocrystalline ZnS and CdS. J. Cryst. Growth **291**(1), 45–51 (2006)
25. Pässler, R.: Alternative analytical descriptions of the temperature dependence of the energy gap in cadmium sulfide. Phys. Stat. Sol. (b) **193**(1), 135–144 (1996)
26. Thomas, D.G., Hopfield, J.J.: Exciton spectrum of cadmium sulfide. Phys. Rev. **116**(3), 573–582 (1959)
27. À la Guillaume, C.B., Debever, J.-M., Salvan, F.: Radiative recombination in highly excited CdS. Phys. Rev. **177**(2), 567–579 (1969)
28. Anedda, A., Fortin, E.: Temperature shift of excitons in CdS. Phys. Stat. Sol. (b) **36**(1), 385–389 (1976)
29. Seiler, D.G., Heiman, D., Fejgenblatt, R., Aggarwal, R.L., Lax, B.: Two-photon magnetospectroscopy of A-exciton states in CdS. Phys. Rev. B. **25**(12), 7666–7677 (1982)

30. Brus, L.E.: Electron-electron and electron-hole interactions in small semiconductors crystallites: The size dependence of the lowest excited electronic state. J. Chem. Phys. **84** (9), 4403–4409 (1984)
31. Ekimov, A.I., Onushchenko, A.A.: Size quantization on electron energy spectrum in microscopic semiconductor crystals. **40**(8), 337–340 (1984) (in Russian) (Engl. Transl.: JETP Lett. **40**(8), 1136–1139 (1984))
32. Berger, L.I.: Semiconductor Materials, p. 472. CRC Press, New York (1997)
33. Kobiakov, I.B.: Elastic, piezoelectric and dielectric properties of ZnO and CdS single crystals in a wide range of temperatures. Solid State Commun. **35**(3), 305–310 (1980)
34. Ninomiya, S., Adachi, S.: Optical properties of wurtzite CdS. J. Appl. Phys. **78**(2), 1183–1190 (1995)
35. Badalyan, A.M., Belyi, V.I., Gel'fond, N.V., Igumenov, I.K., Kosinova, M.L., Morozova, N.B., Rastorguev, A.A., Rumyantsev, Yu.M., Smirnova, T.P., Fainer, N.I., Yakovkina, L.V.: Chemical composition and structure of thin films produced by chemical vapor deposition. Zh. Strukt. Khim. **43**(4), 605–628 (2002) (in Russian) (Engl. Transl.: J. Struct. Chem. **43**(4), 556–580) (2002)
36. Kozhevnikova, N.S., Vorokh, A.S., Uritskaya, A.A.: Cadmium sulfide nanoparticles prepared by chemical bath deposition. Russ. Chem. Rev. **84**(3), 225–250 (2015)
37. Chopra, K.L., Das, S.R.: Thin Film Solar Cells, 607 pp. Plenum Press, New York (1983)
38. Hodes, G.: Semiconductor and ceramic nanoparticle films deposited by chemical bath deposition. Phys. Chem. Chem. Phys. **9**(18), 2181–2196 (2007)
39. Pawar, S.M., Pawar, B.S., Kim, J.H., Joo, O.-S., Lokhande, C.D.: Recent status of chemical bath deposited metal chalcogenide and metal oxide thin films. Curr. Appl. Phys. **11**(2), 117–161 (2011)
40. Kuznetsova, Yu.V., Rempel, A.A.: Size and zeta potential of CdS nanoparticles in stable aqueous solution of EDTA and NaCl. Neorgan. Mater. **51**(3), 262–266 (2015) (in Russian) (Engl. Transl.: Inorg. Mater. **51**(3), 213–219) (2015)
41. Kozhevnikova, N.S., Vorokh, A.S., Rempel, A.A.: Preparation of stable colloidal solution of cadmium sulfide CdS using ethylenediaminetetraacetic acid. Zh. Obsh, Khimii. **80**(3), 365–368 (2010) (in Russian) (Engl. Transl.: Russ. J. Gen. Chem. **80**(3), 391–394 (2010))
42. Kozhevnikova, N.S., Demin, A.M., Krasnov, V.P., Rempel, A.A. The use of 3-mercaptopropyltrimethoxysilane for stabilization of luminescent cadmium sulfide nanoparticles. Dokl. Akad. Nauk. **452**(1), 47–51 (2013) (in Russian) (Engl. Transl.: Dokl. Chem. **452**(1), 215–219 (2013))
43. Rempel, A.A., Kozhevnikova, N.S., Rempel, S.V. Structure of cadmium sulfide nanoparticle micelle in aqueous solutions. Izv. Akad. Nauk. Ser. Khimicheskaya. (2), 400–404 (2013) (in Russian) (Engl. Transl.: Russ. Chem. Bull. Int. Ed. **62**(2), 398–402 (2013))
44. Murray, C.B., Norris, D.J., Bawendi, M.G.: Synthesis and characterization of nearly monodisperse CdE (E = sulfur, selenium, tellurium) semiconductor nanocrystallites. J. Am Chem. Soc. **115**(19), 8706–8715 (1993)
45. Kim, S., Bawendi, M.G.: Oligomeric ligands for luminescent and stable nanocrystal quantum dots. J. Am. Chem. Soc. **125**(48), 14652–14653 (2003)
46. Ma, L., Wu, S.-M., Huang, J., Ding, Y., Pang, D.-W., Li, L.: Fluorescence in situ hybridization (FISH) on maize metaphase chromosomes with quantum dot-labeled DNA conjugates. Chromosoma **117**(2), 181–187 (2008)
47. Gerion, D., Pinaud, F., Williams, S.C., Parak, W.J., Zanchet, D., Weiss, S., Alivisatos, A.P.: Synthesis and properties of biocompatible water-soluble silica-coated CdSe/ZnS semiconductor quantum dots. J. Phys. Chem. B. **105**(37), 8861–8871 (2001)
48. Gao, X., Cui, Y., Levenson, R.M., Chung, L.W.K., Nie, S.: In vivo cancer targeting and imaging with semiconductor quantum dots. Nat. Biotechnol. **22**(8), 969–976 (2004)
49. Chu, Y.-C., Wang, C.-C., Chen, C.-Y.: Synthesis of luminescent and rodlike CdS nanocrystals dispersed in polymer templates. Nanotechnology **16**(1), 58–64 (2005)

50. Liu, S.H., Qian, X.F., Yin, J., Ma, X.D., Yuan, J.Y., Zhu, Z.K.: Preparation and characterization of polymer-capped CdS nanocrystals. J. Phys. Chem. Solids **64**(3), 455–458 (2002)
51. Sooklal, K., Hanus, L.H., Ploehn, H.J., Murphy, C.J.: A blue-emitting CdS/dendrimer nanocomposite. Adv. Mater. **10**(14), 1083–1087 (1998)
52. Barglik-Chory, C., Buchold, D., Schmitt, M., Kiefer, W., Heske, C., Kumpf, C., Fuchs, O., Weinhardt, L., Stahl, A., Umbach, E., Lentze, M., Geurts, J., Mueller, G.: Synthesis, structure and spectroscopic characterization of water-soluble CdS nanoparticles. Chem. Phys. Lett. **379**(5–6), 443–451 (2003)
53. Spoerke, E.D., Voigt, J.A.: Influence of engineered peptides cadmium sulfide nanocrystals. Adv. Funct. Mater. **17**(13), 2031–2037 (2007)
54. Zou, L., Fang, Z., Gu, Z., Zhong, X.: Aqueous phase synthesis of biostabilizer capped CdS nanocrystals with bright emission. J. Lumin. **129**(5), 536–540 (2009)
55. Zhuang, Z., Lu, X., Peng, O., Li, Y.: Direct synthesis of water-soluble ultrathin CdS nanorods and reversible tuning of the solubility by alkalinity. J. Am. Chem. Soc. **132**(6), 1819–1821 (2010)
56. Gabellieri, E., Cioni, P., Balestreri, E., Morelli, E.: Protein structural changes induced by glutathione-coated CdS quantum dots as revealed by Trp phosphorescence. Eur. Biophys. J. **40**(11), 1237–1245 (2011)
57. Nelwamondo, S.M.M., Moloto, M.J., Krause, R.W., Moloto, N.: Direct synthesis of water soluble CuS and CdS nanocrystals with hydrophilic glucuronic and thioglycolic acids. Mater. Res. Bull. **47**(12), 4392–4397 (2012)
58. Huang, P., Jiang, Q., Yu, P., Yang, L., Mao, L.: Alkaline post-treatment of Cd(II)–glutathione coordination polymers: toward green synthesis of water-soluble and cytocompatible CdS quantum dots with tunable optical properties. ACS Appl. Mater. Interfaces. **5**(11), 5239–5246 (2013)
59. Rempel, A.A.: Nanotechnologies. Properties and applications of nanostructured materials. Russ. Chem. Rev. **76**(5), 435–461 (2007)
60. Vollath, D.: Nanomaterials: an Introduction to Synthesis, Properties and Applications, 2nd edn, 386 pp. Wiley-VCH, Weinheim (2013)
61. Uritskaya, A.A., Kitaev, G.A., Belova, N.S.: Kinetics of cadmium sulfide precipitation from aqueous thiourea solutions. Zh. prikl. Khimii. **75**(5), 864–865 (2002) (in Russian) (Engl. Transl.: Russ. J. Appl. Chem. **75**(5), 846–848 (2002))
62. Belova, N.S., Uritskaya, A.A., Kitaev, G.A.: Kinetics of lead sulfide precipitation from citrate solutions of thiourea. Zh. prikl. Khimii. **75**(10), 1598–1601 (2002) (in Russian) (Engl. Transl.: Russ. J. Appl. Chem. **75**(10), 1562–1565 (2002))
63. Kozhevnikova, N.S., Uritskaya, A.A., Rempel, A.A.: Dependence of the size of nanoparticles of lead sulfide PbS on the chemical affinity of its formation reaction. Dokl. Akad. Nauk. **453**(2), 167–171 (2013) (in Russian) (Engl. Transl.: Dokl. Phys. Chem. **453**(1), 270–273 (2013))
64. Vorokh, A.S., Kozhevnikova, N.S.: A Cd(OH)$_2$-CdS heteronanostructure of the core-shell type. Dokl. Akad. Nauk. **419**(1), 58–64 (2008) (in Russian) (Engl. Transl.: Dokl. Phys. Chem. **419**(1), 41–46 (2008))
65. Sugimoto, T., Dirige, G.E., Muramatsu, A.: Synthesis of uniform CdS particles from condensed Cd(OH)$_2$ Suspension. J. Colloid Interface Sci. **173**(1), 257–259 (1995)
66. Sugimoto, T., Dirige, G.E., Muramatsu, A.: Formation mechanism of uniform CdS particles from condensed Cd(OH)$_2$ suspension. J. Colloid Interface Sci. **176**(2), 442–453 (1995)
67. Sugimoto, T., Dirige, G.E., Muramatsu, A.: Formation mechanism of monodisperse CdS particles from concentrated solutions of Cd–EDTA complexes. J. Colloid Interface Sci. **182**(2), 444–456 (1996)
68. O'Brien, P., McAleese, J.: Developing an understanding of the processes controlling the chemical bath deposition of ZnS and CdS. J. Mater. Chem. **8**(11), 2309–2314 (1998)
69. Hodes, G.: Chemical Solution Deposition of Semiconductor Films, 377 pp. Dekker, New York (2002)

70. Katysheva, A.S., Markov, V.F., Maskaeva, L.N.: Mechanism of $PbSe_yS_{1-y}$ film formation in chemical deposition from aqueous solutions. Zh. Neorg. Khim. **58**(7), 940–945 (2013) (in Russian) (Engl. Transl.: Russ. J. Inorg. Chem. **58**(7), 833–838 (2013))
71. Markov, V.F., Maskaeva, L.N., Ivanov, P.N.: Hydrochemical Synthesis of Metal Sulfide Films: Modeling and Experiment, 217 pp. Publications of the Ural Division of the RAS, Ekaterinburg (2006) (in Russian)
72. Kozhevnikova, N.S, Uritskaya, A.A., Vorokh, A.S., Rempel, A.A.: Ionic equilibria in alkaline aqueous solutions of metal complex salts. Zh. Obsh. Khim. **78**(4), 568–574 (2008) (in Russian) (Engl. Transl.: Russ. J. Gen. Chem. **78**(4), 551–556 (2008))
73. Bär, M., Weinhardt, L., Heske, C., Muffler, H.-J., Umbach, E., Lux-Steiner, M.C., Niesen, T. P., Karg, F., Fischer, C.-H.: Chemical insights into the Cd2+/NH3 treatment: An approach to explain the formation of Cd-compounds on Cu(In, Ga)(S, Se)2 absorbers. Solar Energy Mater. Solar Cells. **90**(18–19), 3151–3157 (2006)
74. Uritskaya, A.A., Bol'shchikova, T.P., Kozhevnikova, N.S.: Kinetic relationships of the formation of antimony sulfide. Zh. Obsh. Khim. **77**(5), 717–720 (2007) (in Russian). (Engl. Transl.: Russ. J. Gen. Chem. **77**(5), 818–821 (2007))
75. Dona, J.M., Herrero, J.: Chemical bath deposition of CdS thin films: an approach to the chemical mechanism through study of the film microstructure. J. Electrochem. Soc. **144**(11), 4081–4091 (1997)
76. Ortega-Borges, R., Lincot, D.: Mechanism of chemical bath deposition of cadmium sulfide thin films in the ammonia-thiourea system: In situ kinetic study and modelization. J. Electrochem. Soc. **140**(12), 3464–3473 (1993)
77. Froment, M., Lincot, D.: Phase formation processes in solution at the atomic level: metal chalcogenide semiconductors. Electrochim. Acta **40**(10), 1293–1303 (1995)
78. Vorokh, A.S., Kozhevnikova, N.S., Rempel, A.A., Magerl, A.: Formation of cadmium sulfide CdS nanofilm on a $Cd(OH)_2/SiO_2$ precursor layer. Zh. strukt. Khim. **51**(6), 1206–1210 (2010) (in Russian) (Engl. Transl.: J. Struct. Chem. **51**(6), 1170–1175 (2010))
79. Kozhevnikova, N.S., Rempel, A.A., Hergert, F., Magerl, A.: Structural study of the initial growth of nanocrystalline CdS thin films in a chemical bath. Thin Solid Films **517**(8), 2586–2589 (2009)
80. Potter, B.G., Simmons, J.H.: Quantum size effects in optical properties of CdS-glass composites. Phys. Rev. B. **37**(18), 10838–10847 (1988)
81. Hayes, T.M., Lurio, L.B., Persans, P.D.: Growth and dissolution of CdS nanoparticles in glass. J. Phys. Condens. Matter **13**(3), 425–431 (2001)
82. Kumar, J., Verma, A., Pandey, P.K., Bhatnagar, P.K., Mathur, P.C., Liu, W., Tang, S.H.: Study of optical absorption and photoluminescence of quantum dots of CdS formed in borosilicate glass matrix. Phys. Scr. **79**(6), 065601–065604 (2009)
83. Dey, C., Molla, A.R., Goswami, M., Kothiyal, G.P., Karmakar, S.: Synthesis and optical properties of multifunctional CdS nanostructured dielectric nanocomposites. J. Opt. Soc. Am. B **31**(8), 1761–1770 (2014)
84. Kuznetsova, Yu.V., Rempel, A.A.: Synthesis of cadmium sulfide CdS nanoparticles in a silicate glass matrix. Neorg. Mater. **51**(9), 1013–1018 (2016) (in Russian) (Engl. Transl.: Inorg. Mater. **51**(9), 933–938 (2015))
85. Popov, I.D., Kuznetsova, Yu.V., Vlasova, S.G., Rempel, S.V., Rempel, A.A.: Synthesis and optical properties of glass with cadmium sulfide nanoparticles. Fiz. Khim. Stekla. **42**(1), 58–63 (2016) (in Russian) (Engl. Transl.: Glass Phys. Chem. **42**(1), 38–42 (2016))
86. Kuznetsova, Yu.V., Putyrskii, D.S., Rempel, S.V., Tyurnina, N.G., Tyurnina, Z.G., Rempel, A.A.: Formation of CdS nanoparticles in the matrix of silicate glass and its optical properties. Fiz. Khim. Stekla. **42**(3), 351–359 (2016) (in Russian) (Engl. Transl.: Glass Phys. Chem. **42**(3), 251–256 (2016))
87. Kuznetsova, YuV, Rempel, A.A., Meyer, M., Pipich, V., Gerth, S., Magerl, A.: Small angle X-ray and neutron scattering on cadmium sulfide nanoparticles in silicate glass. J. Cryst. Growth **447**, 13–17 (2016)

88. Liu, L.-C., Risbud, S.H.: Quantum-dot slze-distrlbution analysis and precipitation stages in semiconductor doped glasses. J. Appl. Phys. **68**(1), 28–32 (1990)

89. Simonova, N.B., Tuzikov, F.V., Khramov, R.N., Tuzikova, N.A., Tuzikov, M.F., Vakshtein, M.S.: Study of CdSe/CdS quantum dots in solutions and gels by small-angle X-ray scattering. Poverkhnost (2), 27–34 (2011) (in Russian) (Engl. Transl.: J. Surf. Invest. X-ray, Synchr. Neutr. Techn. **5**(1), 126–133 (2011))

90. Mei, G., Carpenter, S., Felton, L.E., Persans, P.D.: Size effects on optical transition energies in CdS_xSe_{1-x} semiconductor nanocrystal glass composites. J. Opt. Soc. Am. B **9**(8), 1394–1400 (1992)

91. Naik, S.D., Apte, S.K., Sonawane, R.S.: Nanostructured CdS/CdSSe glass composite for photonic application. J. Phys. **65**(4), 707–712 (2005)

92. Sonawane, R.S., Naik, S.D., Apte, S.K.: CdS/CdSSe quantum dots in glass matrix. Bull. Mater. Sci. **3**(3), 495–499 (2008)

93. Ushakov, V.V., Aronin, A.S., Karavanskiĭ, V.A., Gippius, A.A.: Formation and optical properties of CdSSe semiconductor nanocrystals in the silicate glass matrix. Fiz. Tverd. Tela. **51**(10), 2036–2040 (2009) (in Russian) (Engl. Transl.: Phys. Solid State. **51**(10), 2161–2165 (2009))

94. Efros, Al.L., Efros, A.L.: Interband absorption of light in a semiconductor sphere. Fiz. Tekhn. Poluprovodn. **16**(7), 1209–1214 (1982) (in Russian) (Engl. Transl.: Sov. Phys. Semiconductors. **16**(7), 772–775 (1982))

95. Borland Delphi Enterprise.: Version 7.0 Copyright © 1983–2002 Borland Software Corporation

96. Guinier, A., Fournet, G.: Small-Angle Scattering of X-Rays, p. 268. Wiley, New York (1955)

97. Feigin, L.A., Svergun, D.I.: Structure Analysis by Small-Angle X-ray and Neutron Scattering, p. 335. Plenum Press, New York - London (1987)

98. Ilavsky, J., Jemian, P.R.: Irena: tool suite for modeling and analysis of small-angle scattering. J. Appl. Cryst. **42**(2), 347–353 (2009)

99. Le Parc, R., Champagnon, B., Levelut, C., Martinez, V., David, L., Faivre, A., Flammer, I., Hazemann, J.L., Simon, J.P.: Density and concentration fluctuations in SiO_2-GeO_2 optical fiber glass investigated by small angle x-ray scattering. J. Appl. Phys. **103**(9), Paper 094917, 8 pp (2008)

100. Rempel, A.A., Magerl, A.: Non-periodicity in nanoparticles with close-packed structures. Acta Crystallogr. A **A66**(4), 479–483 (2010)

101. Vorokh, A.S., Rempel, A.A.: Atomic structure of cadmium sulfide nanoparticles. Fiz. tverd. Tela. **49**(1), 143–148 (2007) (in Russian) (Engl. Transl.: Phys. Solid State. **49**(1), 148–153 (2007))

102. Vorokh, A.S., Rempel, A.A.: Disordered structure and the shape of nanoparticles of cadmium sulfide CdS. Dokl. Akad. Nauk. **413**(6), 743–746 (2007) (in Russian) (Engl. Transl.: Dokl. Physics. **52**(4), 200–203 (2007))

103. Vorokh, A.S., Kozhevnikova, N.S., Rempel, A.A.: Transition of the CdS disordered structure to the wurtzite structure with an increase in the nanoparticle size. Izv. Akad. Nauk. Seriya fizich. **72**(10), 1472–1475 (2008) (in Russian) (Engl. Transl.: Bull. Russ. Acad. Sci.: Physics. **72**(10), 1395–1398 (2008))

104. Kepleri, J.: Strena seu De nive sexangula. Sacs. Caes. Majest. Mathematici, 24 pp. G. Tambach, Francofurti-Ad-Moenum (1611)

105. Hales, T.C.: A proof of the Kepler conjecture. Ann. Mathem. **162**(3), 1065–1185 (2005)

106. Lagarias, J.C.: Bounds for local density of sphere packings and the Kepler conjecture. Discrete Comput. Geom. **27**(2), 165–193 (2002)

107. Hall, B.D., Zanchet, D., Ugarte, D.: Estimating nanoparticle size from diffraction measurements. J. Appl. Cryst. **33**(6), 1335–1341 (2000)

108. Gusev, A.I.: Nanomaterials, Nanostructures, Nanotechnologies, 3rd edn, 416 pp. Nauka, Moscow (2007) (in Russian)

109. Hoare, M.R., Pal, P.: Physical cluster mechanics. Statics and energy surfaces for monatomic systems. Adv. Phys. **20**(84), 161–196 (1971)
110. Hoare, M.R., Pal, P.: Physical cluster mechanics. Statistical thermodynamics and nucleation theory for monatomic systems. Adv. Phys. **24**(5), 645–678 (1975)
111. Schaffer, P.T.B.: A review of the structure of silicon carbide. Acta Crystallogr. B **B25**(3), 477–488 (1969)
112. Gusev, A.I.: Phase equilibria in M-X-X' and M-Al-X ternary systems (M - transition metal; X, X' = B, C, N, Si) and crystal chemistry of related ternary compounds. Usp. Khim. **65**(5), 407–451 (1996) (in Russian) (Engl. Transl.: Russ. Chem. Rev. **65**(5), 379–419 (1996))
113. Mackay, A.L.: A dense non-crystallographic packing of equal spheres. Acta Crystallogr. **15** (9), 916–918 (1962)
114. International Tables for X-ray Crystallography. Vol. A. Space group symmetry. In: Hahn, T. (ed.). 4th revised edn, 878 pp. Kluwer Academic Publishers, Dordrecht (1995)
115. Vorokh, A.S., Kozhevnikova, N.S., Rempel, A.A., Magerl, J.: Disordering in cadmium sulfide nanoparticles. In: Borisenko, V.E., Gaponenko, S.V., Gurin V.S. (Eds.) Physics, Chemistry and Application of Nanostructures/Proceedings of International Conference on Nanomeeting-2007 (22–25 May 2007, Minsk, Belarus), pp. 312–315. World Scientific Publishing Company, New Jersey (2007)
116. Rempel, A.A., Vorokh, A.S., Neder, R., Magerl, A.: Disordered structure of cadmium sulphide nanoparticles. Poverkhn. (11), 8–11 (2011) (in Russian). (Engl. Transl.: J. Surf. Invest. X-ray, Synchr. Neutron Techn. **5**(6), 1028–1031 (2011))
117. Proffen, Th, Neder, R.B.: DISCUS: a program for diffuse scattering and defect-structure simulation. J. Appl. Crystallogr. **30**(2), 171–175 (1997)
118. Neder, R.B., Proffen, Th.: DISCUS 3.4 Users Guide, 108 pp. (2003)
119. Debye, P.: Zerstreuung von Roentgenstrahlen. Annalen der Physik B **B46**(1), 809–823 (1915)
120. Junkermeier, C.E., Lewis, J.P., Bryant, G.W.: Amorphous nature of small CdS nanoparticles: molecular dynamics simulations. Phys. Rev. B **79**(12), 125323–125328 (2009)
121. Vorokh, A.S., Rempel, A.A.: Direct-space visualization of the short and "average" long-range orders in the noncrystalline structure of a single cadmium sulfide nanoparticle. Pis'ma ZhETP. **91**(2), 106–111 (2010) (in Russian) (Engl. Transl.: JETP Lett. **91**(2), 100–104 (2010))
122. Kovalev, O.V.: Representations of the Crystallographic Space Groups: Irreducible Representations, Induced Representations and Corepresentation, 2nd edn, p. 390. Gordon & Breach Science Publishers, Yverdon (1993)
123. Gusev, A.I., Rempel, A.A., Magerl, A.J.: Disorder and Order in Strongly Nonstoichiometric Compounds: Transition Metal Carbides, Nitrides and Oxides, p. 607. Springer, Berlin (2001)
124. Gusev, A.I.: Nonstoichiometry, Disorder, Short-Range and Long-Range Order in Solids, 856 pp. Nauka-Fizmatlit, Moscow (2007) (in Russian)
125. Pauling, L.: The Nature of the Chemical Bond and the Structure of Molecules and Crystals: An introduction to Modern Structural Chemistry, 3rd edn, 644 pp. Cornell University Press, Ithaca (1960)
126. Vainstein, B.K., Fridkin, V.M., Indenbom, V.L.: Modern Crystallography 2: Structure of Crystals (Modern Crystallography, Vol. 2), 3rd edn, Chapter 1, pp. 97–127. Springer, Berlin (2000)
127. Cowley, J.M.: Diffraction Physics, 3rd revised edn, 481 pp. North-Holland Publishing, Amsterdam (1995)
128. Gusev, A.I.: Short-range order and diffuse scattering in nonstoichiometric compounds. Usp. Fiz. Nauk. **176**(7), 717–743 (2006) (in Russian) (Engl. Transl.: Phys. Uspekhi. **49**(7), 693–718 (2006))
129. Xu, R., Wang, Y., Jia, G., Xu, W., Liang, S., Yin, D.: Zinc blende and wurtzite cadmium sulfide nanocrystals with strong photoluminescence and ultrastability. J. Cryst. Growth **299** (1), 28–33 (2007)

130. Wang, G.Z., Wang, Y.W., Chen, W., Liang, C.H., Li, G.H., Zhang, L.D.: A facile synthesis route to CdS nanocrystals at room temperature. Mater. Lett. **48**(6), 269–272 (2001)

131. Zhang, Y.C., Wang, G.Y., Hu, X.Y.: Solvothermal synthesis of hexagonal CdS nanostructures from a single-source molecular precursor. J. Alloys Comp. **437**(1–2), 47–52 (2007)

132. Wu, X.C., Tao, Y.R.: Growth of CdS nanowires by physical vapor deposition. J. Cryst. Growth **242**(3–4), 309–312 (2002)

133. Oliva, A.I., Castro-Rodrìguez, R., Ceh, O., Bartolo-Perez, P., Caballero-Briones, F., Sosa, V.: First stages of growth of CdS films on different substrates. Appl. Surf. Sci. **148**(1–2), 42–49 (1999)

134. Li, Z., Du, Y., Zhang, Z., Pang, D.: Preparation and characterization of CdS quantum dots chitosan biocomposite. React. Funct. Polym. **55**(1), 35–43 (2003)

135. Ramaiah, K.S., Pilkington, R.D., Hill, A.E., Tomlinson, R.D., Bhatnagar, A.K.: Structural and optical investigations on CdS thin films grown by chemical bath technique. Mater. Chem. Phys. **68**(1–3), 22–30 (2001)

136. Yoshida, T., Yamaguchi, K., Kazitani, T., Sugiura, T., Minoura, H.: Atom-by-atom growth of cadmium sulfide thin films by electroreduction of aqueous Cd^{2+}–SCN^- complex. J. Electroanal. Chem. **473**(1–2), 209–216 (1999)

137. Fainer, N.I., Rumyantsev, YuM, Kosinova, M.L., Zemskova, S.M., Maximovskiy, E.A., Yurjev, G.S., Sivykh, G.F.: The structure study of thin cadmium and copper sulphide films by diffraction of synchrotron radiation. Nucl. Instr. Methods Phys. Res. Sect. A **448**(1–2), 290–293 (2000)

138. Gusev, A.I., Rempel, A.A.: Nanocrysnalline Materials, 351 pp. Cambridge International Science Publishing, Cambridge (2004)

139. Banerjee, R., Jayakrishnan, R., Ayyub, P.: Effect of the size-induced structural transformation on the band gap in CdS nanoparticles. J. Phys. Condens. Matt **12**(50), 10647–10654 (2000)

140. Sapra, S., Sarma, D.D.: Evolution of the electronic structure with size in II-VI semiconductor nanocrystals. Phys. Rev. B. **69**(12), 125304–125307 (2004)

141. Nazzal, A., Fu, H.: Comparative theoretical study of the size dependent electronic and optical properties in CdS and CdSe spherical nanocrystals. J. Comp. Theor. Nanosci. **6**(6), 1277–1289 (2009)

142. Lüning, J., Rockenberger, J., Eisebitt, S., Rubenssona, J.-E., Karl, A., Kornowski, A., Weller, H., Eberhardt, W.: Soft X-ray spectroscopy of single sized CdS nanocrystals: size confinement and electronic structure. Solid State Commun. **112**(1), 5–9 (1999)

143. Schooß, D., Mews, A., Eychmüller, A., Weller, H.: Quantum-dot quantum well CdS/HgS/CdS: theory and experiment. Phys. Rev. B **49**(24), 17072–17078 (1994)

144. Lippens, P.E., Lannoo, M.: Calculation of the band gap for small CdS and ZnS crystallites. Phys. Rev. B **39**(15), 10935–10942 (1989)

145. Pinna, N., Weiss, K., Urban, J., Pileni, M.P.: Triangular CdS nanocrystals: structural and optical studies. Adv. Mater. **13**(4), 261–264 (2001)

146. Rami, M., Benamar, E., Fahoume, M., Chraibi, F., Ennaoui, A.: Effect of the cadmium ion source on the structural and optical properties of chemical bath deposited CdS thin films. Solid State Sci. **1**(4), 179–188 (1999)

147. Klochikhin, A.A., Permogorov, S.A., Reznitskii, A.N.: Exciton luminescence from fluctuation-induced tails in the density of states of disordered solid solutions. Fiz. Tved. Tela. **39**(7), 1170–1182 (1997) (in Russian). (Engl. Transl.: Phys. Solid State **39**(7), 1035–1046 (1997))

148. Kumar, P., Saxena, N., Singh, F., Agarwal, A.: Nanotwinning in CdS quantum dots. Phys. B **407**(17), 3347–3351 (2012)

149. Majumder, M., Chakraborty, A.K., Mallik, B.: Photoluminescence studies on nanostructured cadmium sulfide thin films prepared by chemical bath deposition method and annealed at different temperatures. J. Lumin. **130**(8), 1497–1503 (2010)

150. Wang, Y.R., Duke, C.B.: Cleavage faces of wurtzite CdS and CdSe: surface relaxation and electronic structure. Phys. Rev. B. **37**(11), 6417–6424 (1988)

151. Gorer, S., Ganske, J.A., Hemminger, J.C., Penner, R.M.: Size-selective and epitaxial electrochemical/chemical synthesis of sulfur-passivated cadmium sulfide nanocrystals on graphite. J. Am. Chem. Soc. **120**(37), 9584–9593 (1998)

152. Katsaba, A.V., Ambrozevich, S.A., Vitukhnovsky, A.G., Fedyanin, V.V., Lobanov, A.N., Krivobok, V.S., Vasiliev, R.B., Samatov, G.: Surface states effect on photoluminescence of CdS colloidal nanocrystals. J. Appl. Phys. **113**(18), Paper 184306, 6 pp. (2013)

153. Bagaev, E.A., Zhuravlev, K.S., Sveshnikova, L.L., Badmaeva, I.A., Repinskii, S.M., Voelskow, M.: Photoluminescence from cadmium sulfide nanoclusters formed in the matrix of a Langmuir-Blodgett film. Fiz. Tekhn. Poluprovodnikov. **37**(11), 1358–1362 (2003) (in Russian) (Engl. Transl.: Semiconductors. **37**(11), 3121–1325 (2003))

154. Cragg, G.E., Efros, Al.L: Suppression of Auger processes in confined structures. Nano Lett. **10**(1), 313–317 (2010)

155. Cohn, A.W., Schimpf, A.M., Gunthardt, C.E., Gamelin, D.R.: Size-dependent trap-assisted Auger recombination in semiconductor nanocrystals. Nano Lett. **13**(4), 1810–1815 (2013)

156. Javaux, C., Mahler, B., Dubertret, B., Shabaev, A., Rodina, A.V., Efros, Al.L, Yakovlev, D. R., Liu, F., Bayer, M., Camps, G., Biadala, L., Buil, S., Quelin, X., Hermier, J.-P.: Thermal activation of non-radiative Auger recombination in charged colloidal nanocrystals. Nat. Nanotechnol. **8**(3), 206–212 (2013)

157. Rempel, S.V., Razvodov, A.A., Nebogatikov, M.S., Shishkina, E.V., Shur, V.Ya., Rempel, A.A.: Sizes and fluorescence of cadmium sulfide quantum dots. Fiz. tverd. Tela. **55**(3), 567–571 (2013) (in Russian) (Engl. Transl.: Phys. Solid State **55**(3), 624–628 (2013))

Chapter 4
Nanostructured Silver Sulfide Ag$_2$S

The well known silver sulfide Ag$_2$S is one of the most requisite semiconducting sulfides [1, 2] along with lead, zinc, copper, and cadmium sulfides [3–5].

According to the phase diagram of the system Ag–S [6, 7], silver sulfide Ag$_2$S has three basic polymorphic modifications (Fig. 4.1). Low-temperature semiconducting phase α-Ag$_2$S (acanthite) with monoclinic crystal structure exists at temperatures below $\sim$450 K. Under equilibrium conditions, cubic phase β-Ag$_2$S (argentite) exists in the temperature range 452–859 K, has body centered cubic (bcc) sublattice of sulfur atoms and has a superionic conductivity. High-temperature cubic γ-Ag$_2$S phase with face centered cubic (fcc) sublattice of sulfur atoms is stable from $\sim$860 K up to melting temperature. In the literature, the designations of monoclinic acanthite and cubic argentite are often confounded. For example, in works [8–13] monoclinic phase (acanthite) is designated as β-Ag$_2$S and bcc phase (argentite) as α-Ag$_2$S.

It is thought that monoclinic α-Ag$_2$S phase is stoichiometric, whereas cubic β-Ag$_{2+\delta}$S and γ-Ag$_{2\pm\delta}$S with $\delta \cong 0.002$ are nonstoichiometric phases having either small deficiency or small excess of silver. The homogeneity intervals of cubic allotropic forms of Ag$_2$S have been determined in works [9, 14–19]. According to [20], nonstoichiometric body centered cubic β-Ag$_{2+\delta}$S ($\delta \leq 0.002$) is characterized by high electronic conductivity of about 1.3 $\times$ 10^3 Ω^{-1} cm^{-1} that is 10^6 times higher than in monoclinic α-Ag$_2$S phase. Owing to high electronic conductivity, bcc β-Ag$_2$S can be used in photography [20].

One of the first determinations of the crystal structure of silver sulfide α-Ag$_2$S existing at temperatures below $\sim$450 K was given in work [21] using an acantite mineral sample. According to [21], α-Ag$_2$S phase is monoclinic (space group No. 14 – $B2_1/c$ ($P112_1/b$) (C_{2h}^5)) and its unit cell contains 8 Ag$_2$S formula units. The crystal structure of a single crystal of natural acanthite was carefully determined in study [8], where it was shown that the unit cell of acanthite is primitive monoclinic and belongs to the space group $P2_1/n$ ($P12_1/n1$). The proposed unit cell [8] contains

S.I. Sadovnikov et al., *Nanostructured Lead, Cadmium, and Silver Sulfides*, Springer Series in Materials Science 256, DOI 10.1007/978-3-319-56387-9_4

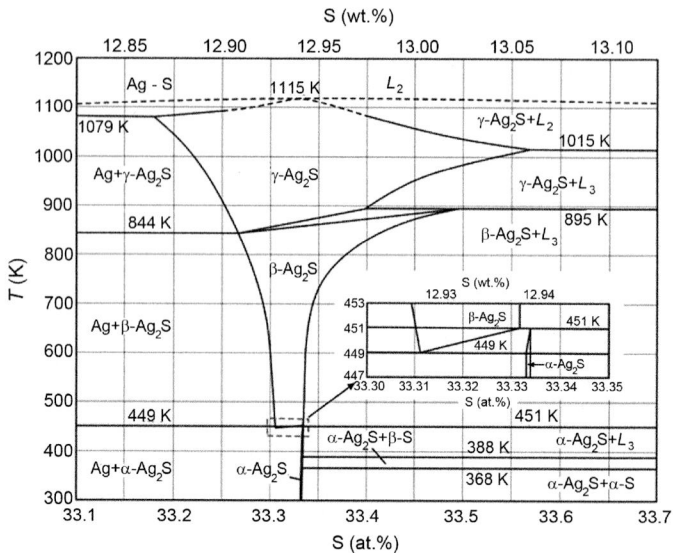

Fig. 4.1 Phase diagram of the Ag–S system in the neighborhood of Ag₂S [6]. Inset shows details of the α-Ag₂S–β-Ag₂S phase equilibria near 450 K

Table 4.1 Monoclinic (space group No 14 – $P2_1/n$ ($P12_1/n1$)) silver sulfide Ag₂S with α-Ag₂S acanthite-type structure at 298 K [8]: $Z = 4$, $a = 0.4229$ nm, $b = 0.6931$ nm, $c = 0.7862$ nm, and $\beta = 99.61°$

Atom	Position and multiplicity	Atomic coordinates			Occupancy
		x/a	y/a	z/a	
Ag1	4(e)	0.758	0.015	0.305	1
Ag2	4(e)	0.285	0.320	0.435	1
S	4(e)	0.359	0.239	0.134	1

4 Ag₂S formula units and has the following parameters: $a = 0.423$ nm, $b = 0.691$ nm, $c = 0.787$ nm and $\beta = 99.6°$ (99°35′). Refined crystal structure of monoclinic (space group $P2_1/n$) α-Ag₂S acanthite [8] is presented in Table 4.1.

Later it was shown [22] that the monoclinic unit cell of natural acantite mineral α-Ag₂S belongs to the space group $P2_1/c$ ($P12_1/c1$) (Table 4.2).

The length and direction of axes $a_{P2_1/c}$ and $b_{P2_1/c}$ of the unit cell [22] coincide with those of axes $a_{P2_1/n}$ and $b_{P2_1/n}$ of the unit cell [8], while $c_{P2_1/c} = (c_{P2_1/n} - a_{P2_1/n}) = (c_{P2_1/n} - a_{P2_1/c})$. According to [22], the axes of the both unit cells of α-Ag₂S acanthite can be represented as a combination of axes a_{bcc}, b_{bcc} and c_{bcc} of the unit cell of bcc argentite.

According to [23], the unit cell of β–Ag₂S argentite has cubic (space group No. 229 – $Im\bar{3}m$ ($I4/m\bar{3}2/m$) (O_h^9)) structure and includes two Ag₂S formula units. Two S atoms occupy crystallographic positions 2(a) and form bcc sublattice.

Table 4.2 Monoclinic (space group No 14 – $P2_1/c$ ($P12_1/c1$)) silver sulfide Ag$_2$S with α-Ag$_2$S acanthite-type structure at 298 K [22]: $Z = 4$, $a = 0.4231$ nm, $b = 0.6930$ nm, $c = 0.9526$ nm, and $\beta = 125.48°$

Atom	Position and multiplicity	Atomic coordinates			Occupancy
		x/a	y/a	z/c	
Ag1	4(e)	0.0712	0.0169	0.3075	1
Ag2	4(e)	0.7259	0.3213	0.4362	1
S	4(e)	0.5000	0.2383	0.1306	1

Table 4.3 Cubic (space group No. 229 – $I m\overline{3}m$ ($I 4/m\,\overline{3}2/m$) (O_h^9)) silver sulfide with β-Ag$_2$S argentite-type structure at 523 K [23]: $Z = 2$, $a = b = c = 0.4889$ nm

Atom	Position and multiplicity	Atomic coordinates			Occupancy
		x	y	z	
Ag1	6(b)	0	0.5	0.5	2/9
Ag2	12(d)	0.25	0	0.5	1/9
Ag3	24(h)	0	0.625	0.625	1/18
S	2(a)	0	0	0	1

Four Ag atoms are statistically distributed in 42 positions 6(b), 12(d) and 24(h) with equal probabilities (Table 4.3); note that in original article [23] stale designations 6 (e), 12(h) and 24(j) for crystallographic positions were used. In other words, 4/3 atoms Ag fill the (b), (d) or (h) positions. Taking into account multiplicity of these positions, the probability of their occupancy by Ag atoms are equal (4/3)/ 6 = 2/9, (4/3)/12 = 1/9 and (4/3)/24 = 1/18, respectively. The model structure of β-Ag$_2$S argentite proposed in work [23] is very similar to that of AgI [24, 25].

The neutron diffraction study [26] of artificial β-Ag$_2$S argentite crystal in the temperature interval 459–598 K confirmed the cubic (space group $I m\overline{3}m$) structure of this phase with bcc sublattice of S atoms, but considerably refined the distribution of Ag atoms. According to [26], 4 Ag atoms are statistically distributed in 18 positions 6(b) and 12(d). The occupation of these positions depends on temperature. At 459 K, 6(b) and 12(d) positions are occupied by 0.81 and 3.19 Ag atoms, and at 473 K these positions are occupied by 0.66 and 3.34 Ag atoms as calculated per unit cell, respectively. At 533 and 598 K, all the 4 silver atoms are statistically distributed only in 12(d) positions. Thus, the probabilities of occupation of the 6 (b) and 12(d) positions by Ag atoms are equal to 0.135 and 0.266 at 459 K, to 0.110 and 0.278 at 473 K and to 0 and 0.333 at 533 and 598 K. The temperature 459 K is only by 9 K higher than the temperature of transition of monoclinic acanthite α-Ag$_2$S to bcc argentite β-Ag$_2$S. This allows us to suggest that the probabilities of occupation of 4(e) positions by Ag atoms in the structure of acanthite should also differ from 1. In this case, the composition of silver sulfide with acanthite-type structure can deviate from stoichiometry.

Table 4.4 Cubic (space group No. 229 – $I\,m\bar{3}m$ $(I\,4/m\,\bar{3}2/m)$ (O_h^9)) silver sulfide with β-Ag$_2$S argentite-type structure at 523 K [27]: $Z = 2$, $a = b = c = 0.48914(4)$ nm

Atom	Position and multiplicity	Atomic coordinates			Occupancy
		x	y	z	
Ag1	6(b)	0	0.5	0.5	0.097(4)
Ag2	48(j)	0	0.319(1)	0.430(1)	0.0715(9)
S	2(a)	0	0	0	1.00

According to high-temperature XRD data [27], as distinct from [26], four silver atoms in β-Ag$_2$S argentite are statistically distributed in 54 positions 6(b) and 48 (j) with the occupation probabilities ~ 0.097 and ~ 0.0715, respectively (Table 4.4).

Structure model of high-temperature cubic (space group $F\,m\bar{3}m$) γ-Ag$_2$S silver sulfide has been proposed in study [28]. According to [28], four S atoms occupy crystallographic positions 4(a) and form fcc sublattice, and eight Ag atoms are statistically distributed in 8(c) and 32(f) positions. Blanton et al. [27], using high-temperature XRD data, have refined model [28] by distribution of Ag atoms on positions 48(i) also (Table 4.5). Model [27] suggests some Ag deficiency in γ-phase, corresponding to nonstoichiometric sulfide Ag$_{1.7}$S. Figure 4.2 shows crystal structures of γ-Ag$_2$S silver sulfide proposed in studies [27, 28].

Careful determination of structures of coarse-crystalline and nanocrystalline acanthite α-Ag$_2$S and argentite β-Ag$_2$S has been performed recently in studies [29–35]. These results will be considered and discussed in detail in Sect. 4.2.

In normal conditions, bulk coarse-crystalline silver sulfide with α-Ag$_2$S acanthite-type structure is a direct semiconductor which possesses a wide band gap E_g and low charge-carrier mobility. The band gap E_g of α-Ag$_2$S depends on temperature. According to [10, 11], band gap of acanthite α-Ag$_2$S at a temperature of 300 K is 0.9 eV, and its temperature coefficient $\partial E_g / \partial T = -(1.2 - 1.5) \times 10^{-3}$ eV K^{-1} [11]. For acanthite α-Ag$_2$S at 100 K, band gap is 1.3 eV [36]. According to [11], band gap for argentite β-Ag$_2$S is 0.3 eV; according to [36], $E_g = 0.44$ eV at $T = 500$ K. For the high-temperature β-Ag$_2$S phase, the value of E_g decreases with temperature slightly: $\partial E_g / \partial T = -4 \times 10^{-5}$ эB K^{-1} for argentite β-Ag$_2$S [36]. Conventional band gap of acanthite α-Ag$_2$S at a temperature of 300 K is 0.9–1.1 eV.

Table 4.5 Crystal structure of cubic (space group No. 225 – $F\,m\bar{3}m$ $(F\,4/m\,\bar{3}2/m)$ (O_h^5)) γ-Ag$_2$S silver sulfide at 923 K [27]: $Z = 4$, $a = b = c = 0.62831(8)$ nm

Atom	Position and multiplicity	Atomic coordinates			Occupancy
		x	y	z	
Ag1	8(c)	0.25	0.25	0.25	0.088(7)
Ag2	32(j)	0.303(4)	0.303(4)	0.303(4)	0.15(1)
Ag3	48(i)	0.5	0.381(4)	0.381(4)	0.027(3)
S	4(a)	0	0	0	1.00

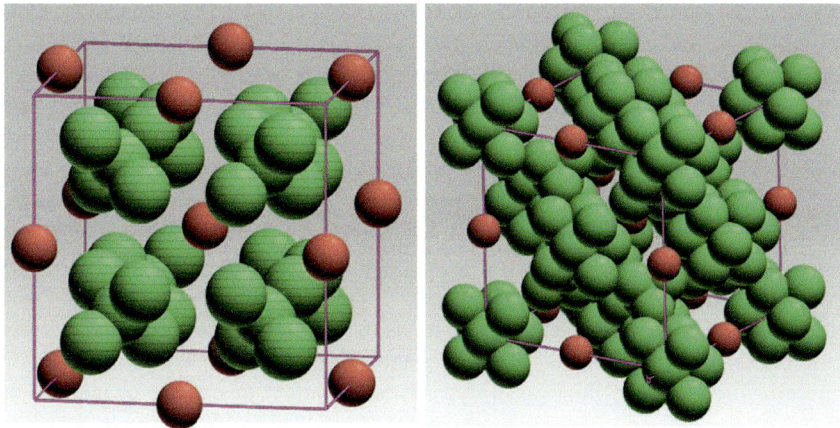

Fig. 4.2 Crystal structures of high-temperature cubic (space group $F\,m\overline{3}m$) γ-Ag₂S silver sulfide reported by Frueh [28] (*left*) and the refined γ-Ag₂S crystal structure from study [27] with an additional Ag positions (*right*)

For bulk coarse-crystalline acanthite α-Ag₂S, the electron and hole effective masses are equal $m_e = 0.286m_0$ and $m_h = 1.096m_0$, respectively [37]. Taking this into account, the reduced exciton mass $\mu_{ex} = m_e m_h/(m_e + m_h)$ for acanthite α-Ag₂S is equal $\sim 0.23m_0 = 2.06 \times 10^{-31}$ kg. For bulk coarse-crystalline argentite β-Ag₂S, the electron and hole effective masses are equal $m_e = 0.23m_0$ and $m_h = 0.23m_0$ [38]. The characteristic size of the Wannier-Mott exciton (or the Bohr radius of the exciton) in the macroscopic semiconductor is $R_{ex} \approx n^2\hbar^2\varepsilon/\mu_{ex}e^2 = (n^2\varepsilon m_0/\mu_{ex})a_B$ where $a_B = \hbar^2/m_0 e^2 = 0.0529$ nm is the Bohr radius. For reduced exciton mass $\mu_{ex} = 0.23m_0$ and dielectric constant $\varepsilon = 5.95$, the radius R_{ex} of the first exciton state with $n = 1$ in coarse-crystalline acanthite α-Ag₂S is ~ 1.4 nm and the diameter D_{ex} is about 2.8 nm.

4.1   Methods of Synthesis of Nanostructured Silver Sulfide Ag₂S

Good chemical stability, ultra-low solubility, the presence of phase transition between a semiconducting acanthite and superionic argentite, which can be induced by external electric field without crystal heating, and unique optical and conductive characteristics of different phases of silver sulfide make it an excellent substance for preparation of nanocrystals, quantum dots, nanostructured films and heteronanostructures with improved properties.

Nanostructured silver sulfide was investigated intensively in recent years due to the possible application in optoelectronics, biosensing and catalysis [39–42]. It is an excellent substance for preparation of heterostructures [43]. Nanostructured silver sulfide Ag$_2$S can be used in photochemical cells [44], infrared detectors [45–47], in resistance-switches and nonvolatile memory devices [48–50]. Silver sulfide is a promising material for conversion of solar energy into electrical energy [51, 52]. Recently, three-dimensional nanoparticle superlattices were built up with Ag$_2$S hollow nanospheres and nanodiscs as building blocks [53]. In this interesting study [53], both types of Ag$_2$S superstructures were formed via microemulsion-based synthesis. Ag$_2$S nanoparticles can be used as photocatalysts in various redox-processes and for producing nanocomposite photocatalysts [54–56]. Ag$_2$S nanoparticles possess antibacterial action [57, 58]. Creation of isolated stable Ag$_2$S quantum dots to be used as biomarkers holds much promise [59, 60]. The intensity (brightness) of emitted signal of such quantum dots exceeds many times that of currently used organic fluorescent dyes. It is important to prepare quantum dots with low toxicity and high emission. Silver sulfide is an ideal semiconducting material for preparing low-toxicity quantum dots.

Nanostructured silver sulfide in the form of nanopowders, quantum dots, and heteronanostructures based on Ag$_2$S has been successfully produced by different methods such as hydrochemical deposition, template method, sol-gel method, synthesis in microemulsions, as well as by sonochemical, hydrothermal, solvothermal, electrochemical, microwave and other techniques. Every method has both advantages and limitations. First, many of these methods are not universal; they allow synthesis of nanostructured silver sulfide only in one form: either quantum dots, or nanopowders, or films, or heteronanostructures. Second, bad control of the size of nanoparticles during synthesis, non-uniform broad size distribution of particles in the prepared nanomaterials, use of multistage (3–4 stages) sequential processes, complicated technique and special defenders, the raised temperatures, rare, expensive and toxic reagents are the main disadvantages of many preparation methods of nanostructured Ag$_2$S. So the preparation methods of nanostructured silver sulfide have still challenges.

Sodium sulfide Na$_2$S [29, 32, 40, 55, 56, 61–64], hydrogen sulfide gas or hydrosulphuric acid H$_2$S [65], the elemental sulfur S dissolved in a concentrated NaOH solution [66], solution of toxic carbon disulfide CS$_2$ in ethanol [67–70], sodium thiosulfate Na$_2$S$_2$O$_3$ [71], solution of 3-thiopropionic (3–7mercaptopropionic) acid C$_3$H$_6$O$_2$S in ethylene glycol as solvent [59], thiocarbamide (thiourea) N$_2$H$_4$CS [52, 53, 72, 73], and thioacetamide CH$_3$C(S)NH$_2$ [74], and their derivatives are used for sulfidizing of soluble complex compounds of silver and other metals.

Recently, much attention is devoted to the production of different hybrid heteronanostructures of the core-shell type which include silver and its compounds (Ag$_2$O@Ag$_2$S, Ag@SiO$_2$, etc.) [75–77]. Particularly familiar core-shell nanostructures are formed by two different semiconductors (CdSe@CdS, CdSe@ZnS, CdS@Ag$_2$S, CdS@ZnS, GaAs@AlS, etc.) [78–82].

The nanosized particles are not stable. The high surface energy will impel the nanoparticles to aggregate. Therefore a creation of core-shell particles can be due to the necessity of fixing of specific groups (organic ligands) on the surface of core, which would prevent agglomeration, growth, and oxidation of particles and provide the production of stable isolated nanoparticles. Using the protective shell, it is possible to control the size of isolated nanoparticles. Such stabilizing and capping agents as trioctylphosphine oxide (TOPO) [83, 84], *L*-cysteine C$_3$H$_7$NO$_2$S [85–87], glutathione C$_{10}$H$_{17}$N$_3$O$_6$S [88], long-chain amine (hexadecylamine, octylamine, dioctylamine, ethylenediamine) [39], ethylenediaminetetraacetic acid (EDTA) and ethylene glycol (EG) [71, 89], are used for creation of a protective shell. However, the majority of the listed stabilizing agents are hazardous to human health and have a serious impact on the environment. One of the most commonly requested non-toxic capping agents which having a high degree of electrostatic stabilization is sodium citrate Na$_3$C$_6$H$_5$O$_7$≡Na$_3$Cit [30, 33, 35, 58, 64, 65, 90–92]. For example, stable isolated silver sulfide nanoparticles with a non-toxic carbon-containing citrate shell (core-shell Ag$_2$S@C heteronanostructures) has been synthesized recently [64]. The spherical Ag nanoparticles has been prepared by the hydrothermal method with using sodium citrate as both a reducing and stabilizing agent [92].

Several investigations [93–100] have shown that extracts or aqueous solutions of natural products (leaves, seeds, fruits, roots, honey, royal jelly, gum of trees, bovine serum albumin, etc.) can be used as stabilizing agents for preparation of isolated nanoparticles of Ag$_2$S, silver Ag and other noble metals (Au, Pt, Pd) with a protective shell. Such exotic stabilizing agents are non-toxic and do not exert harmful impact on environment due to their high antioxidant potential. However, the complete identification of the complex composition of listed natural stabilizing agents is an open and undecided problem till now. Therefore, real application of natural stabilizing agents is rather limited.

The most pronounced role in the formation of the Ag$_2$S nanoparticles has been played by the complexing agents. According to data of "nanoComposix" R@D company (San Diego, US), sodium citrate is the main complexing agent for Ag$^+$ ions. One must point that the sodium citrate is completely harmless standardized food additive E331.

Let us briefly consider some solution methods of synthesis of nanostructured silver sulfide Ag$_2$S.

The solution chemical methods used in the synthesis of semiconductor compounds can be tentatively classified into three main groups [101]: controlled precipitation (deposition), the cluster-build-up approach, and the molecular precursor approach.

Controlled precipitation (deposition) integrates those methods based in traditional precipitative techniques, involving the formation of a stable sol from the mixture of the chemical components of the semiconductor in the form of the respective ionic sources. The stability of the sol is attained either by using stabilizers present in solution (e.g., polymers, surfactants) or by electric double-layer repulsion. The formation of nanoparticles using sol-gel methods is also included in group of controlled precipitation (deposition). The cluster-build-up approach involves the preparation of quantum dots by assembling of inorganic clusters with a definite chemical composition, whose dimensions are generally less than those of synthesized nanostructure. The molecular precursor approach includes methods of the preparation of quantum dots using thermal or chemical treatment applied to a system containing precursor molecular species, which may consist of metal complexes or molecular compounds.

4.1.1   Synthesis by Decomposition of Molecular Precursors

Application of molecular precursors has some attractive features. On the one hand, it provides such important advantages as simplicity, safety and compatibility with metalorganic chemical vapor deposition [101]. On the other hand, the use of molecular precursors may lead to the unusual selectivity in crystal growth or formation of metastable phase of the final products, which are not always achievable by conventional synthesis methods.

Among the solution methods, the injection of an organometallic precursor into a hot solvent provides a simple route to produce particles with desirable properties (e.g. high crystallinity, and uniform shapes and sizes with a high degree of monodispersity).

Lim et al. [39] have discovered that such air-stable precursor as silver thiobenzoate Ag(SCOPh) meets these requirements. The precursor crystals were found to decompose in amine at room temperature to give Ag$_2$S nanoparticles. When a solution of the precursor dissolved in trioctylphosphine is added to a warm solution (353, 373, or 393 K) of a long-chain amine such as hexadecylamine (HDA), well-defined nanocrystals are formed, the size and shape of which can be controlled by the reaction conditions. The most important parametres of synthesis are the reaction temperature, type of amine, relative concentration of the reagents, and reaction time. Lim et al. [39] have found that by increasing the injection temperature to 393 K, cube-shaped Ag$_2$S nanocrystals are obtained exclusively (Fig. 4.3).

The uniform Ag$_2$S nanocubes self-assemble into ordered two-dimensional arrays on the surface of the transmission electron microscope grid (Fig. 4.3a). The average size of these nanocubes is 44 $\pm$ 4 nm. SEM image in Fig. 4.3b illustrates that large quantity of these nanocubes can be obtained by using this approach. HRTEM images (Fig. 4.3c, d) clearly show that Ag$_2$S nanocubes are single crystals. The influence of temperature and amine-to-precursor ratio on the size and morphology of Ag$_2$S nanoparticles is shown in Fig. 4.4.

Later Ag$_2$S nanocrystals were obtained via a modified hot-injection process of the same single-source molecular precursor Ag(SCOPh), which can potentially generate both Ag* and AgS* fragments simultaneously [102]. When the precursor molecules are injected into a preheated reaction system at 433 K, spherical Ag$_2$S nanocrystals are directly obtained even without a molecular activator, such as alkylamines [102]. Wang et al. [103] obtained Ag$_2$S nanocrystallites by heating such molecular precursor as silver diethyldithiocarbamate (Ag-DDTC) in air at 473 K for 3 h, and used this air-stable molecular precursor as the reactant source.

Fig. 4.3 Ag$_2$S nanocubes produced at a temperature of 393 K [39]: **a** TEM image of Ag$_2$S nanocubes; **b** SEM image of clusters formed by Ag$_2$S nanocubes; **c, d** HRTEM images of Ag$_2$S nanocubes

Proposed method is both cost-effective and non-toxic. Monodisperse Ag$_2$S nanoparticles with controlled size were successfully synthesized by thermolysis of harmless silver xanthates as a single-source molecular precursors [104]. In study [104], thermolysis was carried out without the use of surfactants and solvents. In the experiment [104], the diameter of the Ag$_2$S nanoparticles ranges from 8.9 ± 1.2 to 48 ± 4 nm (Fig. 4.5).

Control of the particle size has been achieved by simply changing the alkyl chain length in the precursors. According to [104], Ag$_2$S nanoparticle size decreases with increasing alkyl chain length of precursors. At the same time, when the temperature increasing, the xanthate ligand will be absorbed on the surface of Ag$_2$S nanoparticles limiting their growth.

Figure 4.6 shows a possible growth mechanism of Ag$_2$S nanoparticles. As the reaction temperature increased, C–S or Ag–S bond in precursor may be broken. Ag$_2$S nano-nuclei and surface-capping agent (xanthate ligand) are formed simultaneously. In part of precursors the C–S bonds were broken and AgS* fragments were generated. The AgS* will combine with Ag* fragments which are from the

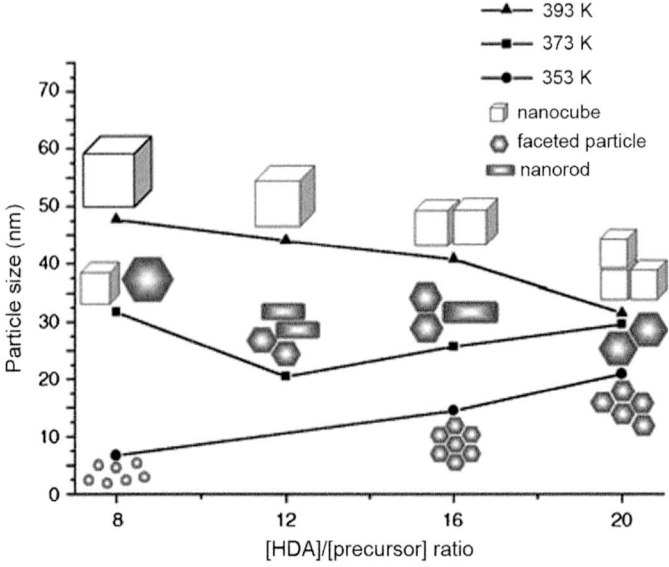

Fig. 4.4 Effect of temperature and hexadecylamine-to-precursor ratio on the shape and size of the Ag$_2$S nanocrystals [39]. The size of the nanorods is not included in the graph

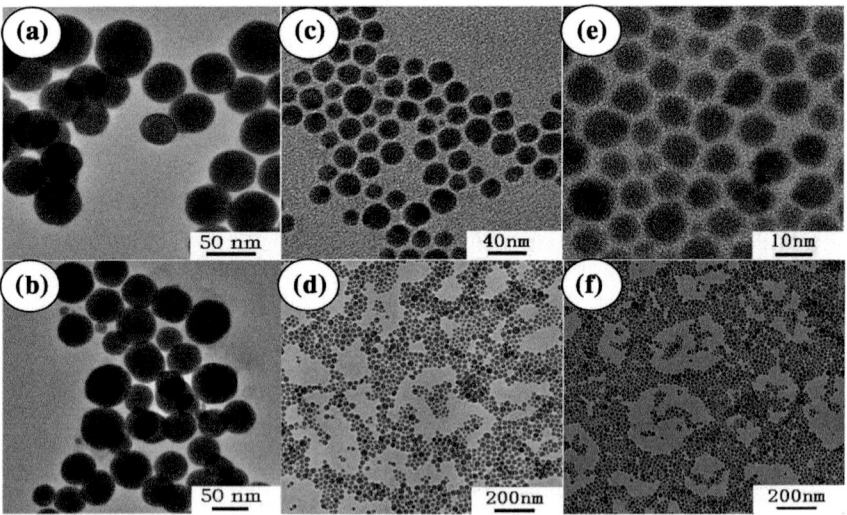

Fig. 4.5 TEM images of Ag$_2$S nanoparticles synthesized by solventless thermolysis of **a**, **b** silver octyl xanthate, **c**, **d** silver hexadecyl xanthate, and **e**, **f** silver carnaubyl xanthate [104]

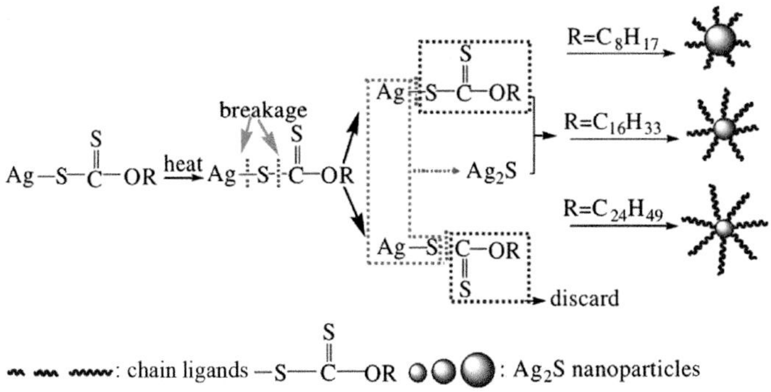

Fig. 4.6 Schematic diagram of a possible growth mechanism of Ag$_2$S nanoparticles [104]. As a result of heating, Ag–S or S–C bonds can be broken in Ag-DDTC precursor

rupture of Ag–S bonds in another part of precursors to form Ag$_2$S nanoparticles. At the same time, when the Ag–S in precursor was broken there will be complete xanthate ligand dissociated from precursor and absorbed to the surface of Ag$_2$S nanoparticles as capping agent to control particle growth. These form xanthate ligand-modified silver sulfide nanoparticles. The long chain ligands more effectively to prevent the growth of nanonuclear. When the kind of precursor changes from octyl xanthate silver to hexadecyl xanthate silver and carnaubyl xanthate silver, the thermal decomposition products show a significant decrease in average particle size at the same reaction conditions.

4.1.2   Synthesis of Nanostructured Silver Sulfide with Different Morphology

Recently, the great efforts have been focused on the preparation of silver sulfide nanoparticles of various morphology and on the morphology control of the semiconductor nanocrystals [105]. Various methods for the preparation of Ag$_2$S nanostructures and their formation mechanism have been studied. For example, Zhao et al. [106] prepared rod-like Ag$_2$S nanocrystals by using sodium thiosulfate Na$_2$S$_2$O$_3$ as a sulfur source via gamma-ray irradiation of aqueous solutions at room temperature. In this experiment, polyvinylpyrrolidone (C$_6$H$_9$NO)$_n$ (PVP) plays an important role in the formation of rod-like Ag$_2$S nanocrystals. The diameter of synthesized Ag$_2$S nanorods varies from 200 to 500 nm. According to [106], the band gap of Ag$_2$S nanorods calculated from the UV-vis absorption spectrum is 2.34 eV. It is very doubtful result because the size of prepared nanorods is large.

An alcohol solution method to synthesize nanostructured Ag$_2$S using carbon bisulfide CS$_2$ as sulfur source has been described in study [107]. All the products

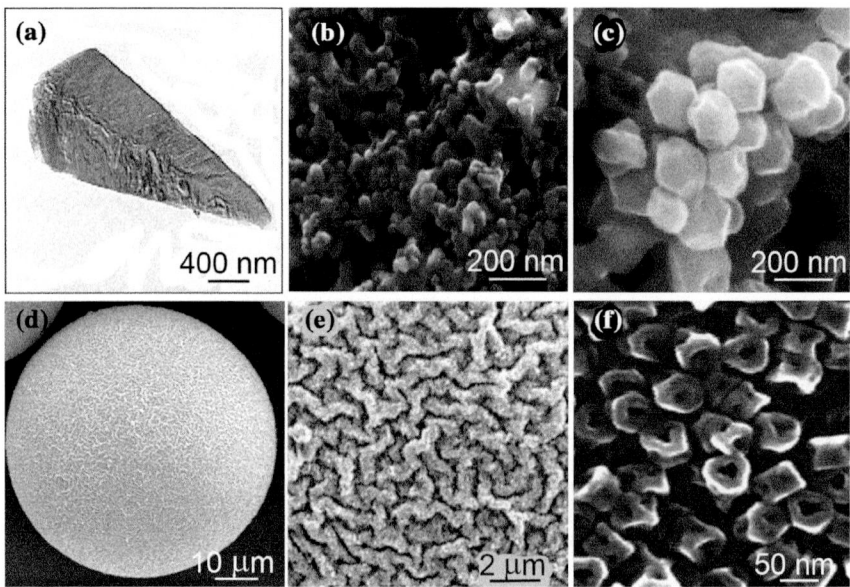

Fig. 4.7 Nanoparticles of nanostructured silver sulfide Ag₂S with different morphology. **a** Typical leaf-like morphology of the Ag₂S nanosheets prepared by hydrothermal treatment of alcohol-water solutions at 433 K for 10 h [108]; **b** worm-like nanoparticles [109], **c** Ag₂S nanopolyhedrons [109]; **d, e** SEM images of Ag₂S-PNIPAM-MAA composite microsphere and of part of the surface of this microsphere at high magnification [112]; **f** SEM image of hollow Ag₂S nanohexagons [113]

were irregular Ag₂S micro- and nanostructures. In the alcohol-water medium, because of the difference in polarity and solubility, the growth of Ag₂S nanocrystals differs from those in water or alcohol. When the reaction medium was changed from water and alcohol-water to alcohol, the morphology of synthesized silver sulfide changed from big irregular nanosheets to leaf-like nanosheets, elliptical, and Y-shaped flake Ag₂S nanoparticles.

Later, Chen et al. [108] reported that the leaf-like Ag₂S nanosheets (Fig. 4.7a) were prepared successfully by a facile hydrothermal method from mixture of alcoholic CS₂ solution with aqueous solution of AgNO₃ and NH₃. Carbon bisulfide, CS₂, has been used as a sulfur source.

When the two solutions were mixed together, the following reactions took place:

$$2NH_3 + CS_2 \rightarrow NH_4NHCSSH, \tag{4.1}$$

$$2Ag(NH_3)^+ + NH_4NHCSSH \rightarrow Ag_2S \downarrow + NH_4SCN + 2NH_4^+. \tag{4.2}$$

Ag₂S micro- and nanostructures with the different morphologies, including micrometer bars, nanowires, and nanopolyhedrons have been synthesized by a facile one-step method at room temperature [109]. In proposed method, no any organic template materials were added into the reaction mixture which contained aqueous solutions of AgNO$_3$, NH$_3$, and N$_2$H$_4$CS. By changing the reactant concentration ratio, the size and morphology of prepared Ag$_2$S particle can be easily tuned (Fig. 4.7b, c).

In last years, polyhedral nanocrystals including face-centered cubic sulfide nanocrystals [39, 110] have been successfully fabricated. Wang et al. [110] have prepared Ag$_2$S nanocrystals by thermolysis of organometallic precursor Ag[S$_2$P (OR)$_2$] (R = C$_n$H$_{2n+1}$). The above-mentioned hydrothermal method has been improved by Dong et al. [111]. The ratio of AgNO$_3$ and N$_2$H$_4$CS concentrations in aqueous solutions was changed with assistance of cetyltrimethyl ammonium bromide C$_{19}$H$_{42}$BrN (CTAB). According to [111], the cooperation effect of CTAB and thiourea N$_2$H$_4$CS is responsible for the formation of the prepared Ag$_2$S nanocrystals. The size of Ag$_2$S particles ranged from 40 to 80 nm. Most of the observed Ag$_2$S nanoparticles looked hexagonal in shape.

Ag$_2$S-poly(N-isopropylacrylamide-co-methacrylic acid) (PNIPAM-MAA) and Ag$_2$S-PNIPAM composite microspheres with patterned surface structures have been synthesized by a polymeric minigel template method [112]. The minigels were prepared by reverse suspension polymerization of the aqueous solution of specified acids initiated by solution of ammonium persulfate in n-heptane. For the preparation of the composite microspheres, the minigels were swelled in aqueous AgNO$_3$ solution. Then, the minigels containing Ag$^+$ ions were re-suspended in n-heptane with mild stirring. 40 min after the addition of metal-containing minigels, hydrosulphuric acid H$_2$S was introduced. The surface structures of "sulfide-polymer" composite microspheres depend not only on the nature of the template, but also on the nature of the sulfide. Figures 4.7d, e show the SEM images of Ag$_2$S-PNIPAM-MAA composite microsphere, and its enlarged surface structures. The surface structure of Ag$_2$S-PNIPAM-MAA microsphere (Fig. 4.7e) looks like flowers.

Single-crystalline Ag$_2$S hollow nanohexagons with narrow size distribution were successfully synthesized in aqueous solutions of AgNO$_3$, N$_2$S$_2$O$_3$, and C$_{19}$H$_{42}$BrN (CTAB) at a temperature of 318 K [113]. Figure 4.7f displays the SEM images of a typical uniform Ag$_2$S nanohexagons: An edge-to-edge distance is 49 ± 2 nm. The formation mechanism of Ag$_2$S hollow nanohexagons is shown in Fig. 4.8.

Tetrahedral colloidal crystals of Ag$_2$S nanoparticles have been synthesized from aqueous solutions of AgNO$_3$, NH$_3$, and dodecanethiol CH$_3$(CH$_2$)$_{11}$SH in autoclave

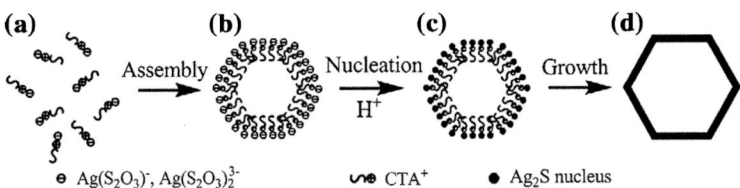

Fig. 4.8 Schema of the formation mechanism of hollow Ag$_2$S nanohexagons [113]: **a** the soluble CTA$^+$–Ag(S$_2$O$_3$) and CTA$^+$–Ag(S$_2$O$_3$)$_2^{3-}$ ion pairs; **b** hexagon-like micellar composites with Ag (S$_2$O$_3$)$^-$ and Ag(S$_2$O$_3$)$_2^{3-}$; **c** Ag$_2$S nuclei; **d** hollow Ag$_2$S nanohexagon

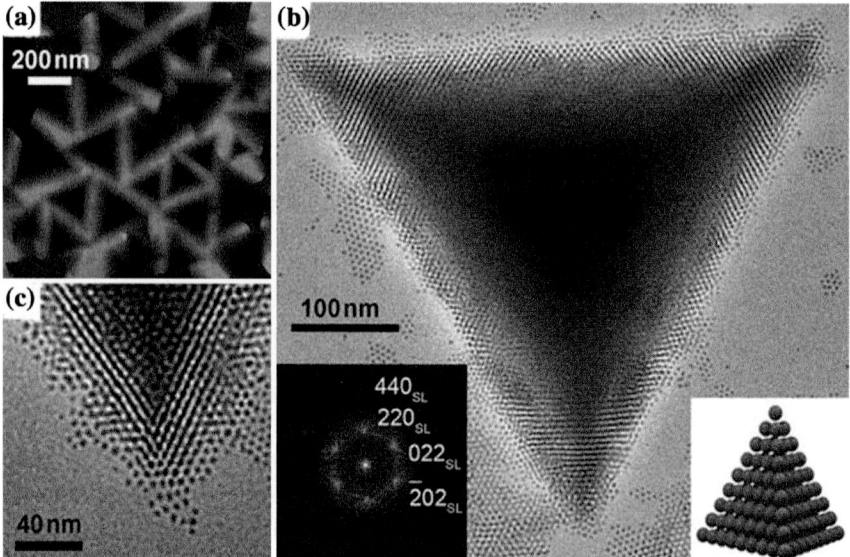

Fig. 4.9 TEM images of the obtained Ag$_2$S tetrahedral superlattice colloidal crystals [114]: **a** TEM image at low magnification; **b** a typical HRTEM image of an individual tetrahedron at high magnification; the *left inset* is the diffraction pattern calculated using Fast Fourier Transform (FFT) of HRTEM image of the tetrahedral superlattice and the *right inset* is a scheme of the tetrahedral superlattice colloidal crystal; **c** a magnified TEM image of a vertex of the tetrahedron

at a temperature of 473 K for 5 h [114]. Due to the high uniformity and Van der Waals interactions, Ag$_2$S nanoparticles spontaneously assemble into tetrahedral colloidal aggregates comprising of a perfectly ordered 3D superlattice structure (Fig. 4.9).

A sacrificial core of sulfur nanoparticles is used to synthesize Ag$_2$S hollow nanospheres via a wet chemical method at room temperature [115]. Sulfur nanoparticles as cores were synthesized from Na$_2$S$_2$O$_3$ sodium thiosulfate in the presence of cetyltrimethyl ammonium bromide aqueous solution. After the completion of the cores formation, the AgNO$_3$ solution was added. Then produced particles were washed by water ethanol mixture and were treated with CS$_2$ for the complete conversion of AgBr to Ag$_2$S, as well as to remove the cores to form Ag$_2$S hollow nanospheres (Fig. 4.10).

Worm-like Ag$_2$S nanofibers with length up to several micrometers and the diameter of 25–50 nm have been prepared in reverse microemulsion in the presence of thioacetamide as a sulfur source and ethylenediaminetetraacetic acid as a chelating ligand [116]. According to [116], as-synthesized Ag$_2$S nanofibers exhibit strong absorption in UV region; the absorption edge which is observed at wavelength λ about 290 nm corresponds to the band gap of 4.3 eV. Thus, appreciable blue shift of the absorption edge of Ag$_2$S nanofibers was observed in comparison

Fig. 4.10 TEM image of
hollow Ag₂S particle
synthesized from reaction
mixture of silver nitrate
AgNO₃ and thiosulfate [115]

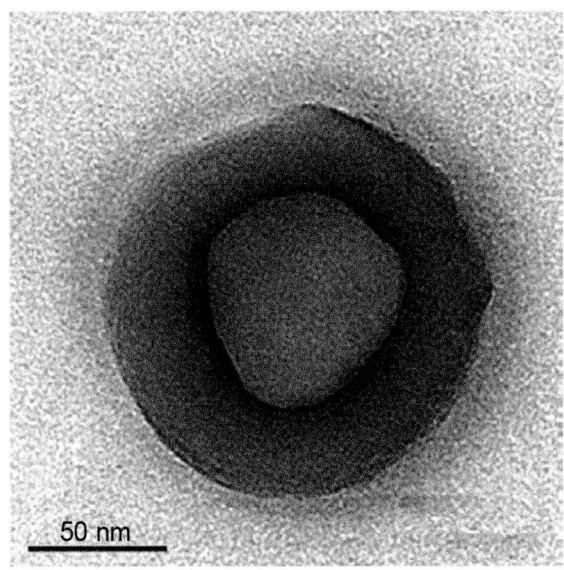

50 nm

with the absorption edge of bulk Ag₂S ($\lambda = 1240$ nm, $E_g = 0.9$–1.0 eV). It is very doubtful result because the diameter of prepared Ag₂S nanofibers is too large for the occurrence of quantum-size effect.

In study [117], rice-shaped Ag₂S nanoparticles were produced by reaction between $Ag(NH_3)_2^+$ and Na_2S in the presence of polyvinylpyrrolidone (PVP) through hydrothermal method. The solution mixture has been subjected to heat treatment in autoclave at a temperature of 433 K for 10 h. According to [117], the formation of the of the rice-shaped Ag₂S nanoparticles depends mainly on the type of silver source, influence of the pyrrolidone rings of PVP, reaction time, and temperature. Well-dispersed Ag₂S nanoparticles have rice-shaped morphology, and length and width of most of the particles are about 70–90 nm and 30–35 nm, respectively (Fig. 4.11a). It was found that the rise-shaped Ag₂S nanoparticles are not formed when AgNO₃ was used instead of $Ag(NH_3)_2^+$ as the silver source. Thus, the pyrrolidone ring plays an important role in the formation of the Ag₂S nanorice by hydrothermal method.

Large rice-shaped Ag₂S particles have been synthesized by hydrochemical bath deposition from aqueous reaction mixture of silver nitrate, sodium sulfide, and sodium citrate with the concentrations of 0.05, 0.4, and 0.005 mol l^{-1}, respectively [29]. Reaction mixture was heated in closed vessel at a temperature of 373 K under the pressure $\sim 2 \times 10^5$ Pa within 2 h. Synthesized Ag₂S powder contained a small amount of separate rice-shaped particles about 2000 nm in length and ~ 400 nm in width (Fig. 4.11b) along with large irregular agglomerates.

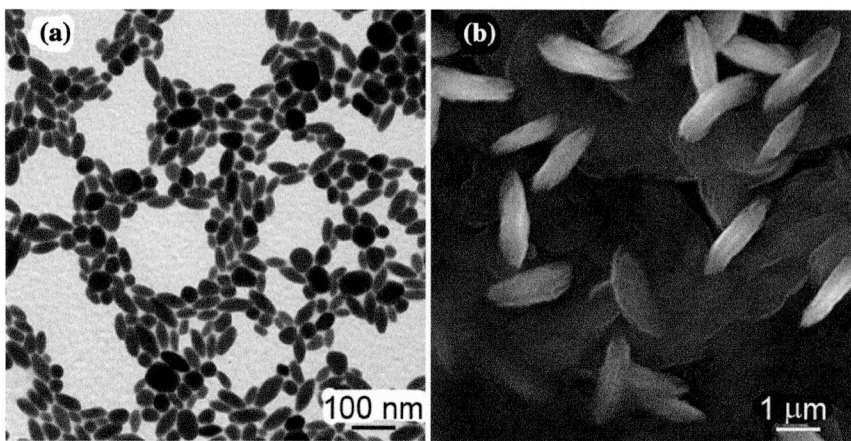

Fig. 4.11 Rice-shaped Ag$_2$S particles: **a** TEM image of Ag$_2$S nanorices about 70–90 nm in length which were synthesized by hydrothermal method at 433 K for 10 h [117]; **b** SEM image of separate rice-shaped Ag$_2$S particles about 2 μm in length and ∼400 nm in width [29]

4.1.3   Biotechnological and Other Methods for Preparing Nanostructured Ag$_2$S Silver Sulfide

Ag$_2$S quantum dots are treated as an ideal optical probe with fluorescence emission from UV to NIR region because these QDs have lower toxicity compared with chalcogenide quantum dots of such heavy metals as Pb, Ca, and Hg [118, 119]. The synthesis of quantum dots which can be used as optical probes for in vitro and in vivo molecular imaging has made great progresses [120]. In study [121], highly monodisperse and water soluble clusters representing Ag$_2$S quantum dots, covered by Ribonuclease-A, have been synthesized in aqueous medium via biomimetic route, i.e. by the method which mimics biochemical processes. Ribonuclease-A serves as not only a stabilizing agent in the formation of Ag$_2$S quantum dots to avoid aggregation, but also is a biomolecule to modify the surface of Ag$_2$S nanocrystals to decrease toxicity. The TEM and HRTEM images of Ag$_2$S quantum dots, covered by Ribonuclease-A, are shown in Fig. 4.12. It is seen that the "Ribonuclease-A–Ag$_2$S" clusters have irregular morphology (Fig. 4.12a), and an individual Ag$_2$S nanoparticles have well-defined crystal lattice and the average nanoparticle size is 5–6 nm (Fig. 4.12b). According to [121], the "Ribonuclease-A–Ag$_2$S" clusters prepared have great potential application for imaging in living cells and tissues.

Siva et al. [122] have noted that the biomolecules assisted the formation of inorganic nanostructures, facilitate electrostatic stabilization, improving optical properties of nanoparticles. In study [122], aqueous solutions of silver nitrate AgNO$_3$ and aurochloric acid HAuCl$_4$, and also solution of L-cysteine in the mixture of water, ethylene glycol, and ethanol were used for synthesis of L-cysteine capped

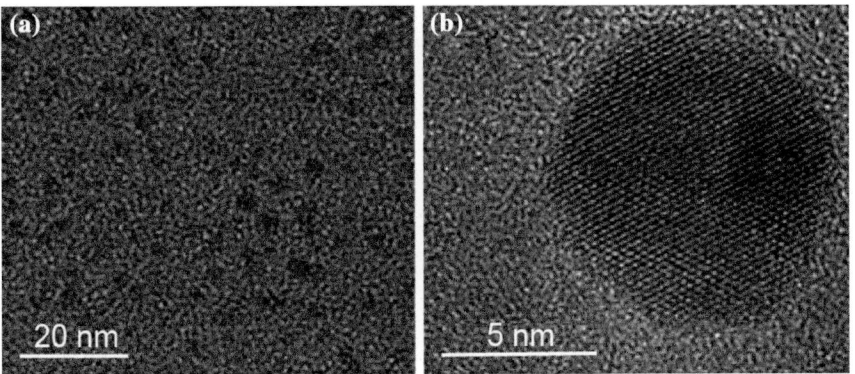

Fig. 4.12 Ag$_2$S quantum dots, covered by Ribonuclease-A [121]: **a** TEM image of the "Ribonuclease-A—Ag$_2$S" clusters; **b** HRTEM image of an individual Ag$_2$S nanoparticles with a size about 6 nm

Ag$_2$S and Ag$_3$AuS$_2$ nanocrystals. Ag$_2$S nanocrystals were prepared using Ag silver nuclei as a core. Biomolecule *L*-cysteine can act as a sulfur source and stabilizing capping agent which prevents the agglomeration of sulfide nanoparticles [122, 123]. Siva et al. [122] proposed the following scheme of formation of particles Ag$_2$S:

$$Ag^+ + L\text{-cysteine} \xrightarrow{298\,K} \left[Ag(L\text{-cysteine})_n \right]^+, \qquad (4.3a)$$

$$\left[Ag(L\text{-cysteine})_n \right]^+ + OH^- \xrightarrow{\text{ethhanol decomposition, 343K}} L\text{-cysteine capped Ag}, \quad (4.3b)$$

$$L\text{-cysteine capped Ag} \xrightarrow{\text{C–S bond rupture, 343 K}} Ag_2S \qquad (4.3c)$$

This scheme is presented in Fig. 4.13a. The mixing of *L*-cysteine and silver nitrate in ethylene glycol results in a transparent solution, which confirms the formation of metal-cysteine complexes owing to the appearance of "ion Ag$^+$–thiol" chains as discussed in earlier study [86]. During heating the Ag$^+$ ions with ethanol, Ag silver nanoparticles coated by *L*-cysteine are formed, due to the ethanol decomposition. Further the C–S bonds are ruptured and result in Ag$_2$S silver sulfide nucleation and growth of Ag$_2$S nanoparticles. The HRTEM image of *L*-cysteine capped Ag$_2$S nanoparticles is shown in Fig. 4.13b. It is seen that the Ag$_2$S nanoparticles are almost uniform in their sizes and are interconnected among themselves. The size distribution indicates that the Ag$_2$S nanoparticle size ranges from 4.3 to 5.8 nm, and average nanoparticle size is 5.2 nm (Fig. 4.13b, inset).

There are many other methods for synthesis of nanostructured silver sulfide. In study [124], Ag$_2$S nanoparticles were prepared by pyrolysis using AgNO$_3$ and S powder as precursors, and oleylamine C$_{18}$H$_{35}$NH$_2$ as a solvent. Oleylamine acts as both reducing agent and stabilizer during the synthesis. The Ag$_2$S nanoparticles

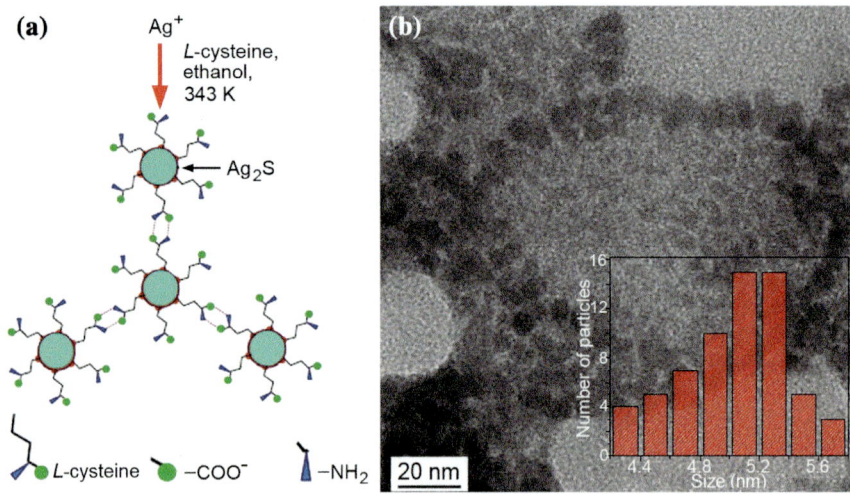

Fig. 4.13 a Schematic representation of plausible formation mechanism of Ag$_2$S nanoparticles and **b** HRTEM image of Ag$_2$S nanoparticles; *inset* shows the histogram of particle size distribution [122]

with uniform size have been prepared by controlling the ratio amounts of AgNO$_3$:S and ripening time.

Shakouri-Arani and Salavati-Niasari [125] have produced the Ag$_2$S nanoparticles by a solvothermal process via reaction of AgNO$_3$ and a new sulfuring agent from class of thio Schiff-base (2-(benzylidene amino) benzenethiol C$_{13}$H$_{11}$NS) in presence of various solvents. In a typical experiment, two solutions A and B were prepared. At first, the benzaldehyde C$_6$H$_5$CHO was dissolved in propylene glycol C$_3$H$_8$O$_2$ (PG) under stirring. In a separate beaker, 2-aminothiophenol C$_6$H$_7$NS was dissolved in absolute propylene glycol and this solution has been added on drops in previous solution under 30 min stirring at a room temperature. Thus the solution A has been prepared. The solution B has been produced by dissolving AgNO$_3$ in dissolved in propylene glycol under stirring. Next, solution A was added into solution B slowly under stirring for 2 h. The color of a mixture became yellow. This solution was transferred to Teflon-line stainless steel autoclave and filled with propylene glycol up to 80% of the total volume. The autoclave was sealed and maintained at 433 K for 12 h.

In study [125], synthesis of Ag$_2$S at temperature of 433 K for 12 h with propylene glycol as solvent has been selected as a basic reaction.

Nanostructured silver sulfide was obtained in absence of surfactant or a presence of such anionic surfactant as sodium dodecyl sulfate C$_{12}$H$_{25}$SO$_4$Na (SDS) or such cationic surfactants as cetyltrimethyl ammonium bromide C$_{19}$H$_{42}$BrN (CTAB) and polyethylene glycol PEG 20000 C$_{2n}$H$_{4n+2}$O$_{n+1}$ (HO–(C$_2$H$_4$O)$_n$–H). Surfactants were dissolved in such solvents as H$_2$O and 1-butanol C$_4$H$_9$OH.

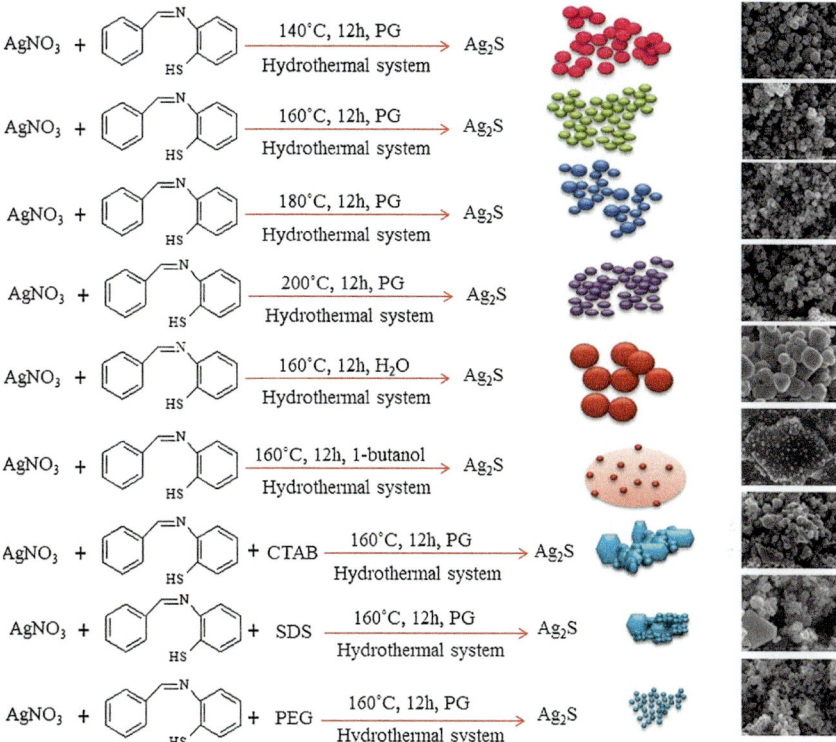

Fig. 4.14 Schematic diagram illustrating the formation of Ag$_2$S nanoparticles at various conditions (PG is propylene glycol C$_3$H$_8$O$_2$, CTAB is cetyltrimethyl ammonium bromide C$_{19}$H$_{42}$BrN, SDS is sodium dodecyl sulfate C$_{12}$H$_{25}$SO$_4$Na, and PEG is polyethylene glycol 20000 C$_{2n}$H$_{4n+2}$O$_{n+1}$) [125]

Experimental results [125] indicate that the reaction temperature, presence of surfactant, and type of solvent has effect on the size of Ag$_2$S nanoparticles. Figure 4.14 illustrates influence of various conditions of synthesis on formation of Ag$_2$S nanoparticles.

Biocompatible Ag$_2$S quantum dots were prepared through thermal decomposition of silver diethyldithiocarbamate (C$_2$H$_5$)$_2$NCS$_2$Ag with 1-dodecanethiol CH$_3$(CH$_2$)$_{11}$SH as covalent ligand and solvent [126]. The reaction mixture was heated to 483 K at a heating rate of 15 K min^{-1} and kept for 1 h under N$_2$ atmosphere. As a result, hydrophobic Ag$_2$S quantum dots with an average diameter from 5.4 to 10.0 nm (Fig. 4.15) coated with dodecanethiol as the surface ligand were obtained. The dodecanethiol molecules might be involved in the reaction as partial sulfur source and the only capping ligand. The Ag$_2$S quantum dots obtained in study [126] can be used as a promising second near-infrared probe with both bright photoluminescence and high biocompatibility.

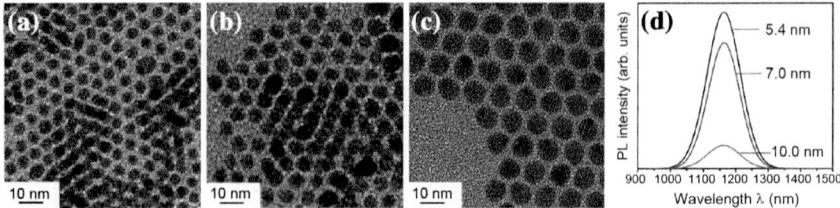

Fig. 4.15 TEM images (**a, b, c**) and photoluminescence spectra (**d**) of Ag$_2$S quantum dots with different sizes [126]: **a** 5.4 nm; **b** 7.0 nm; **c** 10.0 nm

Fig. 4.16 Scheme of aqueous synthesis of 2-MPA coated Ag$_2$S quantum dots [127]

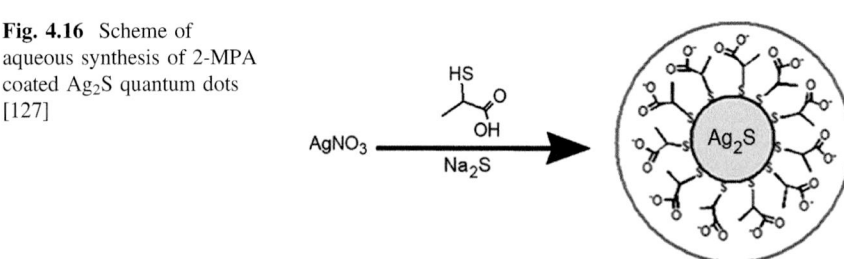

Stable and highly luminescent near-IR emitting Ag$_2$S colloidal quantum dots were prepared by a simple aqueous method using 2-mercaptopropionic acid C$_3$H$_6$O$_2$S (2-MPA) as a coating [127]. At first, 2-mercaptopropionic acid is dissolved in deoxygenated deionized water. Then AgNO$_3$ is added and the solution is brought to the desired temperature (303, 323, or 363 K). Next, aqueous Na$_2$S solution is added slowly to the reaction mixture under mechanical stirring. Nanoparticle size can be tuned between 2.3 and 3.1 nm with an emission maximum between 780 and 950 nm. Scheme of one-step synthesis of Ag$_2$S quantum dots with 2-MPA coating is demonstrated in Fig. 4.16.

4.2   Crystal Structure of Coarse-Crystalline and Nanostructured Ag$_2$S Silver Sulfide

It is noted in the beginning of Chap. 4 that silver sulfide Ag$_2$S has three basic polymorphic modifications: monoclinic α-Ag$_2$S acanthite, cubic β-Ag$_2$S argentite, and high-temperature cubic γ-Ag$_2$S sulfide. Structures of Ag$_2$S different phases have been defined originally in studies [8, 21, 23] and later have been specified in works [22, 26, 27].

Crystal structures of these phases are fairly complex. For this reason, in the majority of experimental works devoted to synthesis and properties of silver sulfide Ag$_2$S, description of the crystal structure of synthesized sulfide is either lacking [60, 128, 129] or it is made by comparing experimental XRD or transmission electron microscopy results [46, 49, 59, 60, 66, 69, 100, 117, 130–132] with stale diffraction data [8, 133, 134]. So, in studies [46, 59, 131, 132, 135], without performing full-profile structure refinement, it is suggested that Ag_2S sulfide synthesized in the form of film, nanocrystalline powder or nanoparticles has crystal structure of natural acanthite, whereas in work [71] it is assumed that synthesized Ag_2S film has a structure of argentite. However the crystal structure of synthetic silver sulfide may have considerable differences that affect the properties of Ag_2S. For example, it is confirmed by the presence of additional reflections, which do not belong to acanthite-type structure, on the diffraction pattern of Ag_2S nanoparticles [131].

It is worth noting that the vast majority of available experimental diffraction data on silver sulfide are unsuitable for the quantitative refinement of the crystal structure since they have been recorded in a narrow 2θ angle range (usually 20°–60°) and with inadequate data acquisition.

In this connection, careful determination of crystal structures of coarse-crystalline and nanocrystalline acanthite α-Ag_2S and argentite β-Ag_2S has been performed recently in studies [29–35].

4.2.1 Crystal Structure of Artificial Coarse-Crystalline Ag₂S Silver Sulfide

According to [8, 22], the structure of acanthite α-Ag_2S can be interpreted as a result of distortion of the β-Ag_2S argentite structure. Indeed, the unit cells of α-Ag_2S acanthite proposed in studies [8, 22] have axes which can be represented as a combination of axes a_{bcc}, b_{bcc} and c_{bcc} of the unit cell of bcc argentite.

Figure 4.17a displays the arrangement of S atoms in bcc sublattice of argentite and the contours of monoclinic unit cells [8, 22] of α-Ag_2S acanthite. It is seen that axes $a \equiv a_{P2_1/c} = a_{P2_1/n} = (a_{bcc} + b_{bcc} - c_{bcc})/2$, $b \equiv b_{P2_1/c} = b_{P2_1/n} = (a_{bcc} - b_{bcc})$, $c_{P2_1/n} = (a_{bcc} + b_{bcc} + 3c_{bcc})/2$ and $c_{P2_1/c} = 2c_{bcc}$. The monoclinic-distorted sub-lattice of S atoms with contours of monoclinic (space group $P2_1/n$ and $P2_1/c$) unit cells is demonstrated in Fig. 4.17b with allowance for found coordinates of sulfur atoms [22].

The diagram of mutual transformation of monoclinic unit cells [20, 21] is shown in Fig. 4.18. Since $a_{P2_1/c} = a_{P2_1/n} \equiv a$, $b_{P2_1/c} = b_{P2_1/n} \equiv b$ and $c_{P2_1/c} = (c_{P2_1/n} - a_{P2_1/n})$, then, taking into account the values of the parameters $a = |a_{P2_1/n}|$, $b = |b_{P2_1/n}|$, $c = |c_{P2_1/n}|$ and β of the unit cell [20] and the above transformations, the unit cell with space group $P2_1/c$ has the parameters $c_{P2_1/c} = |c_{P2_1/c}| = (a^2 + c^2 - 2ac \cdot \cos \beta)^{1/2} = 0.9536$ nm and $\beta_{P2_1/c} = \beta + \beta^* = 125.54°$, where $\beta^* = \arcsin[(a/c_{P2_1/c}) \cdot \sin\beta] = 25.94°$.

The coordinates y and z of the same atoms in the unit cells [8, 22] coincide. The coordinate $x \equiv x_{P2_1/n}$ of arbitrary atom in the unit cell [8] changes to the coordinate $x_{P2_1/c} = (x_{P2_1/n} + \Delta x)$ of the same atom in the unit cell [22]. In accordance with

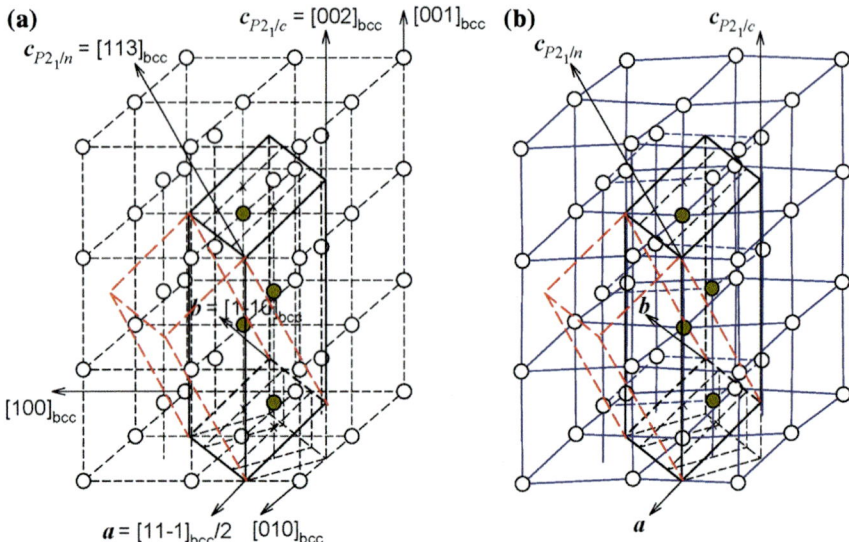

Fig. 4.17 The distortion of sulfur atom sublattice at transition from argentite β-Ag$_2$S to acanthite α-Ag$_2$S [29, 32]: **a** the arrangement of S atoms in bcc sublattice of argentite; **b** displacement of S atoms from bcc sublattice positions and their arrangement in monoclinic sublattice of acanthite. The undistorted bcc sublattice is shown by thin short *dashed lines*, the contours of the unit cells with space groups $P2_1/c$ and $P2_1/n$ are shown by *thick solid lines* and *thick long dashed lines*, respectively. *Open circle* and *filled circle* are S atoms located outside and inside monoclinic (space group $P2_1/c$) unit cell of acanthite α-Ag$_2$S, respectively. Ag atoms are not shown

Fig. 4.18 The diagram of mutual transformation of the unit cells proposed in works [8, 22] for the monoclinic phase α-Ag$_2$S:
$$c_{P21/c} = (c_{P2_1/n} - a_{P2_1/n})$$
relative to $a_{P2_1/n}$ and $c_{P21/n}$ axes of the unit cell proposed in [8]

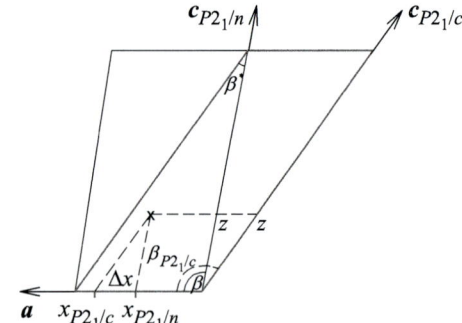

Fig. 4.18 and usual geometric relationships, $a/c = \sin\beta^*/\sin(180 - \beta_{P2_1/c}) \equiv \sin\beta^*/\sin(\beta_{P2_1/c})$ and $a \cdot \Delta x/cz = \sin\beta^*/\sin(180 - \beta_{P2_1/c}) \equiv \sin\beta^*/\sin(\beta_{P2_1/c})$, from which $\Delta x = z \cdot (c/a) \cdot (\sin\beta^*/\sin(\beta_{P2_1/c})) = z \cdot (c/a) \cdot (a/c) \equiv z$. Therefore the coordinate $x_{P2_1/c}$ of arbitrary atom in the unit cell [22] is equal to $x_{P2_1/c} = x_{P2_1/n} + z$.

Table 4.6 The coordinates of atoms in the monoclinic unit cell proposed for acanthite by Frueh [8] and in the monoclinic unit cell transformed in accordance with [22] for the same phase α-Ag$_2$S

Atom	Position and multiplicity	Atomic coordinates [8]			Atomic coordinates [22]		
		$x \equiv x_{P2_1/n}$	y	z	$x_{new} \equiv x_{P2_1/c}$	y	z
Ag 1	4(e)	0.758	0.015	0.305	0.063	0.015	0.305
Ag 2	4(e)	0.285	0.320	0.435	0.721	0.320	0.435
S	4(e)	0.359	0.239	0.134	0.493	0.239	0.134

The coordinates of atoms in the unit cell proposed for acanthite by Frueh [8] and in the unit cell transformed in accordance with Sadanaga et al. [22] for the same phase α-Ag$_2$S are given in Table 4.6.

The unit cells with space groups $P2_1/n$ and $P2_1/c$ proposed in [8, 22] were chosen in one and the same structure and have the same volume. Indeed, the choice of unit cell in the crystal lattice can be made by different methods and is ambiguous [136, 137]. In each lattice, an infinity of unit cells can be chosen. By combining the translational vectors of any of two monoclinic cells, it is possible to obtain several more monoclinic unit cells with the same volume, but with different translational vectors. Because of the ambiguity of the unit cell choice, one and the same crystal in the experimental studies [8, 22] was described differently. Generally, the sole unit cell describing the lattice can be revealed by Delaunay reduction [138–141]. However in [137, 142] it was shown that in the case of lattices with monoclinic symmetry, Delaunay reduction also gives ambiguous results.

Description of the diffraction pattern of monoclinic silver sulfide by any of the two unit cells [8, 22] provides the sets of diffraction reflections with similar positions and intensity that differ only in their Miller indices (hkl). The difference between the two unit cells consists in the length and direction of axis c, and $c_{P2_1/n} = (c_{P2_1/c} + a_{P2_1/c})$ (see Fig. 4.17a). With that in mind, the transition between the reflection indices is very simple: $h_{P2_1/n} = h_{P2_1/c}$, $k_{P2_1/n} = k_{P2_1/c}$ and $l_{P2_1/n} = l_{P2_1/c} + h_{P2_1/n}$ or, vice versa, $l_{P2_1/c} = l_{P2_1/n} - h_{P2_1/n}$.

The important advantage of the unit cell with space group $P2_1/c$ is more convenient description of the α-Ag$_2$S–β-Ag$_2$S transformation [22].

Recently [29, 32] the crystal structure of α-Ag$_2$S acanthite has been refined for the first time on synthesized artificial samples of coarse-crystalline powder of silver sulfide with the use of full-profile analysis of XRD data. Let us consider the results [29, 32] in more detail.

The powders of silver sulfide Ag$_2$S were synthesized by chemical deposition from aqueous solution of silver nitrate AgNO$_3$, sodium sulfide Na$_2$S and sodium citrate Na$_3$C$_6$H$_5$O$_7 \equiv$ Na$_3$Cit. The concentrations of AgNO$_3$, Na$_2$S and Na$_3$Cit in the reaction mixture for synthesis of the coarse-crystalline Ag$_2$S powder with particle size of $\sim$500 nm were 0.05, 0.5 and 0.005 mol l^{-1}, which corresponds to the ratio of concentrations of silver, sulfide and citrate ions, [Ag$^+$]:[S^{2-}]:[Cit^{3-}] = 1:10:0.1. Deposited Ag$_2$S powder was washed with distilled water, filtered and dried in air at 323 K. The average particle size D of coarse-crystalline Ag$_2$S powder was estimated from the value of specific surface area $S_{sp} = 1.6 \pm 0.1$ m^2 g^{-1} measured by

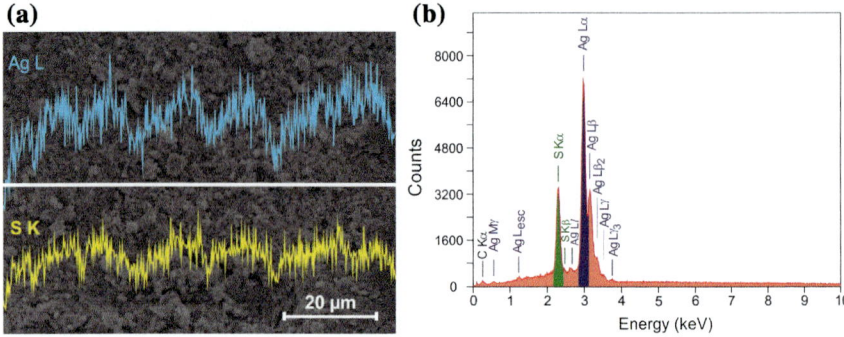

Fig. 4.19 EDX analysis of synthesized coarse-crystalline Ag₂S powder with average particle size of ∼500 nm: **a** the EDX distributions of Ag and S and scanning direction (*white horizontal line*) of the Ag₂S agglomerated powder during energy dispersive X-ray analysis; **b** cumulative elemental EDX pattern

the Brunauer-Emmett-Teller (BET) method. In the approximation that all particles have similar size and spherical shape, the average particle size $D = 6/\rho S_{\mathrm{sp}}$ ($\rho = 7.25$ g cm^{-3} is the density of silver sulfide), and is equal ∼515 nm for coarse-crystalline powder.

According to the Energy Dispersive X-ray Spectroscopy (EDX) results (Fig. 4.19), the content of silver Ag and sulfur S in the synthesized dried silver sulfide powder with average particle size of ∼500 nm is 86.8 ± 0.4 and 12.9 ± 0.1 wt.%, which corresponds to stoichiometric sulfide Ag₂S. Besides silver and sulfur, the EDS spectrum contains a $K\alpha$ line of carbon from the carbon planchet, onto which the examined powder was applied. No other impurity elements have been observed in the synthesized silver sulfide powder.

The XRD pattern of synthesized Ag₂S powder with average particle size of ∼500 nm used for crystal structure refinement is shown in Fig. 4.20. Preliminary analysis revealed that the observed set of diffraction reflections corresponds to one-phase silver sulfide with monoclinic α-Ag₂S acanthite type structure.

The observed I_{obs}, calculated I_{calc} and difference ($I_{\mathrm{obs}} - I_{\mathrm{calc}}$) values corresponding to the refinement of the XRD pattern for synthesized Ag₂S silver sulfide are shown in Fig. 4.20. The refinement of the crystal structure of synthesized Ag₂S silver sulfide using the X'Pert Plus software package [143] provides the following results: synthesized silver sulfide has a crystal structure of α-Ag₂S acanthite type; monoclinic (space group $P2_1/c$) unit cell parameters are equal to $a = 0.42264(2)$ nm, $b = 0.69282(3)$ nm, $c = 0.95317(3)$ nm and $\beta = 125.554(2)°$; the site occupancy factor of all crystallographic positions by Ag and S atoms is equal to 1.0; the Rietveld reliability factor R_I (R_B) is equal to 0.0247. These unit cell parameters are in good agreement with the data [22]. Experimental conditions and characteristics of the refinement for the artificial coarse-crystalline Ag₂S phase with acanthite-type crystal structure are given in Table 4.12 of Appendix.

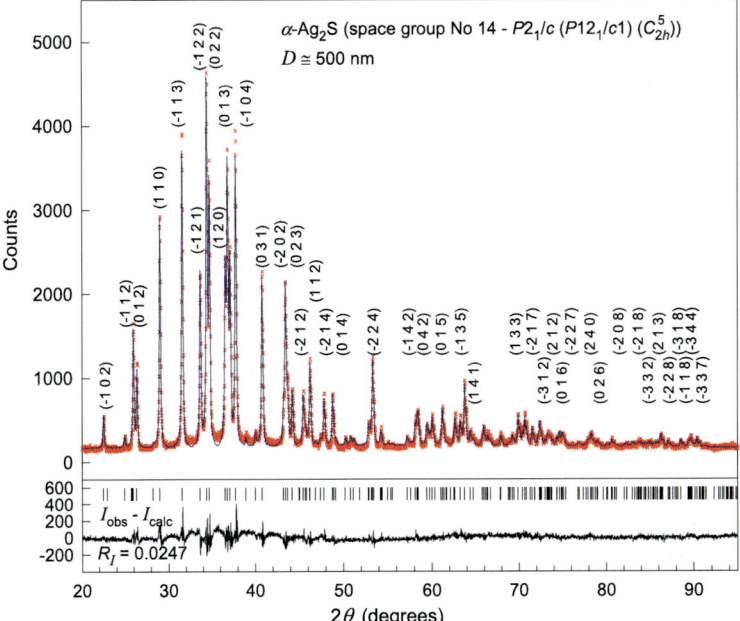

Fig. 4.20 The experimental ($\times$) and calculated (–) XRD patterns of coarse-crystalline Ag_2S powder. The difference between the experimental and calculated XRD patterns ($I_{obs} - I_{calc}$) is shown in the lower part of the figure. Refinement of the structure of artificial silver sulfide showed that it has monoclinic (space group $P2_1/c$ ($P12_1/c1$)) α-Ag_2S acantite-type structure and allowed a high degree of convergence between experiment and calculation to be achieved: the Rietveld reliability factors are equal to $R_p = 0.0719$, $\omega R_p = 0.0974$ and R_I (R_B) = 0.0247. The XRD pattern is recorded in $CuK\alpha_1$ radiation. Reprinted from [29] with permission from Elsevier

Table 4.7 Refined crystal structure of artificial coarse-crystalline monoclinic (space group No 14 – $P2_1/c$ ($P12_1/c1$)) silver sulfide Ag_2S with α-Ag_2S acanthite-type structure [29]: $Z = 4$, $a = 0.42264(2)$ nm, $b = 0.69282(3)$ nm, $c = 0.95317(3)$ nm, and $\beta = 125.554(2)°$

Atom	Position and multiplicity	Atomic coordinates			Occupancy	$B_{iso} \times 10^{-4}$ (pm^2)
		x/a	y/b	z/c		
Ag1	4(e)	0.0715(7)	0.0151(3)	0.3094(3)	1.00	5.59(3)
Ag2	4(e)	0.7264(7)	0.3241(4)	0.4375(3)	1.00	5.25(3)
S	4(e)	0.492(2)	0.234(1)	0.1321(7)	1.00	1.680

The coordinates of Ag and S atoms in the refined crystal structure of artificial silver sulfide Ag_2S are listed in Table 4.7. Artificial silver sulfide Ag_2S is stoichiometric since within the limits of the refinement errors the degree of occupancy of all its crystal lattice sites is 1.0.

However, small deficiency or small excess (less than 0.04) of Ag atoms, i.e. weak nonstoichiometry of monoclinic Ag_2S cannot be excluded. Neutron diffraction study is necessary for determination of such small nonstoichiometry. The

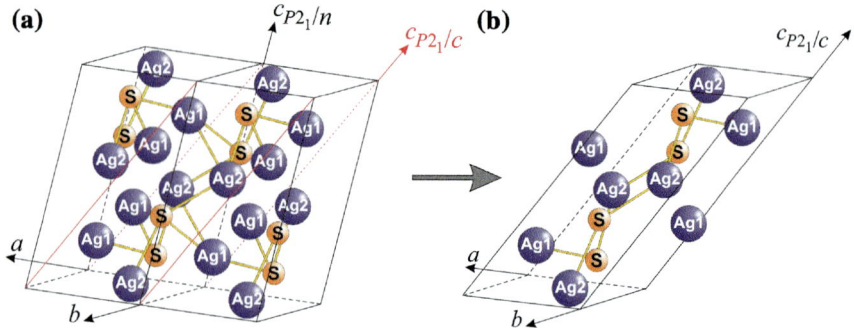

Fig. 4.21 The arrangement of Ag and S atoms in the crystal lattice of synthesized monoclinic (space group $P2_1/c$) silver sulfide Ag$_2$S [29, 32]: **a** the contour of doubled unit cell with space group $P2_1/n$, the position of the $c_{P21/c}$ axis and the contour of monoclinic (space group $P2_1/c$) unit cell in the crystal lattice are shown; **b** monoclinic (space group $P2_1/c$) unit cell of synthesized silver sulfide Ag$_2$S with acanthite structure (shown are only the atoms entering into the unit cell and the nearest bonds between them, Ag1-S and Ag2-S, of length 0.2511 and 0.2548 nm, respectively). Reprinted from [29] with permission from Elsevier

arrangement of Ag and S atoms in the unit cell of artificial monoclinic (space group $P2_1/c$) silver sulfide Ag$_2$S with α-Ag$_2$S acanthite type structure is displayed in Fig. 4.21. The least distances between S and Ag1 atoms in the crystal lattice of artificial silver sulfide Ag$_2$S are equal to 0.25113 and 0.25183 nm, and those between S and Ag2 atoms are 0.25475 and 0.25926 nm. The list of the observed diffraction reflections for the refined crystal structure of synthesized monoclinic (space group $P2_1/c$) silver sulfide Ag$_2$S is given in Table 4.13 of Appendix.

SEM image of the coarse-crystalline Ag$_2$S powder at a magnification of 10000 times is shown in Fig. 4.22. Most of the particles have an elongated dumbbell shape and they present an aggregate of two roundish particles. Elongated particles have a length about 400–500 nm and thickness ∼200 nm. Particles with a rounded shape have a size of 200–250 nm.

4.2.2 Crystal Structure and Nonstoichiometry of Nanostructured α-Ag2S Silver Sulfide

Determination of influence of nanoparticle size on their nonstoichiometry is fundamental scientific problem. Relative content of atoms on the surface of nanoparticles is high. Surface atoms are bonded weakly than atoms inside nanoparticle. This circumstance can lead to nonstoichiometry of a nanoparticle.

The nanocrystalline powder of silver sulfide Ag$_2$S was synthesized by chemical deposition from aqueous solution of silver nitrate AgNO$_3$ and sodium sulfide Na$_2$S containing sodium citrate Na$_3$C$_6$H$_5$O$_7$≡Na$_3$Cit as complexing and stabilizing agent. The concentrations of AgNO$_3$, Na$_2$S and Na$_3$Cit in the reaction mixture for

Fig. 4.22 SEM image of coarse-crystalline Ag_2S powder at a magnification of 10000 times. Most of the particles have an elongated shape and a length about 400–500 nm and thickness $\sim$200 nm. Reprinted from [29] with permission from Elsevier

synthesis of Ag_2S nanopowder with particle size smaller than 45 nm were equal to 0.05, 0.025, and 0.025 mol l^{-1}, respectively [30]. Synthesis was carried out at room temperature. When the reagents are mixed together, silver sulfide deposit is formed almost immediately.

Silver sulfide deposit was examined by XRD method on a Shimadzu XRD-7000 diffractometer. Experimental conditions and characteristics of the refinement for the synthesized Ag_2S nanopowder are given in Table 4.14 of Appendix. The instrumental resolution function $\text{FWHM}_R(2\theta) = (u \tan^2 \theta + v \tan \theta + w)^{1/2}$ of the Shimadzu XRD-7000 X-ray diffractometer was found in a special diffraction experiment on cubic lanthanum hexaboride LaB_6 (NIST Standard Reference Powder 660a) with lattice constant $a = 0.415692$ nm. The resolution function $\text{FWHM}_R(2\theta)$ measured in degrees has the parameters $u = 0.005791$, $v = -0.004627$ and $w = 0.010201$. The determination of the crystal lattice parameters and final refinement of the structure of synthesized silver sulfide nanopowder was carried out with the use of the X'Pert Plus program package [143].

The microstructure, particle size and element chemical composition of silver sulfide nanopowder were studied by the SEM method on a JEOL-JSM LA 6390 microscope coupled with a JED 2300 Energy Dispersive X-ray Analyzer. High-resolution-TEM (HR-TEM) images were recorded on a JEOL JEM-2100 transmission electron microscope with a resolution 0.14 nm.

The average particle size D (to be more precise, the average size of coherent scattering regions (CSR)) in synthesized silver sulfide nanopowder was estimated by XRD method from the diffraction reflection broadening using the dependence of

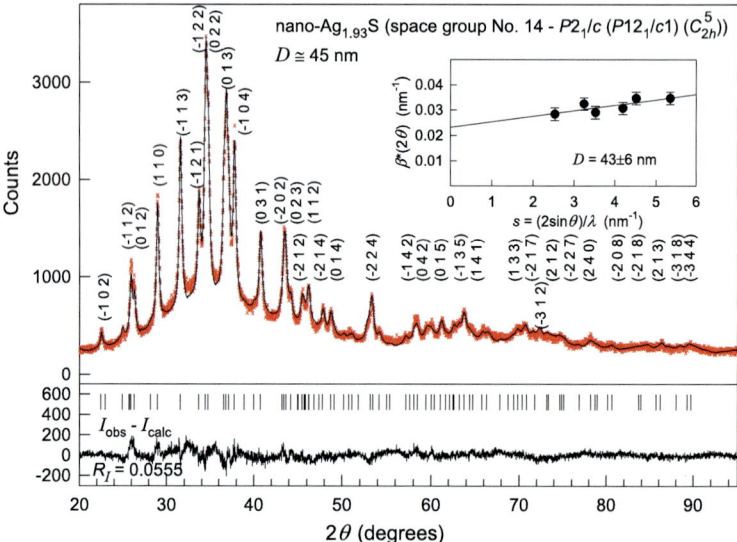

Fig. 4.23 The experimental ($\times$) and calculated ($-$) XRD patterns of Ag$_{1.93}$S nanopowder deposited from the reaction mixture of AgNO$_3$, Na$_2$S and Na$_3$Cit having concentrations 0.05, 0.025, and 0.025 mol l^{-1}, respectively. The difference between the experimental and calculated XRD patterns ($I_{obs} - I_{calc}$) is shown in the lower part of the figure. The *inset* presents the estimate of the average size of the coherent scattering regions from the broadening of non-overlapping diffraction reflections. The XRD pattern is recorded in Cu$K\alpha_1$ radiation. Reproduced from [30] with permission from the PCCP Owner Societies

reduced reflection broadening $\beta^*(2\theta) = [\beta(2\theta)\cos\theta_1/\lambda$ on the scattering vector $s = (2\sin\theta)/\lambda$ [144–147]. Also the average particle size D was estimated from the value of specific surface area S_{sp} of nanopowder. The specific surface area S_{sp} of the silver sulfide nanopowder was found by the BET method.

A nanostructured silver sulfide has been studied extensively in the past two decades. However, until lately there were no experimental works on the determination of the crystal structure of nanocrystalline silver sulfide.

The XRD pattern of synthesized silver sulfide nanopowder is shown in Fig. 4.23. According to the BET data, the particle size of nanopowder is 44 ± 5 nm.

Preliminary analysis revealed that synthesized nanocrystalline powder has monoclinic (space group $P2_1/c$) α-Ag$_2$S acanthite-type structure. The diffraction reflections of the nanopowder are broadened and therefore the reflections located close to each other overlap. The average size D of CSR estimated from broadening of non-overlapping diffraction reflections ($-1\ 0\ 2$), ($1\ 1\ 0$), ($-1\ 1\ 3$), ($-1\ 0\ 4$), ($0\ 3\ 1$) and ($0\ 1\ 4$) is 43 ± 6 nm.

In spite of the large width and overlapping of many diffraction reflections on the XRD pattern of silver sulfide nanopowder (see Fig. 4.23), Sadovnikov et al. [30, 33] have tried to refine the crystal structure of silver sulfide nanopowder. In the first approximation authors of studies [30, 33] supposed that the nanopowder has the same monoclinic (space group $P2_1/c$) crystal structure as a coarse-crystalline silver sulfide with particle size ~500 nm. The observed I_{obs}, calculated I_{calc} and difference ($I_{obs} - I_{calc}$) values corresponding to the refinement of the XRD pattern for silver sulfide nanopowder are shown in Fig. 4.23. The coordinates of Ag and S atoms and unit cell parameters for Ag_2S nanopowder (Table 4.8) are close to those for coarse-crystalline Ag_2S. However, the occupancy of crystallographic positions 4 (e) by Ag1 and Ag2 atoms is somewhat smaller than 1 and is equal to ~0.97 and ~0.96, respectively (Table 4.8). This means that silver sulfide nanoparticles with the size of less than ~50 nm are nonstoichiometric, have a composition of ~$Ag_{1.93}S$ and contain vacant sites in metal sublattice. The list of the observed diffraction reflections for the refined crystal structure of nanocrystalline monoclinic (space group $P2_1/c$) silver sulfide $Ag_{1.93}S$ is given in Table 4.15 of Appendix.

Figure 4.24 displays a SEM image of $Ag_{1.93}S$ nanopowder deposited from the reaction mixture of silver nitrate, sodium sulfide and sodium citrate with the concentrations 0.05, 0.025 and 0.025 mol l^{-1}, respectively. According to SEM data, this nanopowder consists of large irregular agglomerates (Fig. 4.24a) ranging in size from 6 to 10 m. At a magnification of 15000 times, it is seen that these

Table 4.8 Refined crystal structure of monoclinic (space group No 14 – $P2_1/c$ ($P12_1/c1$)) $Ag_{1.93}S$ nanopowder with α-Ag_2S acanthite-type structure and particle size ~45 nm [30]: $Z = 4$, $a = 0.4234(3)$ nm, $b = 0.6949(3)$ nm, $c = 0.9549(5)$ nm, and $\beta = 125.43(6)°$

Atom	Position and multiplicity	Atomic coordinates			Occupancy	$B_{iso} \times 10^{-4}$ (pm^2)
		x/a	y/b	z/c		
Ag1	4(e)	0.0715	0.0151(0)	0.3093(9)	0.97	10.05(5)
Ag2	4(e)	0.7264	0.3240(9)	0.4375(0)	0.96	7.44(6)
S	4(e)	0.4920	0.2339(8)	0.1321(1)	1.00	1.960

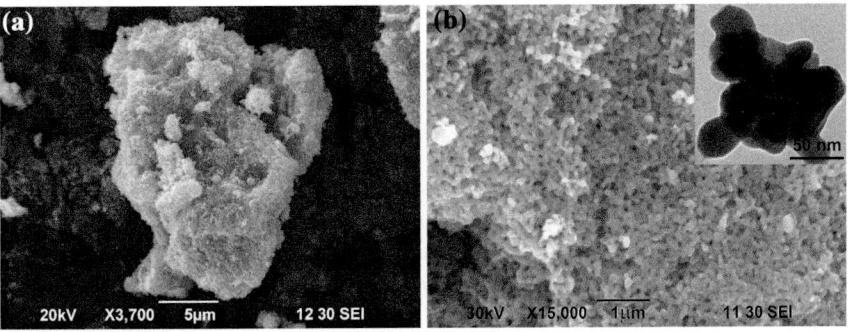

Fig. 4.24 SEM image of $Ag_{1.93}S$ nanopowder: **a** powder consists of large irregular agglomerates with a size from 6 to 10 μm; **b** at a magnification of 15000 times, it is seen that these agglomerates have a loose microstructure and are formed by small rounded particles with a size of ~100–150 nm and elongated dumbbell-shaped particles of length to ~250–300 nm. The *inset* shows the HRTEM image of disoriented crystalline blocks of ~$Ag_{1.93}S$ particles from 20 to 50 nm in size

agglomerates have a loose microstructure and are formed by small particles (Fig. 4.24b). Rounded silver sulfide particles have a size of $\sim$100–150 nm, and elongated silver sulfide particles are up to $\sim$250 nm in length and $\sim$100 nm in thickness. According to HRTEM data (Fig. 4.24b, inset), agglomerates consist of disoriented crystalline blocks from 20 to 50 nm in size. It should be noted that the morphology of coarse-crystalline and nanostructured Ag$_2$S powders is identical, difference consists in particle size only (see Figs. 4.22 and 4.24b).

4.2.3   In Situ High-Temperature Study of "Acanthite α-Ag$_2$S—Argentite β-Ag$_2$S" Phase Transformation

For the first time, complex in situ high-temperature XRD and scanning electronic microscopy study of the "α-Ag$_2$S (acanthite) – β-Ag$_2$S (argentite)" phase transformation in nanocrystalline and coarse-crystalline powders of silver sulfide has been carried out by Sadovnikov et al. in works [31, 34, 35].

The low-temperature monoclinic acanthite α-Ag$_2$S exists at temperatures below $\sim$450 K. The cubic argentite β-Ag$_2$S exists in the temperature interval 452–859 K.

When cubic argentite β-Ag$_2$S is cooled below 450 K, a polymorphous phase transformation takes place giving rise to monoclinic acanthite α-Ag$_2$S. This transformation is accompanied by distortion of the body-centered cubic (bcc) sublattice of S atoms to monoclinic sublattice. The Ag atoms statistically distributed in positions 6(b) and 12(d) (or 6(b) and 48(j)) of bcc structure of argentite are concentrated in the positions of the monoclinic lattice of acanthite and occupy these positions with probability which is close to 1.

The data of differential thermal and differential thermogravimetric analyses (DTA-DTG) [61, 132] show that the acanthite α-Ag$_2$S – argentite β-Ag$_2$S phase transformation takes place at $\sim$448–453 K. According to [148–152], the temperature T_{trans} and enthalpy ΔH_{trans} of the α-Ag$_2$S–β-Ag$_2$S phase transformation are 449.3–451.3 K and 4.0 ± 0.5 kJ mol^{-1}.

Until lately, few data on acanthite–argentite phase transformation were obtained only on bulk coarse-crystalline silver sulfide samples.

Nanocrystalline Ag$_2$S powder was synthesized in studies [31, 34, 35] by chemical deposition from aqueous solutions of silver nitrate AgNO$_3$, sodium sulfide Na$_2$S and sodium citrate Na$_3$C$_6$H$_5$O$_7 \equiv$Na$_3$Cit with concentrations 0.05, 0.025, and 0.0125 mol l^{-1}, respectively. Coarse-crystalline Ag$_2$S powder was produced via hydrothermal synthesis in the closed vessel at 453 K for 4 h from AgNO$_3$ water solution and (NH$_2$)$_2$CS thiocarbamide as the sulfur source. The pressure of saturated vapor above solution was $\sim$1 $\times$ 10^6 Pa.

In situ high-temperature XRD (HT-XRD) measurements were performed using a X'Pert PRO MPD (Panalytical) diffractometer which was combined with a Anton Paar HTK-1200 Oven furnace. High temperature XRD patterns were collected in the angle interval 2θ = 20–67.5° with a step of $\Delta(2\theta)$ = 0.026° and scanning time

200 s in each point. Characteristics of the refined crystal structure of the artificial coarse-crystalline acanthite α-Ag_2S are presented in Table 4.16 of Appendix.

The most important part of a X'Pert PRO MPD diffractometer is a position-sensitive fast sector detector, which simultaneously measures the intensity of reflection not at a separate point as an ordinary proportional counter, but in a 2θ range with a width of 3.154°. This detector has enabled to decrease considerably the time of the X-ray profile recording without losing quality. Also the synthesized powders of silver sulfide were examined by X-ray diffraction on Shimadzu XRD-7000 and STADI-P (STOE) diffractometers in $CuK\alpha_1$ radiation. X-ray measurements were performed in the angle interval $2\theta = 20$–$95°$ with a step of $\Delta(2\theta) = 0.02°$ and scanning time 10 s in each point. Qualitative and semi-quantitative phase composition of the powders was estimated using the Match! Version 1.10 software suite [153]. The determination of the crystal lattice parameters and final refinement of the structure of the synthesized powders of silver sulfide were carried out with the use of the X'Pert Plus software suite [143].

The temperature T_{trans} and enthalpy ΔH_{trans} of the acanthite—argentite transformation were determined by DTA-DTG method on a Setaram SETSYS Evolution 1750 thermal analyzer. DTA and TGA experiments were carried out in argon Ar flow of 20 ml min^{-1} in the temperature range 293–500 K with a heating rate of 5 K min^{-1}.

The phase transformation α-Ag_2S (acanthite)–β-Ag_2S (argentite) in silver sulfide powders was directly observed by the SEM method on a JEOL-JSM LA 6390 microscope with a JED 2300 Energy Dispersive X-ray Analyzer. The elemental chemical composition of Ag_2S powders was studied on the same microscope via EDX analysis. The powders to be examined were deposited on a carbon substrate. In all the experiments, the microscope operating distance was 10 mm, and the electron beam spot size was 30 or 60. The silver sulfide powders were intensively heated in the microscope via surface scanning with an electron beam with enhanced spot size up to 60 and the counting rate increased until ~ 2000 cps (counts per second). Taking into account the instrumental microscope parameters and heating of investigated Ag_2S powders a depth of ~ 400–500 nm, the increase of temperature at the radiation heating was about 250–400 K, i.e., the silver sulfide powders were heated to ~ 550–700 K. The heating temperature of the powder was thus superior to the T_{trans} of the α-Ag_2S–β-Ag_2S phase transition, which equals ~ 450 K.

The synthesized powders of silver sulfide Ag_2S were studied also by high-resolution transmission electron microscopy (HRTEM) on a JEOL JEM-2100 microscope with 0.14 nm resolution.

The X-ray diffraction patterns of the coarse- and nanocrystalline powders of Ag_2S are shown in Fig. 4.25. According to the BET and SEM data, the average particle size in coarse-crystalline powder is ~ 850 nm and the maximal size reaches ~ 3–5 μm. The quantitative refinement of the XRD pattern (Fig. 4.25a) showed that the coarse-crystalline powder synthesized by hydrothermal method is stoichiometric and contains one phase with monoclinic (space group $P2_1/c$) structure of α-Ag_2S acanthite type. This structure of artificial coarse-crystalline silver sulfide was

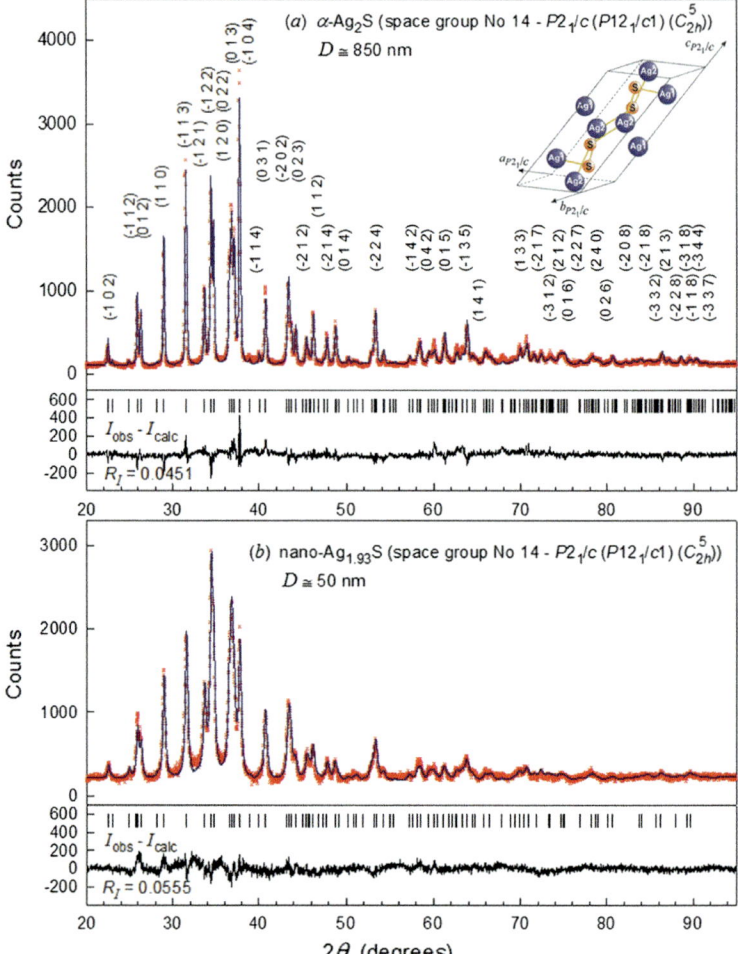

Fig. 4.25 The experimental ($\times$) and calculated ($-$) XRD patterns of silver sulfide powders with particle size **a** $\sim$850 nm and **b** $\sim$50 nm, respectively. The difference ($I_{obs} - I_{calc}$) between the experimental and calculated XRD patterns is shown in the lower part of the figure. The *inset* shows the unit cell of coarse-crystalline monoclinic (space group $P2_1/c$) silver sulfide. Reproduced from [31] with permission from the PCCP Owner Societies

determined in study [29] (see Sect. 4.2.1). Nanocrystalline silver sulfide powder contains one monoclinic (space group $P2_1/c$) phase too (Fig. 4.25b) but this phase is nonstoichiometric and has the composition $\sim$Ag$_{1.93}$S. Detailed refinement of crystal structure of nanocrystalline silver sulfide is given in works [30, 33] (see Sect. 4.2.2).

To precisely determine the phase transformation temperature, the silver sulfide powders were studied via DTA-DTG in both heating and cooling [34] (Fig. 4.26).

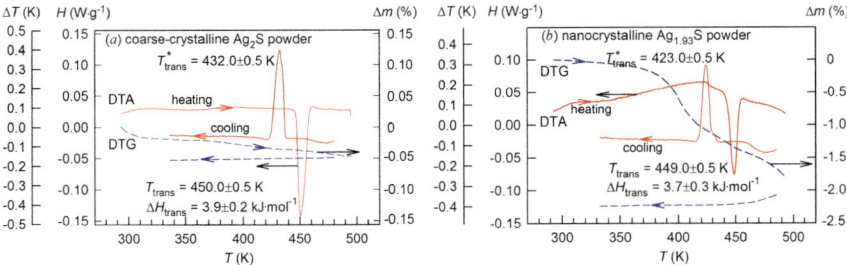

Fig. 4.26 The DTA and DTG curves measured during heating and cooling of **a** coarse-crystalline Ag_2S and **b** nanocrystalline $Ag_{1.93}S$ silver sulfide powders [34]

During heating, the DTA curves have one endothermal peak at $\sim$449–450 K, corresponding to the α-Ag_2S (acanthite)–β-Ag_2S (argentite) phase transformation. In cooling from 500 K to room temperature, the DTA dependencies exhibit one exothermal peak that is related to the argentite-to-acanthite phase transformation toward the lower temperature range by $\sim$20 K. The presence of the T_{trans} temperature hysteresis means this is first-order reversible acanthite–argentite transformation. Found enthalpy ΔH_{trans} of phase transformation is equal $\sim$3.7–3.9 kJ mol^{-1}, which is very close to the value $\Delta H_{trans} = 4.0 \pm 0.5$ kJ mol^{-1} determined in works [148–152].

In order to detect the presence of any chemically adsorbed impurities, the DTG measurements were made. The mass change during heating of the coarse-crystalline powder is negligibly small and is within the measurement error (Fig. 4.26a). Small mass loss of the nanocrystalline powder by $\sim$2 wt.% which is observed at heating (Fig. 4.26b) is due to evaporation of water adsorbed by the nanopowder surface.

No chemical reactions during radiation heating of silver sulfide occurred. Indeed, according to the DTA-DTG data, during heating of silver sulfide to $\sim$500 K, only acanthite-argentite transformation is observed at $\sim$450 K (see Fig. 4.26).

The XRD patterns for coarse-crystalline silver sulfide powder at 300, 433, 453 and 503 K, and the XRD patterns for nanocrystalline silver sulfide powder at 300, 398 and 463 K are shown in Fig. 4.27. The XRD patterns collected at $T < 450$ K (Fig. 4.27a, c) contain the diffraction reflections of monoclinic (space group $P2_1/c$) α-Ag_2S acanthite. According to data [29, 30], monoclinic coarse-crystalline and nanocrystalline silver sulfide powders have the composition Ag_2S and $Ag_{1.93}S$, respectively (see Sects. 4.2.1 and 4.2.2). The XRD patterns collected at $T \geq 453$ K contain the diffraction reflections of cubic β-Ag_2S argentite only. According to the DTA data (see Fig. 4.26), the acanthite α-Ag_2S–argentite β-Ag_2S transformation occurs at a temperature about 449–450 K.

The structure of the high-temperature silver sulfide phase was determined in study [31] using the XRD patterns collected at 463 and 503 K. For the initial refinement of the structure, two models [23, 26] were used. According to the model [23], 4 Ag atoms are distributed in 6(b), 12(d), and 24(h) positions with probabilities equal to 2/9, 1/9, and 1/18, respectively (see Table 4.3). In the model [26],

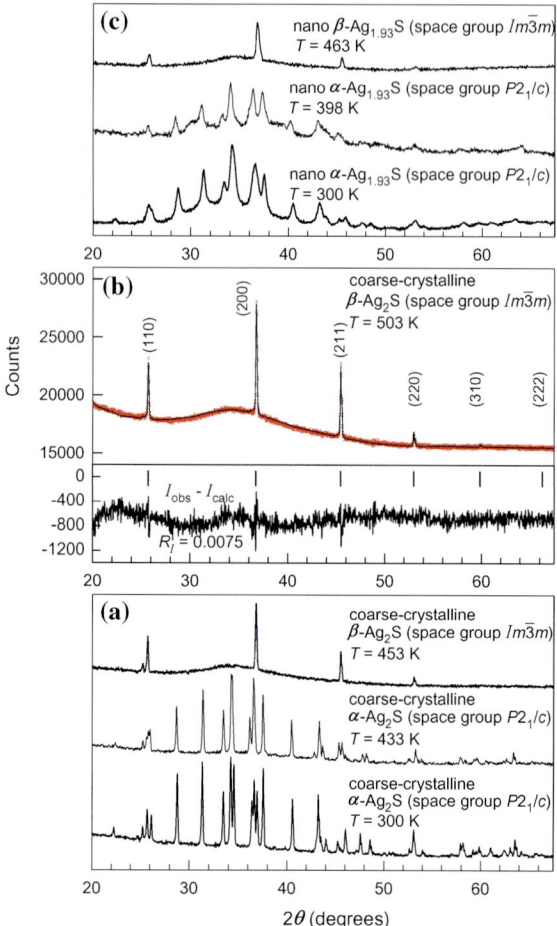

Fig. 4.27 Evolution of XRD patterns of silver sulfide at heating. **a** XRD pattern of coarse-crystalline silver sulfide with monoclinic (space group $P2_1/c$) α-Ag$_2$S acanthite type structure at 300 and 433 K, and with cubic (space group $Im\overline{3}m$) β-Ag$_2$S argentite-type structure at 453 K. **b** Experimental ($\times$) and calculated ($-$) XRD patterns of coarse-crystalline silver sulfide with cubic (space group $Im\overline{3}m$) β-Ag$_2$S argentite-type structure at 503 K and the difference ($I_{obs} - I_{calc}$) between the experimental and calculated XRD patterns; the ticks correspond to reflections of cubic argentite β-Ag$_2$S. **c** XRD patterns of nanocrystalline silver sulfide Ag$_{1.93}$S with monoclinic (space group $P2_1/c$) acanthite type structure at 300 and 398 K, and with cubic (space group $Im\overline{3}m$) β-Ag$_2$S argentite-type structure at 463 K, respectively

four silver atoms are statistically distributed in 6(b) and 12(d) positions. The quantitative analysis of the experimental HT-XRD data showed that the both models [23, 26] are not realistic. Improved crystallographic data were obtained by the redistribution of Ag atoms from the 12(d) positions into the 48(j) sites, and by the refinement of the degrees of occupation of the 6(b) and 48(j) positions (Table 4.9). Thus, at temperatures above 433 K silver sulfide contains one phase with bcc (space group No 229 – $Im\bar{3}m$ ($I4/m\bar{3}2/m$) (O_h^9)) structure of β-Ag$_2$S argentite type. The unit cell of β-Ag$_2$S argentite includes two Ag$_2$S formula units. Two sulfur atoms occupy 2(a) crystallographic positions of the considered cubic structure and form non-metal sublattice, and four silver atoms are distributed in 54 positions 6(b) and 48(j) with occupation degrees 0.0978(7) and 0.0711(0), respectively.

The unit cell of cubic (space group $Im\bar{3}m$) β-Ag$_2$S argentite is shown in Fig. 4.28.

Table 4.9 Refined crystal structure of cubic (space group No. 229 – $Im\bar{3}m$ ($I4/m\bar{3}2/m$) (O_h^9)) coarse-crystalline silver sulfide β-Ag$_2$S (β-Ag$_{2.01}$S) with argentite-type structure at 503 K [31, 34, 35]: $Z = 2$, $a = b = c = 0.4874(1)$ nm

Atom	Position and multiplicity	Atomic coordinates			Occupancy	$B_{iso} \times 10^{-4}$ (pm^2)
		x	y	z		
Ag1	6(b)	0	0.5	0.5	0.0978(7)	0.50
Ag2	48(j)	0	0.3306 (5)	0.4122 (7)	0.0711(0)	0.50
S	2(a)	0	0	0	1.00(0)	0.50

Fig. 4.28 The arrangement of Ag and S atoms in the unit cell of cubic (space group $Im\bar{3}m$) β-Ag$_2$S argentite. Positions 6(b) and 48(j) of silver sublattice on which four Ag atoms are statistically distributed are shown

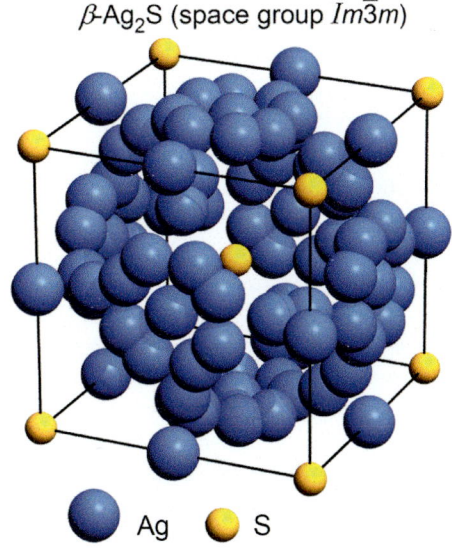

β-Ag$_2$S (space group $Im\bar{3}m$)

Ag   S

The list of the observed diffraction reflections for the refined crystal structure of coarse-crystalline cubic (space group $Im\overline{3}m$) β-Ag$_2$S argentite is given in Table 4.17 of Appendix. Detailed crystallographic information on β-Ag$_2$S argentite presented as "Crystal structure data" in the CIF-file (Crystallographic Information File), attached to the article [31]. This CIF No. 1062400 which placed on the Cambridge Crystallographic Data Centre (CCDC) website, can be found at the following electronic address [154].

Using the temperature dependences of the crystal lattice parameters of monoclinic acanthite and cubic argentite [35, 155] it is possible to estimate the interatomic distances in these phases at comparable temperatures close to the "acanthite–argentite" transition temperature T_{trans}. The least distance between the Ag1 and Ag1 atoms in the crystal lattice of monoclinic α-Ag$_2$S acanthite at 433 K is equal to 0.3351 nm, and the least distance between the Ag1 and Ag2 atoms is in the interval from 0.3085 to 0.3200 nm [35, 155]. In cubic β-Ag$_2$S argentite, possible distances between silver atoms at a similar temperature of 443 K are much smaller. According to [35, 155], in the crystal lattice of cubic β-Ag$_2$S argentite at 443 K the least possible distances between the Ag1 and Ag1 atoms are equal to 0.2428 nm, between the Ag1 and Ag2 atoms the least distances lie in the range from 0.0927 to 0.2971 nm, and between the Ag2 and Ag2 atoms the least distances are from 0.0988 to 0.2998 nm. The covalent radius of Ag atom is $\sim$0.146 nm. With that in mind, it is clear that silver atoms in monoclinic acanthite are at rather large distances from each other and therefore they occupy their crystallographic sites with probability 1. In cubic β-Ag$_2$S argentite, the possible distances between silver atoms are too small for the 6(b) and 48(j) positions to be occupied by Ag atoms with probability 1. For this reason, the occupancies of the 6(b) and 48 (j) positions by Ag atoms (put it otherwise, the probabilities of finding Ag atoms in the 6(b) and 48(j) sites) are very small, less than 0.1 (see Table 4.9). Physically, this means that 4 silver atoms in the cubic argentite lattice are in constant motion over 54 possible crystallographic positions. It is this constant motion of Ag atoms that provides the crystal lattice stability of cubic argentite and its superionic conductivity.

Nanocrystalline silver sulfide at a temperature of 463 K contains only the same cubic (space group $Im\overline{3}m$) argentite β-Ag$_2$S (Fig. 4.27d) with broadened diffraction reflections. The average particle size D (to be more precise, the average size of CSR) in the nanopowder estimated from broadening of diffraction reflections is $\sim$60 nm.

The in situ X-ray study of silver sulfide in the temperature range from 300 to 503 K revealed only acanthite phase at $T < 430$ K and only argentite phase at $T > 433$ K (see Fig. 4.27). No other phases were found.

Thus, heating of monoclinic α-Ag$_2$S acanthite up to temperature $\sim$449–450 K leads to polymorphic phase transition with the formation of bcc β-Ag$_2$S argentite.

An unique electron-microscopic experiment was performed to directly observe the dynamics of the transformation of acanthite into argentite. The SEM images of silver sulfide powders were obtained with electron beam spot size 30 and counting rate 514 cps. The α-Ag$_2$S acanthite powders were heated intensively in a JEOL-JSM LA 6390 microscope by scanning the surface with the electron beam with increased spot size 60 and with increased counting rate $\sim$2000 cps. The temperature change, ΔT, during radiation heating can be estimated as $\Delta T = nE/mC_p = nE/V\rho C_p$, where nE is the electron beam energy, m is the mass of the heated part of Ag$_2$S sample, V is the volume of the heated part of the sample, ρ is the density of silver sulfide, and C_p is the specific heat capacity of acanthite α–Ag$_2$S (for 300 K C_p = 312 J kg^{-1} K^{-1} [150, 152]). The depth of heating was estimated from the SEM images of those samples, where the transformation of acanthite into argentite ended. The amount of the argentite formed is equal to the amount of heated acanthite since both phases have virtually the same density. The maximal length of argentite whiskers is from 500 to 1200 nm and more (see for example Fig. 4.29f). Since the relative volume density of argentite layer is not greater than 30–40%, the thickness of heated acanthite layer (depth of heating) is $\sim$300–500 nm. Considering the microscope instrumental parameters and the depth of heating, the temperature change ΔT during radiation heating was approximately 250–300 K, i.e. the silver sulfide powders were heated up to $\sim$550–600 K.

Thus, the temperature of heating the sample exceeded the temperature $T_{\text{trans}} \cong$ 450 K of the α-Ag$_2$S–β-Ag$_2$S phase transformation. This allowed one to observe the $\cong$of transformation of acanthite into argentite in real time (see Supplementary movie file which is attached to the article [31]). At first, argentite nuclei appear on the surface of acanthite particles as a result of heating (Fig. 4.29). Emerged argentite nuclei continue to grow and take a shape of whiskers. During $\sim$5 min these whiskers reach a length of 500 nm to 1 μm and more.

Some stages of α-Ag$_2$S–β-Ag$_2$S phase transformation which takes place in silver sulfide powders are shown in SEM images (Fig. 4.29).

In the upper row (Fig. 4.29a–c), the initial nanocrystalline silver sulfide powder with separated part of surface, for which the elemental chemical composition is determined by the EDX method, as well as the cumulative elemental EDX pattern of this powder are shown. Upon drying, the nanoparticles formed closely spaced agglomerates. The contents of Ag and S in the silver sulfide nanopowder determined from ratio of the integral intensities of AgL and SK lines of EDX pattern (Fig. 4.29c) are equal to 86.8 $\pm$ 0.4 and 13.1 $\pm$ 0.1 wt.%, respectively. It corresponds to silver sulfide which is close to stoichiometric composition $\sim$Ag$_2$S but has a small deficiency of silver. The EDX spectrum contains also small $K\alpha$ line of C from a carbon planchet, on which the examined powder was applied. No other impurity elements have been detected in the silver sulfide nanopowder. SEM images of part of the heated surface and of the same surface (side view) with grown argentite crystals are shown in the second row of Fig. 4.29.

The transformation of coarse-crystalline acanthite powder into argentite is presented in the 3rd and 4th rows of Fig. 4.29. As distinct from the nanocrystalline powder, the agglomeration in the coarse-crystalline powder is small, but certain intergrowth of particles takes place. In the lower row (Fig. 4.29j, k), the SEM

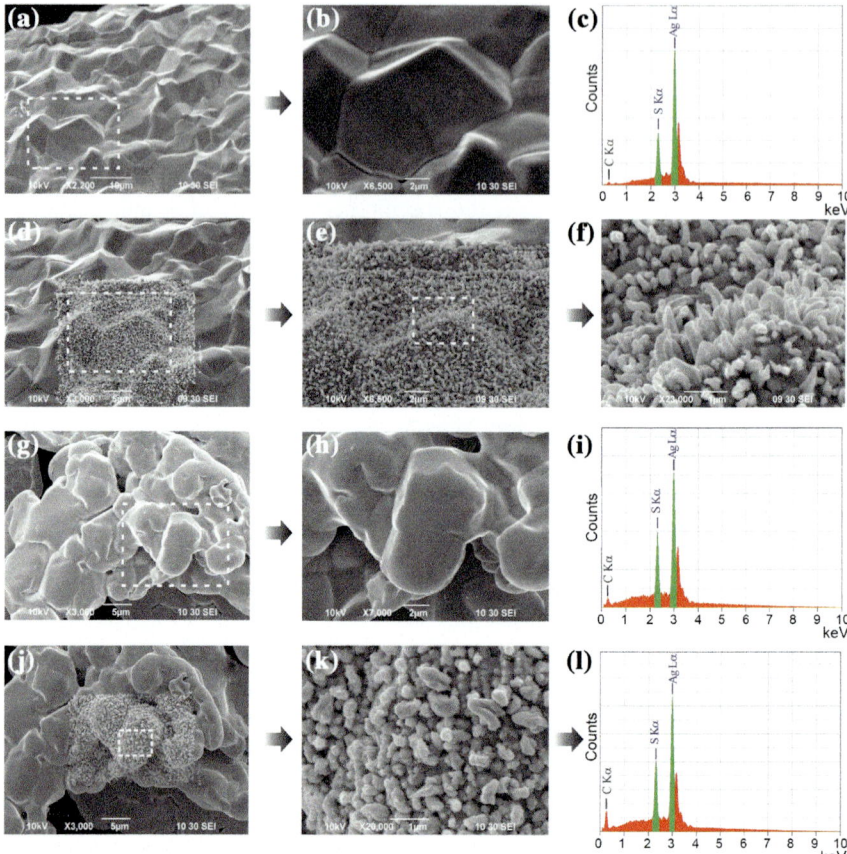

Fig. 4.29 The transformation of acanthite α-Ag$_2$S into argentite β-Ag$_2$S. *The upper row* **a**, **b** the initial nanopowder and (**c**) its cumulative elemental EDX pattern. *The second row* **d–f** the argentite particles are growing on the electron beam heated surface area. *The third row* **g**, **h** coarse-crystalline powder of silver sulfide and **i** its cumulative elemental EDX pattern. *The lower row* **j**, **k** the growth of argentite particles; **l** cumulative elemental EDX pattern of the surface layer of argentite. The *white dotted lines* show the surface areas of Ag$_2$S powders heated by electron beam. Reproduced from [31] with permission from the PCCP Owner Societies

images of acanthite particle (top view), on which argentite particles are growing, are shown. A large particle with a face lying in the plane of image (Fig. 4.29j), was chosen for observation of argentite growth.

Short pyramidal nuclei of β-Ag$_2$S with a base of about 50–80 nm appear on the surface of particle of silver sulfide powder in $\sim$30 s after the beginning of electron beam heating. As a result of heating, the argentite nuclei grow quickly and take the shape of whiskers. Formed argentite is covered all the surface of acanthite particle in $\sim$5 min. During the growth, the length of the whiskers reaches 500–800 nm or

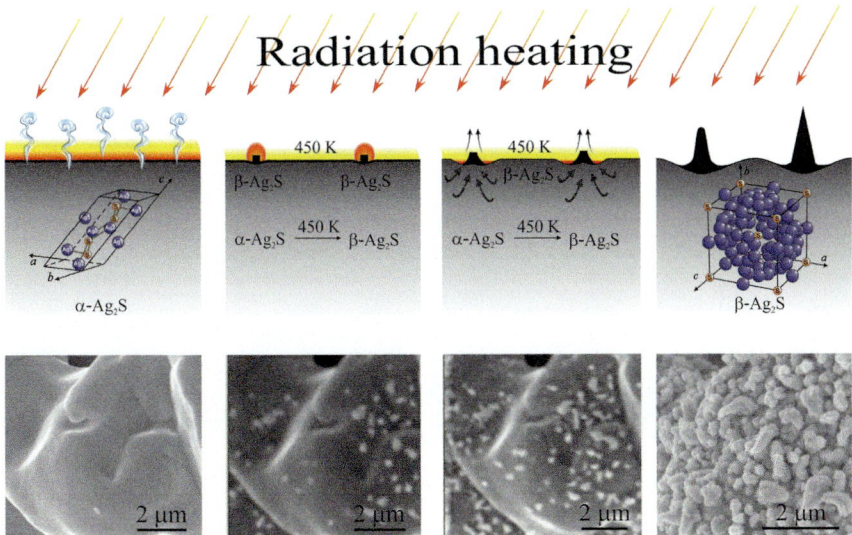

Fig. 4.30 The sequence of transformation of acanthite α-Ag_2S into argentite β-Ag_2S as a result of electron-beam heating. Reproduced from [31] with permission from the PCCP Owner Societies

more, while their thickness increases slightly, to 100–150 nm. According to the EDX data (see Fig. 4.29*l*), Ag and S contents in the formed surface layer are equal to 87.3 ± 0.5 and 12.7 ± 0.2 wt.%, respectively, and correspond to sulfide $\sim Ag_2S$.

The sequence of formation and growth of argentite nanoparticles and the SEM images of silver sulfide surface variation during acanthite–argentite phase transformation are shown in Fig. 4.30. Sadovnikov et al. conventionally divided this sequence into four stages. First stage is heating of acanthite particles by electron-beam irradiation. Then the phase transformation leads to the appearance of argentite and growth of argentite nuclei on the surface of acanthite particles. In the third stage, the β-Ag_2S argentite nuclei interact with the low-temperature acanthite phase acting as a donor. The growth of argentite particles takes place at the interfaces. In the fourth stage argentite growth is ended due to no low-temperature acanthite remains in the surface layer, as the electron beam intensity is not sufficient for heating a thicker acanthite layer up to the transformation temperature ~ 450 K.

HRTEM and TEM study of silver sulfide nanopowder before and after irradiation confirms the formation of argentite (Fig. 4.31).

In the HRTEM image of silver sulfide before heating, the interplanar distance 0.309 nm is observed. This distance coincides with the spacing $d_{(110)}$ between atomic planes (110) of monoclinic (space group $P2_1/c$) α-Ag_2S acanthite (Fig. 4.31a). Upon radiation heating, the interplanar distance 0.245 nm which is equal to the spacing between the (200) atomic planes of cubic (space group $I\,m\overline{3}m$) β-Ag_2S argentite is well visible in the HRTEM image (Fig. 4.31b).

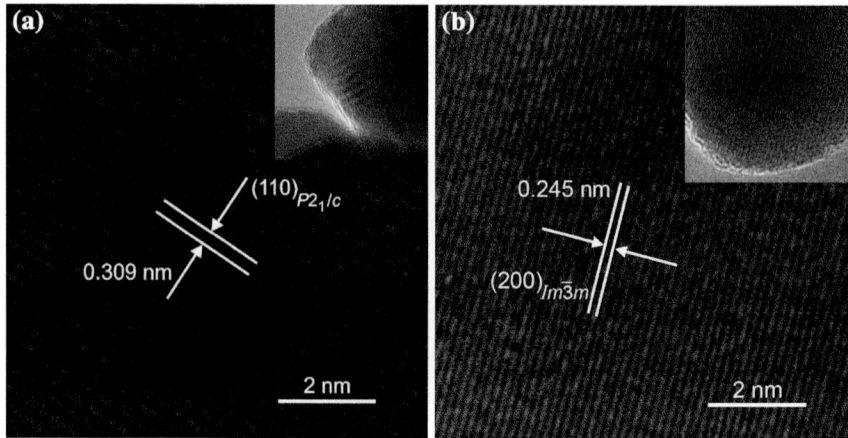

Fig. 4.31 HRTEM images of silver sulfide nanoparticles: **a** the interplanar distance 0.309 nm observed before heating corresponds to monoclinic silver sulfide with α-Ag_2S acanthite type structure; **b** the interplanar distance 0.245 nm observed after heating corresponds to cubic silver sulfide with β-Ag_2S argentite type structure. The *insets* show the TEM images of nanoparticles at smaller magnification. Reproduced from [31] with permission from the PCCP Owner Societies

4.2.4 Thermal Expansion and Heat Capacity of Acanthite α-Ag_2S, Argentite β-Ag_2S, and Cubic γ-Ag_2S Phase

While Ag_2S nanoparticles display electronic properties different from bulk silver sulfide due to the confinement of electrons and holes on the nanometer scale, the thermal properties of nanocrystalline silver sulfide are less affected by the small particle size. Nevertheless the small particle size influences the thermal expansion of silver sulfide. The α-Ag_2S acanthite and β-Ag_2S argentite phases are of most interest for engineering applications. The information about thermal expansion of these silver sulfide phases is very limited, although such data are necessary for application of Ag_2S at elevated temperatures and for the creation of resistance-switches and nonvolatile memory devices, whose operation is based on the acanthite—argentite transformation.

According to [151], the linear thermal expansion coefficient α_{ac} of acanthite is equal to $\sim 20 \times 10^{-6}$ K^{-1}. The temperature range for this coefficient, and the method of its measurement in work [151] are not given. According to [156], the relative thermal expansions $\Delta L/L$ of bulk acanthite in the temperature range 293–450 K and bulk argentite at a temperature from ~ 460 to 570 K linearly depend on temperature. This implies that the linear thermal expansion coefficients of acanthite and argentite in the mentioned temperature ranges are independent of temperature. Indeed, in the temperature range 293–450 K the coefficient α_{ac} of bulk acanthite is equal to 16.8×10^{-6} K^{-1}, and the coefficient α_{arg} of bulk argentite is equal to

$45.8 \times 10^{-6} \, K^{-1}$ at a temperature from ~ 460 to 570 K [156]. Until 2016, the data on the thermal expansion coefficient of nanocrystalline α-Ag_2S acanthite were absent in the literature.

A systematic in situ study of the thermal expansion of coarse-crystalline and nanocrystalline powders of silver sulfide in the region of existence of monoclinic α-Ag_2S acanthite and cubic β-Ag_2S argentite has been performed for the first time in works [35, 155] via the high-temperature X-ray diffraction method.

The synthesis conditions of silver sulfide powders are described in works [31, 35, 155] (see also Sect. 4.2.3).

In situ high-temperature XRD investigation of silver sulfide was performed using a X'Pert PRO MPD (Panalytical) diffractometer with a fast-acting position-sensitive sector detector PIXCEL. Diffractometer was equipped also with a Anton Paar HTK-1200 Oven furnace. The features of the high-temperature XRD experiment with the use of a position-sensitive sector detector PIXCEL are described in Sect. 4.2.3. Reducing the time of recording the XRD patterns due to the use of a detector PIXCEL was also an essential condition, since silver sulfide nanopowder is very hygroscopic, contains adsorbed water, and therefore is quickly oxidized during heating. The HT-XRD measurements were performed in a temperature range 295–623 K with a step of ~ 25–30 K. Each XRD pattern was recorded during 30 min at assigned heating temperature.

The final refinement of the crystal structure of silver sulfide powders was carried out with the use of the X'Pert Plus [143] and X'Pert HighScore Plus [157] software packages.

According to the BET and SEM data, the average particle size D for coarse-crystalline silver sulfide powder is ~ 850 nm and the maximal size reaches ~ 1–2 μm (Fig. 4.32a).

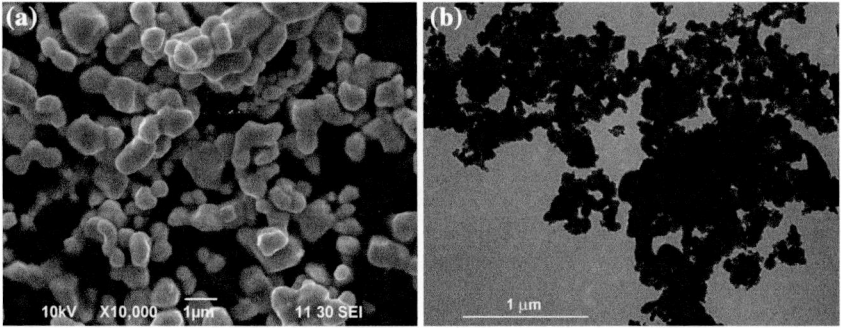

Fig. 4.32 Synthesized coarse-crystalline and nanocrystalline Ag_2S powders: **a** coarse-crystalline Ag_2S powder prepared by hydrothermal synthesis contains a rounded irregular shape particles with a size about 1000 nm; **b** nanocrystalline Ag_2S powder deposited from aqueous solutions of $AgNO_3$, Na_2S and Na_3Cit contains small particles with a size of ~ 45–50 nm. Reproduced from [35] with permission from the PCCP Owner Societies

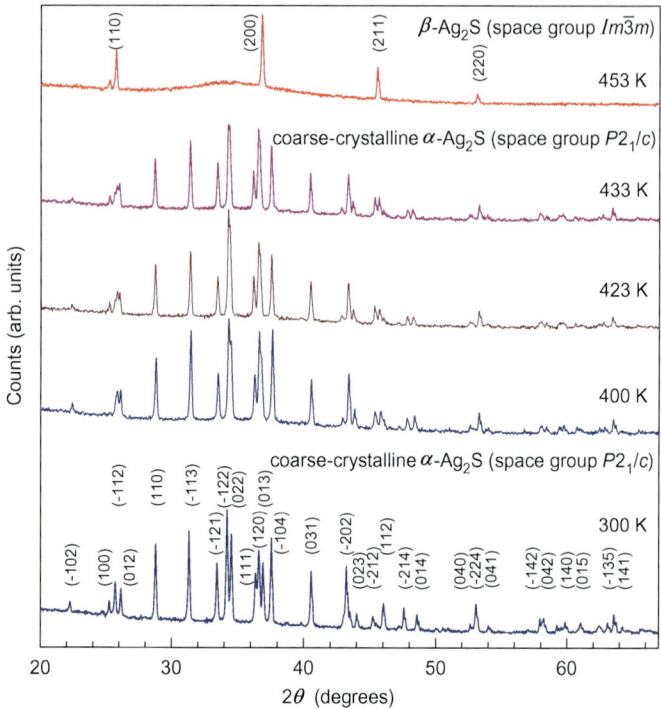

Fig. 4.33 XRD patterns of coarse-crystalline silver sulfide with the average particle size of ~500–800 nm at heating from 300 to 453 K. Transition from α-Ag$_2$S to β-Ag$_2$S is observed at a temperature between 443 and 453 K

The influence of temperature T on the evolution of XRD patterns and on the unit cell parameters of coarse- and nanocrystalline silver sulfide is shown in Figs. 4.33, 4.34 and 4.35.

The XRD patterns of coarse-crystalline silver sulfide recorded in the temperature range 300–433 K (Fig. 4.33) contain the diffraction reflections of only monoclinic (space group $P2_1/c$) stoichiometric acanthite α-Ag$_2$S (see Appendix, Table 4.13). The intensities and positions of the acanthite XRD reflections gradually change during heating up 433 K. A polymorphous phase transformation between monoclinic α-Ag$_2$S acanthite and cubic β-Ag$_2$S argentite occurs at a temperature about 448–453 K (see Sect. 4.2.3). Indeed, XRD pattern at a temperature of 453 K contains already the diffraction reflections of cubic (space group $I\,m\bar{3}m$) argentite β-Ag$_2$S.

Continuous change of XRD patterns of nanocrystalline silver sulfide at heating from 295 to 443 K is shown in Fig. 4.34. The XRD patterns of nanocrystalline silver sulfide recorded at 295, 323, 348, 373, and 398 K contain the same set of broadened diffraction reflections. Qualitative and quantitative analysis of the XRD patterns and also comparison with data [30] reveal that the observed diffraction

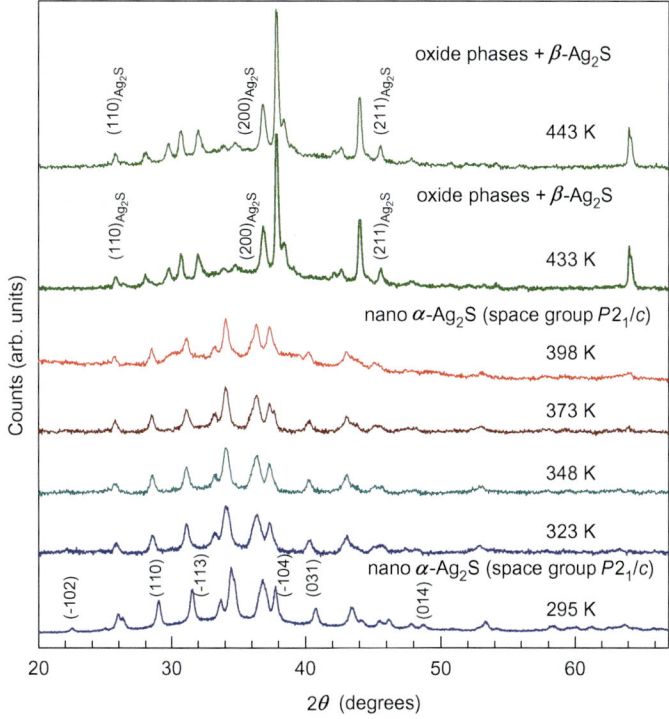

Fig. 4.34 Evolution of XRD patterns of nanocrystalline acanthite α-Ag$_2$S with the average particle size of ~45–50 nm at heating from 300 to 398 K. Nanopowder was oxidized at a temperature ≥433 K, and its XRD patterns contains diffraction reflections of oxide and oxide-sulfate phases along with some silver sulfide reflections

reflections correspond to nanocrystalline nonstoichiometric monoclinic (space group $P2_1/c$) acanthite α-Ag$_{1.93-1.96}$S. The average nanoparticle size D was calculated from the broadening of non-overlapping diffraction reflections (−1 0 2), (1 1 0), (−1 1 3), (−1 0 4), (0 3 1), and (0 1 4). According to these estimates, the average nanoparticle size D in the examined acanthite nanopowders is ~45–50 nm, in good agreement with the TEM data (see Fig. 4.32b).

Nanocrystalline silver sulfide powder is very hygroscopic and even after long-continued drying it contains adsorbed water in an amount of up to 2 wt.%. Therefore its partial oxidation (see Fig. 4.34) by adsorbed water takes place at heating of the nanopowder up to 433 K. The XRD patterns of the nanopowder recorded at ≥433 K contain the diffraction reflections of different oxide and oxide-sulfate phases of silver along with some reflections of cubic argentite β-Ag$_2$S.

The quantitative analysis of the XRD patterns showed that the rise in temperature T leads to gradual variation of all the unit cell parameters (a, b, c, β) of coarse-crystalline and nanocrystalline acanthite (Fig. 4.35) and to an increase in the unit cell volume V.

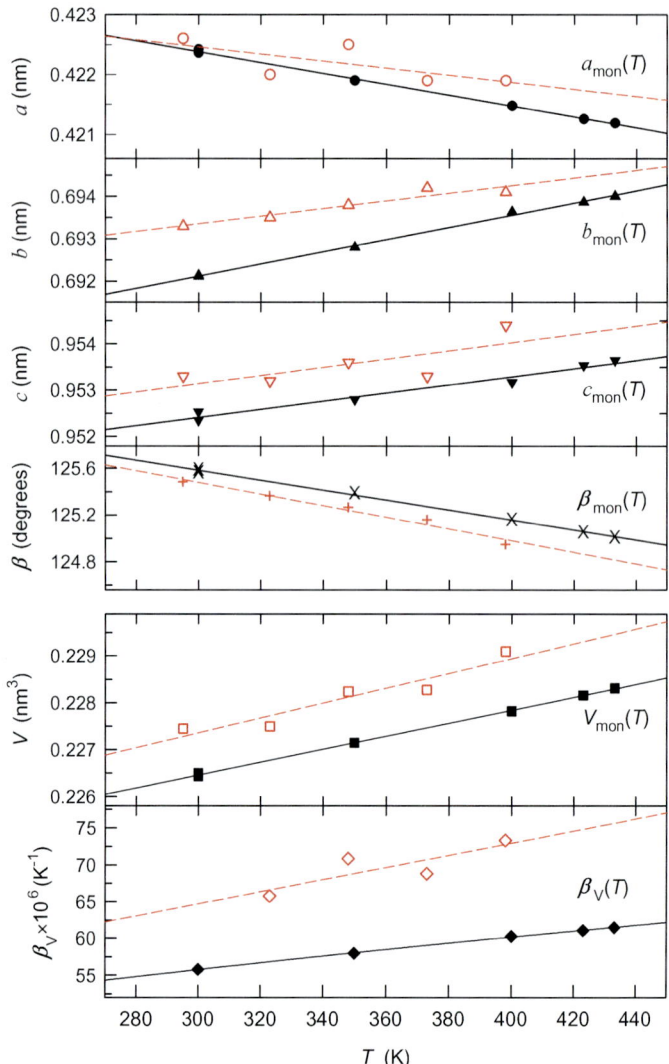

Fig. 4.35 The effect of temperature T on the unit cell parameters a, b, c, β, and volume V, and on the volumetric thermal expansion coefficient β_V of coarse- and nanocrystalline acanthite [35, 155]. The approximation of the experimental data by the solid line, and the closed symbols *filled circle*, *filled triangle*, *inverted filled triangle*, *multiplication sign*, *filled square*, and *filled diamond* correspond to coarse-crystalline acanthite. The approximation of the experimental data by dotted line, and the open symbols *open red circle*, *open red triangle*, *inverted open red triangle*, *red plus sign*, *open red square*, and *open red diamond* correspond to nanocrystalline acanthite. Reproduced from [35] with permission from the PCCP Owner Societies

The volumetric thermal expansion coefficient was determined as

$$\beta_V(T) = \frac{1}{V_{300K}} \frac{\Delta V}{\Delta T} = \frac{V(T) - V_{300K}}{V_{300K}(T - 300)}. \tag{4.4}$$

The temperature dependence of the volumetric thermal expansion coefficient β_V for coarse-crystalline acanthite α-Ag$_2$S is described by the quadratic function

$$\beta_{V_{ac}}(T) = 34.81 \times 10^{-6} + 8.87 \times 10^{-8}T - 62.57 \times 10^{-12}T^2 \pm 2 \times 10^{-6} \left[K^{-1}\right]. \tag{4.5}$$

The isotropic linear thermal expansion coefficient of anisotropic crystal, α_{isotr}, is averaging the thermal expansion over all the crystallographic directions. In this case, the following approximate relationship can be used

$$\alpha_{isotr} \cong \beta_V/3. \tag{4.6}$$

With taking into account (4.5) and (4.6), the isotropic linear thermal expansion coefficient $\alpha_{ac\ isotr}$ of acanthite is

$$\alpha_{ac\ isotr}(T) \cong \beta_{Vac}/3$$
$$= 11.6 \times 10^{-6} + 2.9 \times 10^{-8}T - 20.8 \times 10^{-12}T^2 \pm 1 \times 10^{-6} \left[K^{-1}\right]. \tag{4.7}$$

When the temperature rises from 300 to 433 K, the value of this coefficient $\alpha_{ac\ isotr}$ for coarse-crystalline acanthite increases from $\sim 18.4 \times 10^{-6}$ to $\sim 24.0 \times 10^{-6}$ K^{-1}. This result agrees with the data [151, 156], according to which the linear expansion coefficient of bulk acanthite at a room temperature is equal to 20×10^{-6} K^{-1} or 16.8×10^{-6} K^{-1}, respectively.

Heating leads to the change of all the unit cell parameters (a, b, c, β) and increasing the unit cell volume V of nanocrystalline acanthite. Note that the determination accuracy of the unit cell parameters for nanocrystalline acanthite is much smaller than that for coarse-crystalline acanthite. It follows from the quantitative analysis of the data on the unit cell volume for nanocrystalline acanthite that the volumetric thermal expansion coefficient $\beta_{V\ ac-nano}$ and the isotropic linear thermal expansion coefficient $\alpha_{ac-nano\ isotr}$ in the temperature range 300–400 K can be described by the linear functions

$$\beta_{Vac-nano}(T) = 40.08 \times 10^{-6} + 8.21 \times 10^{-8}T \pm 4 \times 10^{-6} \left[K^{-1}\right] \tag{4.8}$$

and

$$\alpha_{\text{ac-nanoisotr}}(T) = 13.4 \times 10^{-6} + 2.7 \times 10^{-8} T \pm 2 \times 10^{-6} \left[\text{K}^{-1} \right]. \qquad (4.9)$$

The isotropic linear thermal expansion coefficient $\alpha_{\text{ac-nano isotr}}$ of nanocrystalline acanthite is by $\sim 25\%$ larger than the analogous coefficient $\alpha_{\text{ac isotr}}$ of coarse-crystalline acanthite. Such large difference in the $\alpha_{\text{ac-nano isotr}}$ and $\alpha_{\text{ac isotr}}$ coefficients is due to the small particle size in acanthite. Small particle size leads to the variations of the phonon spectrum and its boundaries, and to the change of all thermal (expansion, heat capacity, etc.) of a substance. For example, a considerable difference in the thermal expansion coefficients of nanocrystalline film and coarse-grained sample was observed for lead sulfide PbS [158–160] (see also Sect. 2.3.2).

In studies [159, 160], using the approach [161], it was shown that thermal expansion coefficient of a nanocrystalline substance molecules of which contain n atoms can be presented as

$$\alpha(T, D) = \alpha_{\text{bulk}}(T) + n \frac{\gamma}{3B} \left(\frac{12 k_1 T}{D^2} + \frac{6 k_2 T^2}{D} \right), \qquad (4.10)$$

where $\alpha_{\text{bulk}}(T)$ is the thermal expansion coefficient of bulk coarse-grained substance, γ is the Grüneisen constant, B is the bulk modulus, and D is the particle size. The values k_1 and k_2 are the positive constants, which are associated with the effective propagation velocities of elastic vibrations determined through the velocities of longitudinal and transverse vibrations, c_ℓ and c_t (see details in Sect. 2.3.2).

In the literature, there are no data on the propagation velocities of longitudinal and transverse elastic vibrations, c_ℓ and c_t, for acanthite. Therefore, even a semi-quantitative estimation of the small particles contribution to the acanthite thermal expansion is not possible. Nevertheless, it follows from (4.9) that the small size particles make a positive contribution to the thermal expansion coefficient. The presence of this contribution is experimentally observed in the comparison of thermal expansion of nano- and coarse-crystalline acanthite α-Ag$_2$S.

The thermal expansion coefficient $\alpha(T)$ is related to the coefficient of atomic vibrations anharmonicity, β_{anh}, as $\alpha(T) = k_B \beta_{\text{anh}} / (A_h^2 a_{293K})$, where $a_{293\text{ K}}$ is the crystal lattice constant and A_h is a constant. In the first approximation, the averaged lattice constant of monoclinic acanthite can be calculated as $V^{1/3}$, where V is a volume of the unit cell of acanthite. According to the formula (2.56), deduced in study [159], the anharmonicity coefficient β_{anh} depends on the particle size D (see Sect. 2.3.2).

It follows from (2.56) that reduction of the particle size D in Ag$_2$S should be accompanied by some enhancement of the anharmonicity. Earlier an essential increase of the role of anharmonicity of atomic vibrations for nanomaterials as compared with microstructure was noted in works [145, 159].

The crystal lattice parameters of acanthite remain unchanged when size of particles decreases. If the coefficient A_h remains unchanged too when the particle size D decreases, then the ratio of anharmonicity coefficients for nano- and coarse-crystalline acanthite α-Ag_2S is $\beta_{anh\text{-}nano}/\beta_{anh} = \alpha_{ac\text{-}nano\ isotr}/\alpha_{ac\ isotr}$, hence $\beta_{anh\text{-}nano} = (\alpha_{ac\text{-}nano\ isotr}/\alpha_{ac\ isotr})\beta_{anh}$. At room temperature, the coefficients $\alpha_{ac\ isotr}$ and $\alpha_{ac\text{-}nano\ isotr}$ for coarse-crystalline and nanocrystalline acanthite are 18.6×10^{-6} and 22.6×10^{-6} K^{-1}, respectively. Therefore $\beta_{anh\text{-}nano} \approx 1.2\beta_{anh}$. It means that decreasing α-Ag_2S particle size leads to the enhancement of the anharmonicity.

Let us consider now the thermal expansion of argentite β-Ag_2S.

Heating coarse-crystalline silver sulfide up to 453 K leads to acanthite–argentite phase transformation.

The change of XRD patterns and the unit cell parameter of cubic β-Ag_2S argentite with the temperature variation are presented in Figs. 4.36 and 4.37.

The XRD patterns recorded at heating from 453 K and above contain the reflections of main cubic (space group $Im\bar{3}m$) argentite β-Ag_2S (Fig. 4.36). The XRD patterns collected at 443, 453, 463, and 473 K contain very weak trace of reflection $(100)_{mon}$ of monoclinic α-Ag_2S acanthite to the left of reflection (110) of cubic argentite. The temperature increasing leads to displacement of the XRD reflections of cubic argentite into the region of smaller 2θ angles and larger interplanar distances (Fig. 4.36, inset), i.e. to gradual increasing the lattice constant a_{arg} of argentite.

The temperature dependence of the lattice constant of argentite β-Ag_2S, $a_{arg}(T)$, is not linear (Fig. 4.37) and in the temperature region 443–623 K is described by quadratic function

$$a_{arg}(T) = a_0 + a_1T + a_2T^2, \qquad (4.11)$$

with $a_0 = 0.46747$ nm, $a_1 = 5.6086 \times 10^{-5}$ nm K^{-1},   and $a_2 = -3.3873 \times 10^{-8}$ nm K^{-2}.

The linear thermal expansion coefficient α was determined as an average coefficient in the temperature interval between the temperature T of measurement and the initial temperature of argentite appearance which is equal 443 K, i.e.

$$\alpha(T) = \frac{1}{a_{443K}}\frac{\Delta a}{\Delta T} = \frac{a(T) - a_{443K}}{a_{443K}(T - 443)}, \qquad (4.12)$$

where $a(T)$ and $a_{443\ K}$ are the lattice constants of cubic argentite at temperature T and at the initial temperature 443 K, respectively.

The temperature dependence of the thermal expansion coefficient of argentite, $\alpha_{arg}(T)$, in the interval 443–623 K can be represented through the coefficients of the polynomial (4.9) and has the form

$$\alpha_{arg} = 84.5 \times 10^{-6} - 6.9 \times 10^{-8}T \pm 3 \times 10^{-6}\left[K^{-1}\right]. \qquad (4.13)$$

At a temperature of 523 K, the thermal expansion coefficient α_{arg} of bulk argentite is 45.8×10^{-6} K^{-1} [156]. The thermal expansion coefficient α_{arg} also can

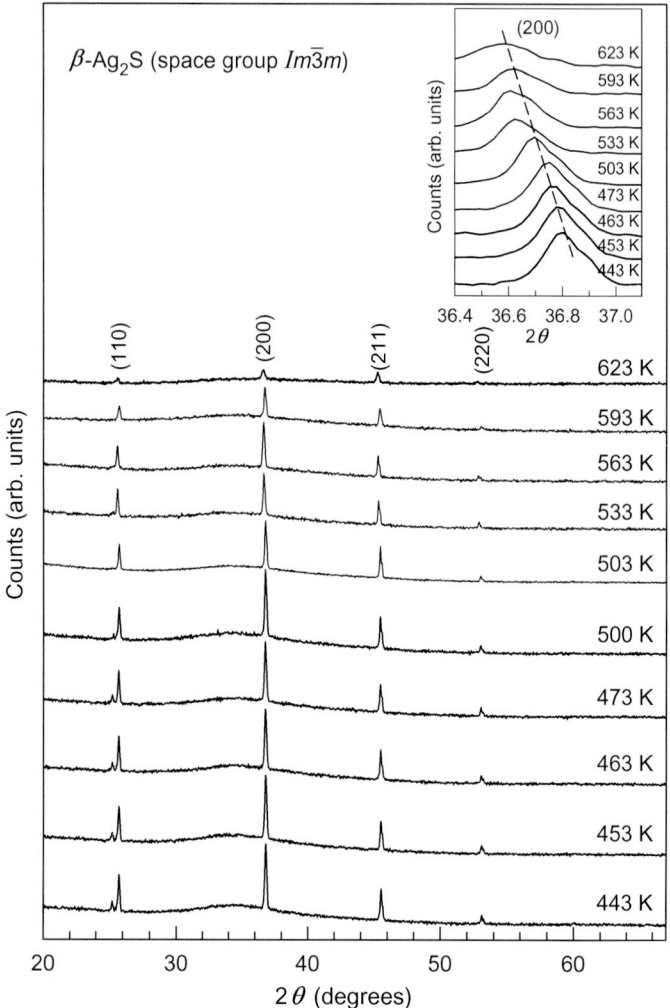

Fig. 4.36 Evolution of XRD patterns of coarse-crystalline argentite β-Ag$_2$S at heating from 446 to 623 K. The *inset* shows a systematic displacement of the (200) diffraction reflection of cubic argentite with increase of measuring temperature. Reproduced from [35] with permission from the PCCP Owner Societies

be estimated from the data [26, 148] on the argentite lattice constant for different temperatures. Such estimations give the coefficients α_{arg} which are equal to $\sim 43 \times 10^{-6}$ and $\sim 37 \times 10^{-6}$ K^{-1}, respectively. These α_{arg} values coincide within the limits of measurement error with the values of α_{arg} found for temperature interval from 443 to 623 K in study [31].

Observed small decrease in thermal expansion coefficient α_{arg} is caused by the peculiarities of crystal structure of argentite β-Ag$_2$S, in which silver atoms are

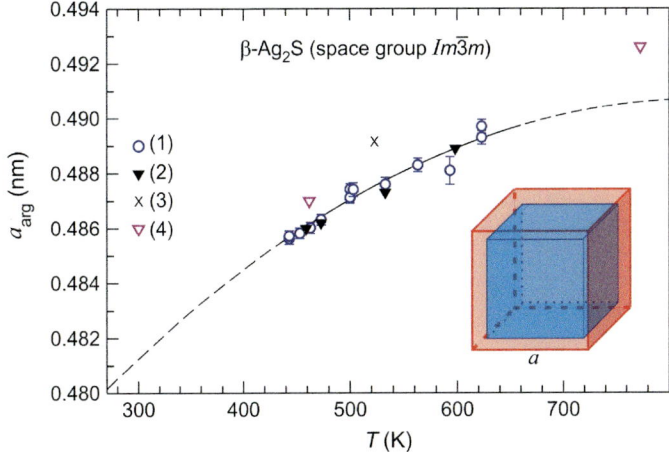

Fig. 4.37 Dependence of the lattice constant a_{arg} of argentite β-Ag$_2$S on the temperature T: (*1*) data [35]; (*2*) data [26], (*3*) data [27], and (*4*) data [152], respectively. The approximation of measured lattice constant $a_{arg}(T)$ by the function (4.11) in the temperature range 440–660 K is shown by *solid lines*. The *inset* shows increase in a cubic argentite unit cell as a result of temperature growth. Reproduced from [35] with permission from the PCCP Owner Societies

statistically distributed in 6(*b*) and 48(*j*) positions with the probabilities less than 0.1 (see Table 4.9). The amount of Ag atoms in argentite β-Ag$_2$S is much smaller than the number of sites of metal sublattice, therefore significant positional disorder in an Ag arrangement and gigantic (more than 92%) concentration of vacant sites facilitate jumping of silver cations and provide superionic conductivity of β-Ag$_2$S phase.

Also, small decreasing coefficient α_{arg} can be due to a large measurement step (25–30 K) of the thermal expansion in works [35, 155]; as a result, an overestimated value of $\alpha_{aver}(443\ K)$ has been registered immediately in the region of the phase transformation, i.e. in the region of the expansion coefficient discontinuity.

Additionally, direct dilatometric measurements of the thermal expansion of coarse- and nanocrystalline silver sulfide at temperature from 290 to 970 K in the region of existence of acanthite α-Ag$_2$S, argentite β-Ag$_2$S and γ-Ag$_2$S phase have been carried out. The thermal expansion coefficient was measured on cylindrical samples 5 and 10 mm in diameter pressed from Ag$_2$S powders under a pressure of $\sim$260 MPa. The dilatometric measurements were made by means of a Linseis L75/1250 dilatometer under vacuum 0.0026 Pa in the temperature range 293–985 K with a step 1 K and also on a NETZSCH DIL 402C dilatometer in He atmosphere at a temperature from 293 to 923 K with a step of 0.5 K. The heating rate was 4 K min^{-1}. Also heat capacity C_p of nanocrystalline silver sulfide Ag$_2$S in the temperature range 300–930 K was measured on a STA 449 thermal analyzer in argon Ar atmosphere with a step of 0.5 K.

The temperature dependences $\alpha_{aver}(T)$ for coarse- and nanocrystalline Ag$_2$S samples measured by dilatometric method are presented in Fig. 4.38.

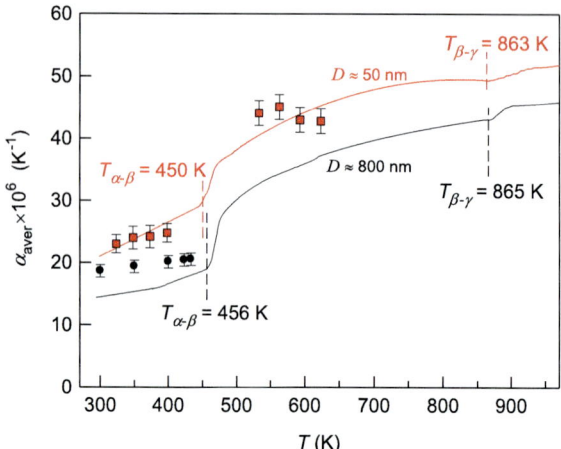

Fig. 4.38 The average thermal expansion coefficients α_{aver} of coarse- and nanocrystalline silver sulfide in the temperature range 293–970 K. The thermal expansion coefficient α_{aver} of coarse-crystalline (*filled circle*) and nanocrystalline (*filled red square*) silver sulfides measured in works [35, 155] by the high-temperature XRD method are shown for comparison. Temperatures $T_{\alpha-\beta}$ and $T_{\beta-\gamma}$ of α-Ag$_2$S–β-Ag$_2$S и β-Ag$_2$S–γ-Ag$_2$S phase transitions are marked by *vertical dashed lines*

Nanocrystalline Ag$_2$S sample produced from the powder with an average nanoparticle size of ~50 nm has the largest coefficient $\alpha_{aver}(T)$ in the examined temperature range 293–970 K. Explicit discontinuities are observed on the $\alpha(T)$ dependences near α-Ag$_2$S–β-Ag$_2$S and β-Ag$_2$S–γ-Ag$_2$S phase transformations. This allowed us to find more accurately the transition temperatures $T_{\alpha-\beta} = 456 \pm 5$ K and $T_{\beta-\gamma} = 865 \pm 10$ K. For comparison, the average linear thermal expansion coefficients α_{aver} of coarse-crystalline and nanocrystalline acanthite α-Ag$_2$S and argentite β-Ag$_2$S measured in studies [35, 155] by the high-temperature XRD method are shown in Fig. 4.38. With taking into account the experimental errors and different measurement methods, all the results on α_{aver} satisfactorily agree with each other.

The heat capacity C_p of Ag$_2$S nanopowder changes rather monotonically with increasing temperature, except for the transition regions (Fig. 4.39). In the temperature range from 300 to 450 K, the heat capacity increases and then, near the transition temperature $T_{\alpha-\beta}$, it experiences a discontinuity. In the following temperature interval from ~470 to ~840 K, where β-Ag$_2$S phase is exist, the heat capacity first decreases slightly to ~670 K and then grows slightly to the transition temperature $T_{\beta-\gamma}$, where C_p experiences second discontinuity. As the temperature increases further to ~890 K, a small reduction of C_p is observed, and at $T > 890$ K the heat capacity grows slightly. According to the C_p measurements, the transition temperatures $T_{\alpha-\beta}$ and $T_{\beta-\gamma}$ of the $\alpha_{-\beta}$ and $\beta_{-\gamma}$ transformations are equal to 451 and 858 K, which agree with the data on thermal expansion for silver sulfide.

Fig. 4.39 The heat capacity of silver sulfide nanopowder. The *inset* shows spasmodic change of C_p (*filled circle*) for Ag$_2$-S nanopowder in the regions of α-Ag$_2$S–β-Ag$_2$S and β-Ag$_2$S–γ-Ag$_2$S transformations. For comparison, heat capacity (*open circle*) of coarse-grained Ag$_2$S [152] is shown

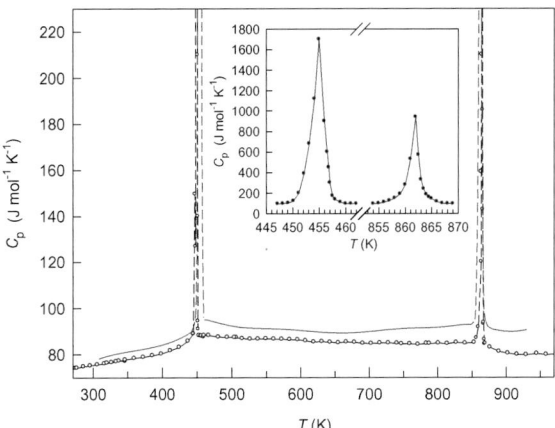

The experimental data on the heat capacity of nanocrystalline silver sulfide Ag$_2$S in the temperature range 298–450 K, where the equilibrium phase is monoclinic acanthite α-Ag$_2$S, are approximated by the following function

$$C_p(\alpha - Ag_2S) = 161.812 - 0.154T + 0.136 \times 10^{-3}T^2 - 15161.76/T \ (J\,mol^{-1}\,K^{-1}).$$

The heat capacity of nanocrystalline silver sulfide Ag$_2$S in the region of existence of argentite β-Ag$_2$S at temperatures from 470 to 840 K is described by the function

$$C_p(\beta - Ag_2S) = 318.777 - 0.420T + 0.247 \times 10^{-3}T^2 - 39419.86/T \ (J\,mol^{-1}\,K^{-1}).$$

The heat capacity of nanocrystalline silver sulfide Ag$_2$S in the region of existence of the cubic phase γ-Ag$_2$S at temperatures from 890 to 930 K is described by the function

$$C_p(\gamma - Ag_2S) = 66.395 + 0.027T \ (J\,mol^{-1}\,K^{-1})(890 - 930\ K).$$

The heat capacity C_p peaks for Ag$_2$S nanopowder in the α–β and β–γ transformation regions are symmetric rather than λ-shaped (Fig. 4.39, inset). Observed symmetric shape of C_p peaks is more characteristic of the phase transformations of first order. The width of C_p peaks base is about 8 K, i.e. peaks are very narrow that also confirms the first order of phase transformations.

The data obtained show that decreasing the silver sulfide particle to the nanosized scale leads to enhanced values of heat capacity C_p at comparable temperatures. The presence of discontinuities on the $C_p(T)$ temperature dependence in the vicinity of the α-Ag$_2$S–β-Ag$_2$S and β-Ag$_2$S–γ-Ag$_2$S phase transformation temperatures suggests these transformations should take place by the first-order phase transition mechanism.

The enthalpies of the phase transformations α-Ag$_2$S–β-Ag$_2$S and β-Ag$_2$S–γ-Ag$_2$S were estimated to be $\Delta H_{\alpha-\beta} = 4.2 \pm 0.4$ and $\Delta H_{\beta-\gamma} = 1.2 \pm 0.3$ kJ mol^{-1}, respectively. According to [148, 149, 152] and [31, 34, 35], the enthalpy $\Delta H_{\alpha-\beta}$ is 3.98, 3.93, 4.06 and 3.7–3.9 kJ mol^{-1}, respectively. As reported in [149, 152], the enthalpy $\Delta H_{\beta-\gamma}$ of the transformation of argentite β-Ag$_2$S into γ-Ag$_2$S phase is ~ 0.50 and ~ 0.78 kJ mol^{-1}, respectively. The found values of enthalpies $\Delta H_{\alpha-\beta}$ and $\Delta H_{\beta-\gamma}$ of these phase transformation are rather close to the literature data.

For comparison, Fig. 4.39 presents experimental heat capacity C_p of nano-crystalline silver sulfide Ag$_2$S and the most reliable literature experimental data on the C_p of coarse-grained Ag$_2$S [152]. The heat capacity C_p of nanocrystalline silver sulfide is by 1–4% larger than that of coarse-crystalline sulfide. The observed difference in the thermal expansion coefficient α and heat capacity C_p is due to a small size of Ag$_2$S particles in nanocrystalline silver sulfide leading to the restriction of the phonon spectrum on the side of low and high frequencies.

4.3 Universal Approach to Synthesis of Silver Sulfide in the Form of Nanopowders, Quantum Dots, Core-Shell Nanoparticles, and Heteronanostructures

Preparation of substances and materials in nanocrystalline state is one of the main trends in modern materials science. Nanostructured sulfides for different applications were synthesized by various methods including hydrochemical deposition, microwave, sonochemical, hydrothermal, solvothermal and electrochemical methods and so on.

The chemical deposition from aqueous solutions which is named also chemical condensation method, *one-pot* synthesis in aqueous solutions and hydrochemical bath deposition is the most popular method used for the synthesis of nanocrystalline sulfide powders.

Hydrochemical bath deposition is a well-known method which allows preparing colloidal solutions of Ag$_2$S nanoparticles, nanocrystalline and coarse-crystalline silver sulfide powders, isolated Ag$_2$S nanoparticles, and different heteronanostructures with Ag$_2$S.

Usually nanostructured silver sulfide synthesizes by hydrochemical deposition from aqueous solutions of silver nitrate AgNO$_3$ and sodium sulfide Na$_2$S. Solutions of sodium citrate Na$_3$Cit or disodium salt of ethylenediaminetetraacetic acid (Trilon B) use as the complexing agents. Coarse-crystalline Ag$_2$S powder prepares by hydrothermal synthesis from aqueous reaction mixture of AgNO$_3$, Na$_2$S and Na$_3$Cit with the subsequent heating of a matrix solution with the precipitated powder in the closed vessel at the elevated temperature and pressure.

Hydrochemical deposition of Ag_2S nanopowders with using of Trilon B, and also hydrothermal synthesis of Ag_2S powder are described in study [63].

Silver sulfide deposition in the presence of Trilon B has occurred according to the following reaction scheme:

$$2AgNO_3 + Na_2S \xrightarrow{EDTA-H_2Na_2} Ag_2S \downarrow + 2NaNO_3. \tag{4.14}$$

The concentrations of $AgNO_3$ and Na_2S in different reaction mixtures were 50 and 25 mmol l^{-1}, respectively, and the concentration of Trilon B was varied from 35 to 50 mmol l^{-1}. Trilon B was added with constant stirring to a silver nitrate solution, and the prepared solution was then mixed with sodium sulfide solution. During the mixing of the solutions, a sulfide formation reaction occurs instantaneously. All nanoparticles deposited during two days.

A coarse-crystalline Ag_2S powder was prepared by hydrothermal synthesis using a reaction mixture with $AgNO_3$, Na_2S, and Na_3Cit concentrations of 50, 400, and 5 mmol l^{-1}, respectively, followed by heating the matrix solution with the deposited powder in a closed vessel at a temperature of 373 K for 2 h. The air and saturated vapor pressure over the solution reached $\sim 2.1\ 10^5$ Pa.

According to [63], all the synthesized silver sulfide powders have monoclinic (space group $P2_1/c$) α-Ag_2S acanthite-type structure.

The average size D of silver sulfide nanoparticles in the deposited Ag_2S nanopowders estimated from broadening of non-overlapping diffraction reflections is equal to 58 ± 8 and 59 ± 8 nm. According to the BET data, the silver sulfide nanoparticles sizes which were deposited from the reaction mixtures with Trilon B concentrations equal 35 and 50 mmol l^{-1} are 60 ± 5 and 63 ± 5 nm, respectively. Thus, increasing the Trilon B concentration in the reaction mixture led to deposition of slightly larger Ag_2S particles. According to the BET data, the particle size of coarse-crystalline Ag_2S powder is 975 ± 15 nm.

The EDX analysis has shown that within the limits of measurement error, silver sulfide nanopowders deposited with Trilon B contain 89.4 ± 0.4 wt.% Ag and 10.6 ± 0.1 wt.% S. The content of silver in these powders exceeds the possible concentration of silver in stoichiometric monoclinic sulfide Ag_2S. This means that excessive silver is present in synthesized Ag_2S nanopowders as impurity phase. Indeed, on the XRD patterns of nanopowders deposited with Trilon B, the weak diffraction reflections of silver were observed. The growth of Trilon B concentration in the reaction mixtures from 35 to 50 mmol l^{-1} is accompanied by increasing the amount of Ag impurity in deposited nanopowders up to 3.0 and 4.5 wt.%, respectively. No metallic silver impurity was detected in the coarse-crystalline silver sulfide powder prepared by hydrothermal synthesis.

On the whole, disadvantages of the chemical deposition of silver sulfide from aqueous $AgNO_3$, Na_2S, and Trilon B solutions are the large size of prepared Ag_2S nanoparticles and the presence of a considerable amount of metallic silver impurity.

Hydrochemical bath deposition with using sodium citrate Na_3Cit is a well-known, simple and reliable universal green method which allows preparing non-toxic colloidal solutions of Ag$_2$S nanoparticles, isolated Ag$_2$S nanoparticles with protective shell, Ag$_2$S/Ag heteronanostructures, nanocrystalline and coarse-crystalline powders of silver sulfide. Weak aqueous solution of AgNO$_3$ silver nitrate, which is widely applied in pharmacology and medicine and possesses antibacterial action, is usually used as a source of silver ions for Ag$_2$S synthesis. It is noteworthy that all forms of nanostructured silver sulfide can be prepared from the same chemical reagents by varying only their concentrations in solution and the conditions of synthesis also. The important advantage of hydrochemical deposition in comparison with other methods is a reproducibility of obtained results. Study [163] has generalized and continued a cycle of systematic studies of nanostructured silver sulfide [29–35, 63, 64, 155, 164–168] beginning with synthesis conditions and ending with crystal structure and properties of different forms of nanostructured Ag$_2$S, and has opened previously unknown possibilities of hydrochemical bath deposition for preparing nanostructured Ag$_2$S.

Hydrochemical deposition of nanostructured Ag$_2$S with using sodium citrate Na$_3$Cit is an example of green chemistry as the design of chemical products and processes that reduce or eliminate the use and generation of hazardous substances [169]. Indeed, hydrochemical bath deposition allows one to obtain valuable products from harmless substances using environmentally friendly methods.

4.3.1 Green Synthesis of Nanostructured Ag$_2$S Without Hazardous Substances

In study [163], different forms of nanostructured silver sulfide Ag$_2$S has been synthesized by hydrochemical bath deposition from aqueous solutions of silver nitrate AgNO$_3$ and sodium sulfide Na$_2$S used as sources of silver Ag$^+$ and sulfide S^{2-} ions. Trisodium citrate dihydrate (sodium citrate) Na$_3$C$_6$H$_5$O$_7$·2H$_2$O≡Na$_3$Cit was used as a complexing agent and electrostatic stabilizer. All reagents with ACS highest purity used without any further purification. All aqueous solutions have been prepared using high quality pure deionized water.

The zeta potential ζ and size (hydrodynamic diameter D_{dls}) of Ag$_2$S nanoparticles in the colloidal solutions were determined by non-invasive Dynamic Light Scattering (DLS) on a Zetasizer Nano ZS facility at 298 K. The He–Ne laser with radiation wavelength 633 nm was used as a radiation source.

The turbidity of colloidal solutions was measured by a HI 93703 Turbidity Meter (Hanna Instruments) in NTU (Nephelometric Turbidity Unit) at 298 K.

All the samples synthesized were examined by X-ray diffraction (XRD) , Energy Dispersive X-ray (EDX) analysis, X-ray photoelectron spectroscopy (XPS), Raman scattering spectroscopy, UV-vis spectroscopy, scanning electron microscopy (SEM), high-resolution transmission electron microscopy (HRTEM), and Brunauer-Emmett-Teller (BET) methods. The average particle size D (to be more precise, the average size of coherent scattering regions (CSR)) in deposited silver sulfide powders was estimated by XRD method from the diffraction reflection broadening [145, 147].

The microstructure, size and element chemical composition of Ag_2S particle were studied by the scanning electron microscopy (SEM) method on a JEOL-JSM LA 6390 microscope coupled with a JED 2300 Energy Dispersive X-ray Analyzer. Nanostructured silver sulfide was examined by transmission electron microscopy (TEM) method. The high-resolution TEM images were recorded on a JEOL JEM-2010 transmission electron microscope with 0.14 nm (1.4 Å) lattice resolution.

Photoluminescence (PL) emission spectra of colloidal silver sulfide solutions were collected on a fluorescence spectrometer (HITACHI, F-4600) at 298 K. UV-vis absorption spectra of colloidal Ag_2S nanoparticles and optical properties of the Ag_2S deposits were recorded on a Shimadzu UV-2401 PC spectrophotometer at a temperature of 298 K.

The solubility product K_{sp} of silver sulfide Ag_2S is very small (at 298 K $K_{sp} = 6.3 \times 10^{-50}$ [170]), and silver sulfide Ag_2S is formed from aqueous solution of $AgNO_3$ and Na_2S on simple reaction

$$2AgNO_3 + Na_2S = Ag_2S \downarrow + 2NaNO_3, \qquad (4.15)$$

where concentrations of sulfide S^{2-} and silver Ag^+ ions are related by the relation $C_{S^{2-}} = C_{Ag^+}/2$. Small variation in the ratio of reagent concentrations as compared with perfect reaction (4.15), introduction of a complexing agent, and control over synthesis condition allow to prepare different forms of silver sulfide—from coarse-crystalline powder to colloidal nanoparticles.

A generalized scheme of synthesis of different types of nanostructured Ag_2S silver sulfide and Ag_2S/Ag heteronanostructures is shown in Fig. 4.40.

Coarse-crystalline silver sulfide Ag_2S is deposited almost instantly from an aqueous solution of $AgNO_3$ and Na_2S with excessive concentration of sulfide S^{2-} ions ($C_{S^{2-}} \geq C_{Ag^+}$) and without addition of a complexing agent (Fig. 4.40a). If the concentration of sulfide S^{2-} ions is sufficient or differs slightly from the concentration required for the chemical bonding all silver Ag^+ ions, i.e. $C_{S^{2-}} \approx C_{Ag^+}/2$, then addition of sodium citrate to the solution promotes the formation of silver sulfide nanoparticles and quantum dots (Fig. 4.40b). Deposition of silver sulfide takes place in neutral medium at pH ≈ 7 by the following reaction scheme

$$2AgNO_3 + (1 \pm \delta)\, Na_2S \xrightarrow{Na_3C_6H_5O_7} Ag_2S \downarrow + 2NaNO_3, \qquad (4.16)$$

where $\delta \geq 0$.

In aqueous solutions of $AgNO_3$ and Na_2S, sodium citrate plays a triple role.

First, it is a complexing and stabilizing agent during deposition of Ag_2S nanoparticles, which occurs both in the light and in the dark (Fig. 4.40b). Second, during deposition in the dark, Na_3Cit is adsorbed on silver sulfide nanoparticles impeding their agglomeration. In this case, increased duration of deposition and the

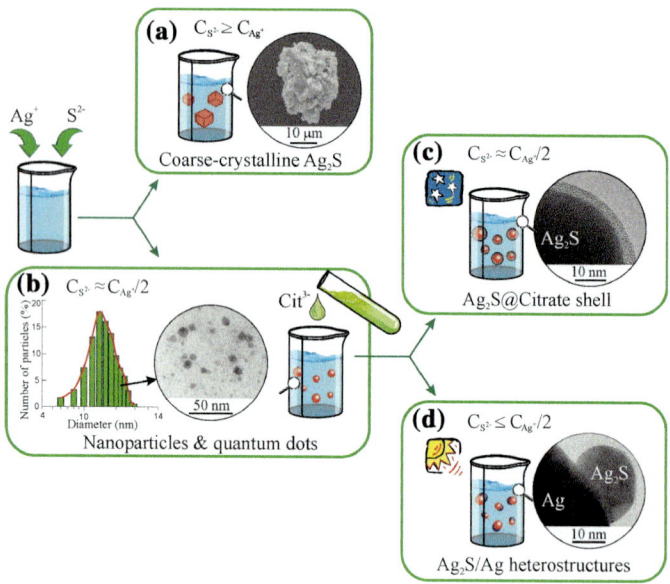

Fig. 4.40 Generalized scheme of synthesis of various types of nanostructured Ag₂S silver sulfide and Ag₂S/Ag heteronanostructures: **a** coarse-crystalline Ag₂S; **b** Ag₂S quantum dots; **c** Ag₂S@C core-shell nanoparticle with carbon-containing citrate shell; **d** Ag₂S/Ag heteronanostructures. Reproduced from [163] with permission from Wiley

use of reaction mixtures with enhanced concentration of Na₃Cit lead to the formation of a protective citrate shell on the surface of Ag₂S nanoparticles (Fig. 4.40c). Third, during deposition in the light in aqueous solutions with lowered content of S^{2-} ions, sodium citrate can reduce Ag^+ ions to metallic silver [171]. During deposition in the light, sodium citrate, as a reducing agent, takes part in the photochemical reaction

$$C_6H_5O_7^{3-} + 2Ag^+ \xrightarrow{h\nu} C_5H_4O_5^{2-} + CO_2 + H^+ + 2Ag \downarrow, \qquad (4.17)$$

in which $C_6H_5O_7^{3-}$ citrate ions reduce the Ag^+ ions present in the solution to metallic Ag nanoparticles and transform to acetone-1,3-dicarboxylate $C_5H_4O_5^{2-}$ ions. In this case, selection of the concentrations of AgNO₃, Na₂S, and Na₃Cit in the initial solution meeting the condition $C_{S^{2-}} \leq C_{Ag^+}/2$ makes it possible to deposit Ag nanoparticles along with Ag₂S sulfide nanoparticles and to synthesize Ag₂S/Ag heteronanostructures (Fig. 4.40d).

In principle, citric acid (harmless standardized food additive E330) can be used as a complexing agent for Ag^+ ions, but its application in silver sulfide synthesis displaces the equilibrium into the acid region, which is not desirable.

Let us consider the peculiarities of synthesis of each type of silver sulfide in greater detail.

4.3.2 Coarse-Crystalline Ag₂S Powders

Coarse-crystalline silver sulfide powders have been synthesized from aqueous solutions of $AgNO_3$ and Na_2S with a large excess concentration of sodium sulfide both without and with addition of sodium citrate (Table 4.10, reaction mixtures 1–4) at room temperature in the dark. XRD patterns of synthesized coarse-crystalline silver sulfide powders are shown in Fig. 4.41a. According to BET data, an average particle size of Ag_2S powders deposited from the reaction mixtures 1, 2, 3, and 4 is equal from ~ 1000 to ~ 200 nm (Table 4.10).

SEM images of coarse-crystalline Ag_2S powder at different magnification of 2000, 10000, and 20000 times are shown in Fig. 4.41b. This powder is deposited from the reaction mixture 2 with concentrations of $AgNO_3$, Na_2S and Na_3Cit which are equal to 50, 500 and 5 mmol l^{-1}, respectively. Most of the particles have an elongated dumbbell shape and they present an aggregate of two roundish particles. Elongated particles have a length about 400–500 nm and thickness about 200 nm. Particles with a rounded shape have a size of 200–250 nm.

The crystal structure of the low-temperature monoclinic α-Ag_2S phase is very complex. The crystal structure of a single crystal of natural acanthite was determined in study [8], where it was shown that the unit cell of acanthite is monoclinic and belongs to the space group $P2_1/n$ ($P12_1/n1$). Later it was shown that the monoclinic unit cell of natural acanthite mineral α-Ag_2S belongs to the space group $P2_1/c$ ($P12_1/c1$) [22].

Recently, the crystal structure of artificial coarse-crystalline silver sulfide was carefully determined in study [29]. According to [29], artificial coarse-crystalline silver sulfide powder with average particle size of ~ 500 nm has monoclinic (space group $P2_1/c$) α-Ag_2S acanthite-type structure and is stoichiometric.

Analysis of the XRD patterns of synthesized coarse-crystalline silver sulfide powders (Fig. 4.41a) has shown that these powders are one-phase and have the same monoclinic (space group $P2_1/c$) structure of α-Ag_2S acanthite-type (see Sect. 4.2.1 and Table 4.7). The degrees of occupancy of all crystal lattice sites by Ag and S atoms are equal to 1.0. The unit cell of artificial monoclinic (space group $P2_1/c$) silver sulfide with α-Ag_2S acanthite-type structure is displayed in Fig. 4.21b.

Synthesis from reaction mixture 1 with excessive concentration of Na_2S and without addition of sodium citrate led to the deposition of Ag_2S powder with particle size of ~ 1000 nm. When Na_3Cit was added to reaction mixtures 2, 3, and 4 having excessive concentration of Na_2S, the average size of Ag_2S particles decreased, but did not reach the nanometer scale (Table 4.10).

The Energy Dispersive X-ray Spectroscopy (EDXS) has shown that the content of silver Ag and sulfur S in synthesized coarse-crystalline silver sulfide powders corresponds to stoichiometric sulfide Ag_2S. As an example, EDX analysis of Ag and S distributions in Ag_2S powder deposited from the reaction mixture 2 (see Table 4.10) is shown in Fig. 4.41b. The content of Ag and S in this powder with average particle size of ~ 500 nm is equal to $\sim 86.8 \pm 0.4$ and $\sim 12.9 \pm 0.1$ wt.

Table 4.10 Composition of the reaction mixtures, average particle size D (nm) of silver sulfide Ag$_2$S in the deposited powders, synthesized colloidal solutions, core-shell nanoparticles, and heteronanostructures

| Type of nano-structured Ag$_2$S | No. | Concentration of reagents in the reaction mixture (mmol l^{-1}) | | | Dark/light[a] | D (nm) in deposited powders | | D and h_{shell} (nm) in colloidal solutions | | |
| | | AgNO$_3$ | Na$_2$S | Na$_3$Cit | | BET | XRD | DLS | TEM | |
								D	D	h_{shell}
Coarse-crystalline particles	1	50	200	0	Dark	1008	–	10 ± 3	7–15	–
	2	50	500	5	Dark	515	–	–	–	–
	3	50	100	25	Dark	430	–	–	–	–
	4	50	50	100	Dark	163	85 ± 7	–	–	–
Nano-particles	5	50	25 + δ_2^b	12.5	Dark	56 ± 5	46 ± 8	55 ± 10	–	–
	6	50	25 + δ_2^b	25	Dark	44 ± 5	43 ± 6	60 ± 10	–	–
	7	50	25 + δ_2^b	100	Dark	53 ± 5	49 ± 8	66 ± 10	–	–
Quantum dots	8	0.3125	0.16 + δ_1^b	5	Dark	No deposits	–	2.3 ± 1	–	–
	9	0.3125	0.16 + δ_1^b	2.5	Dark	No deposits	–	2.7 ± 1	–	–
	10	0.3125	0.16 + δ_1^b	1	Dark	No deposits	–	3.1 ± 1	–	–
	11	0.625	0.3125 + δ_1^b	5	Dark	No deposits	–	4.2 ± 2	–	–
	12	0.625	0.3125 + δ_1^b	3.75	Dark	No deposits	–	5.6 ± 2	–	–
	13	2.5	1.25 + δ_1^b	1	Dark	No deposits	–	8.0 ± 2	–	–
	14	1.25	0.625 + δ_1^b	1.25	Dark	No deposits	–	8.2 ± 2	–	–
	15	0.625	0.3125 + δ_1^b	2.5	Dark	No deposits	–	9.2 ± 2	–	–
	16	0.625	0.3125 + δ_1^b	1.25	Dark	No deposits	–	10.0 ± 2	–	–
	17	2.5	1.25 + δ_1^b	2.5	Dark	No deposits	–	15.0 ± 3	–	–
	18	0.625	0.3125 + δ_1^b	15	Dark	No deposits	–	16.0 ± 4	–	–
	19	1.25	0.625 + δ_1^b	7.5	Dark	–	–	17.0 ± 5	–	–

(continued)

Table 4.10 (continued)

Type of nano-structured Ag_2S	No.	Concentration of reagents in the reaction mixture (mmol l^{-1})			Dark/light[a]	D (nm) in deposited powders		D and h_{shell} (nm) in colloidal solutions		
		$AgNO_3$	Na_2S	Na_3Cit		BET	XRD	DLS D	TEM D	h_{shell}
$Ag_2S@C$ core-shell nano-particles	20	5	2.5	5	Dark	–	–	–	40^c	8^d
	21	50	25	100	Dark	–	–	–	36^c	12^e
	22	50	25	50	Dark	40 ± 5	45 ± 5	–	27^c	10^e
	23	50	25	25	Dark	–	–	16 ± 2	18^c	5^f
	24	50	25	10	Dark	–	–	–	13^c	2^f
	25	50	25	5	Dark	–	–	–	12^c	1^f
Ag_2S/Ag heteronano-structures	26	50	$25 - \delta_2^b$	100	Light	56	48 ± 6	28 ± 6	36^g	–
	27	50	$25 - \delta_2^b$	25	Light	84	52 ± 8	32 ± 6	38^g	–
	28	50	$25 - \delta_2^b$	12.5	Light	56	46 ± 6	34 ± 6	42^g	–

[a]Synthesis in dark or in light (under illumination)

[b]$\delta_1 = 0.01$ and $\delta_2 = 0.5$ mmol l^{-1} (small excess of sodium sulfide Na_2S is necessary for synthesis of silver sulfide colloidal solutions and nanopowders without an impurity of metallic Ag; small deficiency of Na_2S is necessary for the synthesis of Ag_2S/Ag heterostructures)

[c]Core size D_{core}

[d]3200 min

[e]1200 min,

[f]420 min are the dwell times t of Ag_2S nanoparticles in colloidal solution which is required for the formation of citrate shell with the specified thickness h_{shell}

[g]Size of Ag_2S nanoparticle; the size of Ag nanoparticles is 2–3 times smaller

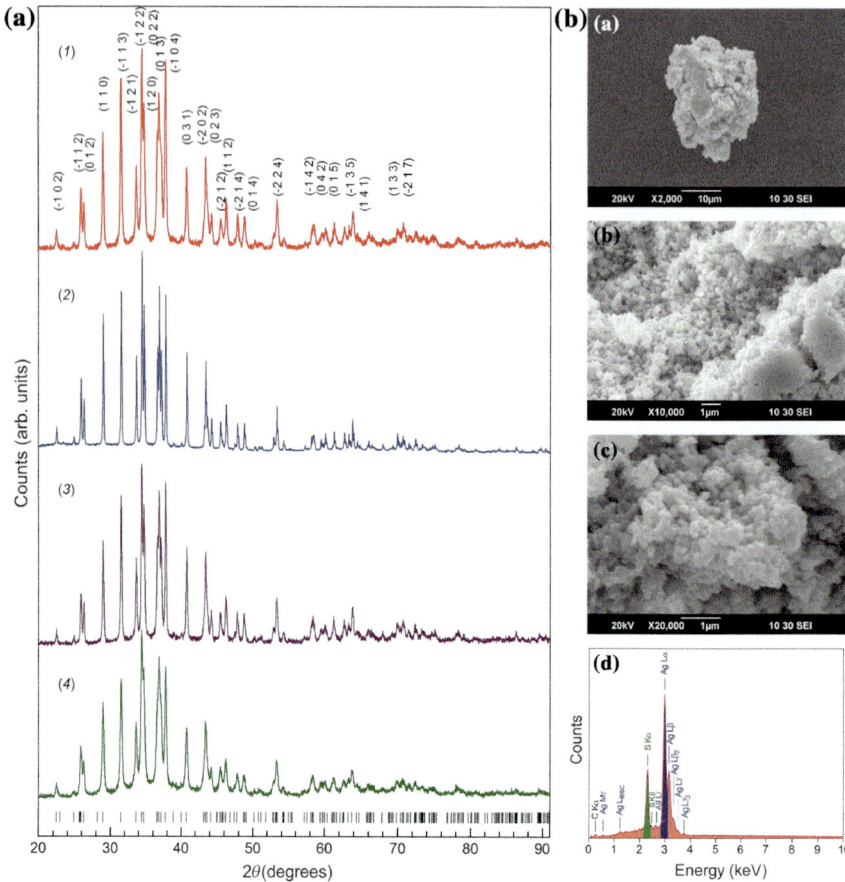

Fig. 4.41 **a** The XRD patterns of synthesized coarse-crystalline monoclinic (space group $P2_1/c$) α-Ag$_2$S silver sulfide powders deposited from the reaction mixtures 1, 2, 3, and 4 with large excess concentration of sodium sulfide Na$_2$S (see Table 4.10). *Vertical marks* indicate the positions of diffraction reflections of stoichiometric monoclinic α-Ag$_2$S phase. **b** SEM images of coarse-crystalline Ag$_2$S powder at a magnification of **a** 2000, **b** 10000, and **c** 20000 times, and **d** cumulative elemental EDX pattern of Ag$_2$S. Powder is deposited from the reaction mixture 2. Reproduced from [163] with permission from Wiley

%, respectively, and corresponds to silver sulfide Ag$_2$S with stoichiometric composition.

4.3.3 Deposited Ag$_2$S Nanopowders

Nanosized silver sulfide powders were prepared from aqueous solutions of AgNO$_3$ and Na$_2$S containing sodium citrate Na$_3$Cit. Concentration of sodium sulfide Na$_2$S in the initial reaction mixtures was slightly over half of the AgNO$_3$ concentration, i.e., $C_{Na_2S} = (C_{AgNO_3}/2) + \delta_2$ with $\delta_2 = 0.5$ mmol l^{-1} (see Table 4.10). Some excess of Na$_2$S is necessary for synthesis of silver sulfide without an impurity of metallic Ag nanoparticles. Synthesis was carried out in a dark room.

The use of reaction mixtures 5, 6, and 7 with sodium citrate Na$_3$Cit and with concentrations of AgNO$_3$ and Na$_2$S equal to 50 and $(25 + \delta_2)$ mmol l^{-1}, respectively, resulted in synthesis of unstable colloidal solutions, from which Ag$_2$S nanopowders were deposited.

To estimate the stability of colloidal solutions, the turbidity index was used. Turbidity is the cloudiness of a fluid caused by large numbers of individual small particles that are generally invisible to the naked eye. If the turbidity of synthesized solution differs from zero and changes with time, then the colloidal solution is unstable. For example, the turbidity of solution 5 immediately after synthesis and 60 and 100 days later was equal to 25 ± 0.5, 29 ± 0.5, and 28 ± 0.5 NTU, respectively. According to the DLS data, the size of Ag$_2$S nanoparticles in these solutions ranges from 45 to 65 nm (see Table 4.10), and silver sulfide nanopowders are precipitated from them very quickly.

XRD patterns of deposited silver sulfide nanopowders are shown in Fig. 4.42a. The quantitative analysis of the XRD patterns (see Table 4.8 and Sect. 4.2.2) and comparison with data [30] shown that the observed set of diffraction reflections corresponds to nanocrystalline nonstoichiometric monoclinic (space group $P2_1/c$) acanthite α-Ag$_{1.93}$S. The average size D of coherent scattering regions (CSR) estimated from broadening of non-overlapping diffraction reflections (−1 0 2), (1 1 0), (−1 1 3), (−1 0 4), (0 3 1) and (0 1 4) is presented in the inset to the XRD patterns. From these estimates, the average size D in the examined silver sulfide nanopowders, deposited from the initial reaction mixtures 5, 6, and 7 with Na$_3$Cit is 43 ± 6, 46 ± 8, and 49 ± 8 nm, respectively (Table 4.10). It was impossible to estimate the particles size in silver sulfide nanopowder 19 because of large broadening and overlapping of XRD reflections. Within the limits of experiment error, the size D of coherent scattering regions determined by the XRD method agrees with the particle size estimated by the BET and DLS methods (see Table 4.10).

TEM images of Ag$_2$S nanoparticles deposited from the reaction mixture 5, and experimental electron diffraction reflections are shown in Fig. 4.42b. Crystal structure and interplanar distances for Ag$_2$S nanopowders were determined by HRTEM method also. In study [163], electron diffraction pattern [selected area of electron diffraction (SAED)] was obtained by standard Fast Fourier Transform (FFT) of selected area of HRTEM image. Then authors of study [163] carried out inverse FFT of the selected diffraction reflections in the HRTEM image using the Gatan Microscopy Suite software [172], and determined the interplanar distances corresponding to these diffraction reflections. The indices (hkl) of diffraction reflections were determined by comparison of found interplanar distances with diffraction data [30] for nanocrystalline monoclinic (space group $P2_1/c$) silver sulfide. The scheme of sequence of operation for determination of the interplanar distances for the diffraction reflections, observed on SAED, is shown in Fig. 4.43.

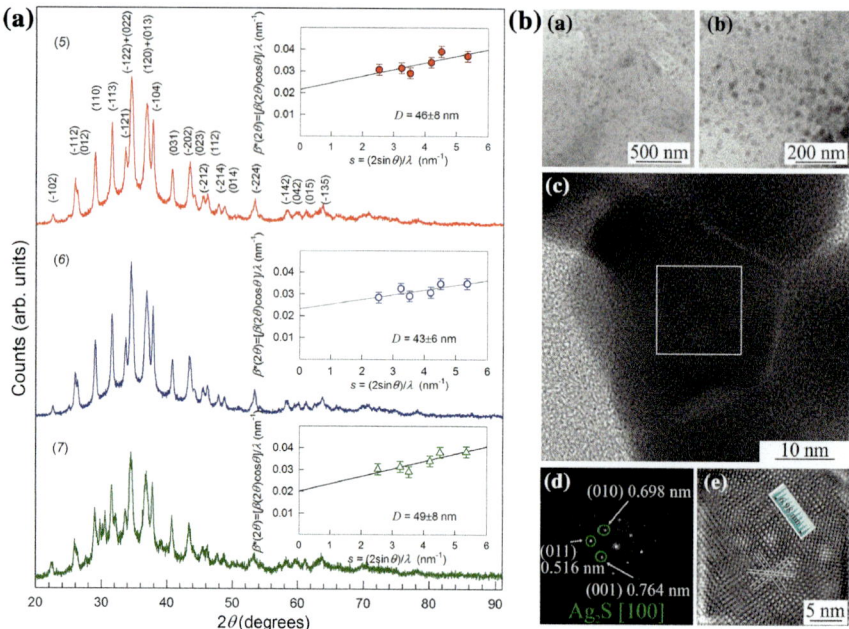

Fig. 4.42 a XRD patterns of synthesized monoclinic (space group $P2_1/c$) Ag$_2$S nanopowders deposited from the initial reaction mixtures 5, 6, 7, and 19 with Na$_3$Cit (see Table 4.10). The *insets* present the estimate of the average size of the coherent scattering regions from the broadening of non-overlapping diffraction reflections. The average CSR size in nanopowder 19 cannot be estimated because of large broadening and overlapping of XRD reflections. *Vertical marks* indicate the positions of diffraction reflections of nonstoichiometric monoclinic α-Ag$_{1.93}$S phase. **b** TEM images (*a*), (*b*), and (*c*) of Ag$_2$S nanoparticles deposited from the reaction mixture 5; (*d*) experimental electron diffraction reflections and (*e*) filtered image of the area in white rectangle of TEM image (*c*). Reproduced from [163] with permission from Wiley

Detailed description of determination of interplanar distances with the use of the Gatan Microscopy software [172] is given on site [173].

All silver sulfide nanopowders were studied using Raman and X-ray photoelectron spectroscopy. The Raman spectrum of silver sulfide nanopowder in the low-frequency region is smeared and broadened. Broadening of the peaks in the Raman spectrum indicates on the formation of nanoparticles. Weak overtone and strong fundamental modes which are observed at ~215 and ~430 cm^{-1} correspond to the first-order and second-order longitudinal optical phonon modes in Ag$_2$S, respectively.

According to the EDX results, the content of silver Ag and sulfur S in nanocrystalline silver sulfide powders corresponds to sulfide ~Ag$_2$S. Besides silver and sulfur, the EDX spectra contain a $K\alpha$ line of carbon C and a $K\alpha$ line of copper Cu. These lines appear from a carbon-containing collodium-glue film which applied to copper grid onto which examined particles were applied. No other elements have been found in the silver sulfide nanoparticles.

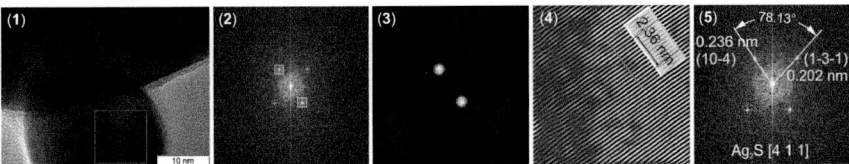

Fig. 4.43 Sequence of operation for determination of the interplanar distances includes following steps: (*1*) uploading of HRTEM image in Gatan Microscopy Suite and selection of area in HRTEM image for FFT (white square); (*2*) creation of FFT from the selected area (SAED) and selection of a spot on SAED; (*3*) creation of a mask from FTT; (*4*) creation of inverse FFT using mask, creation of line scale profile on FFT, and calculation of interplanar distance; (*5*) determination of (*hkl*) indices. Then steps *2–5* are repeated for another spots. Reprinted from [164] with permission from Elsevier

The XPS analysis of the Ag_2S nanopowders revealed the peaks at 367.5 and 373.8 eV, which could be respectively attributed to Ag 3d5/2 and Ag 3d3/2 binding energies, and weak peaks located at around 161.2 and 163.0 eV which could be assigned to sulfur atoms S in the lattice of Ag_2S. No oxygen peaks were found.

Recently, the technology for the production of Ag_2S nanopowders with preset nanoparticle size by hydrochemical deposition has been patented [174].

The optical properties of the Ag_2S powders have been studied from the reflectance spectra. The optical reflectance was measured for the coarse-crystalline and nanocrystalline Ag_2S powders 2, 4, 7 and 19 with average particle size of ~ 500, ~ 90, ~ 60, and 17 nm. Band gap E_g of the Ag_2S powders was determined from the optical reflectance spectra.

The spectral region most informative for determining the band gap from the optical spectra is the region, wherein the reflectance exhibits a noticeable variation depending on the wavelength λ. This region is in the wavelength range 700–800 to 1700–2100 nm corresponding to photon energies from ~ 1.6 to 0.6 eV.

According to the band theory of semiconductors, the band absorption edge is determined by the lowest band gap energy. To calculate the band gap energy of the Ag_2S nanopowders, the following equation was used

$$[\hbar\omega F(R_\infty)]^{1/n} = A(\hbar\omega - E_g), \tag{4.18}$$

where $\omega = 2\pi c/\lambda$ is the incident light frequency, $\hbar w = 2\pi\hbar c/ = hc/\lambda$ is the photon energy, and E_g is the band gap. Function $F(R_\infty) = (1 - R)^2/2R$ is determined from the reflectivity R, measured in relative units. For direct transition, value n is equal to 1/2 and

$$[\hbar\omega F(R_\infty)]^2 = A(\hbar\omega - E_g). \tag{4.19}$$

Equation (4.19) is convenient for quantitative processing of experimental data on the optical properties and for estimation of the band gap. In the ideal case, the experimental results represented in the coordinates "$[\hbar \omega F(R_\infty)]^2 \leftrightarrow \hbar\omega$" should be described by a linear dependence. In real experiments, the band is smeared, and therefore, the dependence $[\hbar\omega F(R_\infty)]^2 = f(\hbar\omega)$ in the vicinity of the band edge is nonlinear. In this case, the band gap E_g is determined as the value of the intercept on the $\hbar\omega$ axis, cut off by the tangent to the linear part of the experimental curve.

Figure 4.44 shows the reflectance spectra of the Ag$_2$S powders constructed on the energy scale, i.e., in the $[\hbar\omega F(R_\infty)]^2 \leftrightarrow \hbar\omega$ coordinates. In all of the spectra, the band edge is smeared at low energies.

The quantitative minimization of the experimental data on the reflectance R of the Ag$_2$S powders with function (4.19) allowed us to determine the values of E_g for these powders. The minimization was performed for nearly linear portions of the dependences $[\hbar\omega F(R_\infty)]^2 = f(\hbar\omega)$, i.e., in the wavelength range from 650–700 to 1400–1500 nm. The calculation shows that the band gap E_g for the coarse-crystalline Ag$_2$S powder is equal to ~ 0.88 eV and coincides with $E_g \cong 0.9$ eV of bulk α-Ag$_2$S silver sulfide [11]. For the Ag$_2$S nanopowders with particle size ~ 90, ~ 60, and ~ 17 nm, the value of E_g is equal to ~ 0.96, ~ 1.21, and ~ 1.67 eV, respectively. The band gap energy of the Ag$_2$S nanopowders is some higher than that of coarse-crystalline Ag$_2$S, indicating that studied Ag$_2$S nanopowders exhibit a blue shift of the optical absorption edge from the infrared region of the spectrum. This agrees with the results [65], where the band gap energy E_g for the Ag$_2$S nanoparticles with a size of ~ 8 nm is equal to ~ 2.85 eV.

Note that the blue shift of the optical absorption edge with decreasing nanoparticle size implies a corresponding blue shift of the luminescence band.

According to [175], for the ground state ($n = 1$) of a semiconductor, the effective band gap is

$$E_g(R) = E_b + (\pi^2\hbar^2/2\mu_{ex}R^2) - (1.78e^2/\varepsilon R), \qquad (4.20)$$

where E_b and ε are the band gap and dielectric constant for bulk crystal, R is the radius of particle, $\mu_{ex} = m_e m_h/(m_e + m_h)$ is the reduced exciton mass, m_e and m_h are the effective masses of the electron and hole, respectively. From (4.20) it follows that a decrease in the nanoparticle size D (or R) should be accompanied by an increase in the effective band gap E_g. Using the data obtained in studies [164, 165] and reported previously [65], we plotted the band gap of Ag$_2$S as a function of D^{-1}, where D is the Ag$_2$S particle size (Fig. 4.44, inset). It is seen from Fig. 4.44 (inset) that the band gap energy of the Ag$_2$S nanopowders is higher than that of coarse-crystalline Ag$_2$S.

4.3.4 Colloidal Solutions of Ag$_2$S Quantum Dots

Stable colloidal solutions of Ag$_2$S quantum dots were prepared from the reaction mixtures 8–19 with silver nitrate concentrations C_{AgNO_3} from 0.3125 to 2.5 mmol l^{-1} (Table 4.10). Concentration C_{Na_2S} of sodium sulfide in the reaction mixtures was slightly over half of the AgNO$_3$ concentration, i.e.,

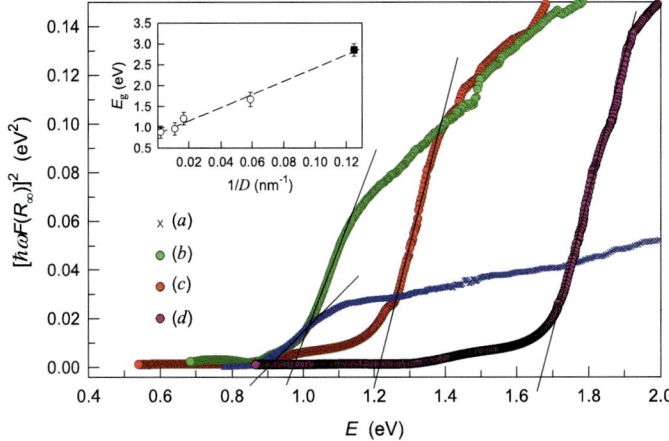

Fig. 4.44 Plots of $[\hbar\omega F(R_\infty)]^2$ with photon energy, $\hbar\omega$, for the Ag_2S powders with different average particle size: **a** coarse-crystalline powder with particle size of ~ 500 nm, **b**, **c** and **d** nanocrystalline powders with particle size of ~ 90, ~ 60, and ~ 17 nm, respectively. Inset shows the dependence of the band gap E_g on the Ag_2S nanoparticle size D, plotted as the function $E_g(D) \propto 1/D$: *Open circle* the results [164, 165], *filled square* data [65]

$C_{Na_2S} = (C_{AgNO_3}/2) + \delta_1$, where $\delta_1 = 0.01$ mmol l^{-1} (see Table 4.10). Small excess of Na_2S is necessary for synthesis of silver sulfide colloidal solutions without an impurity of Ag nanoparticles. The stability of these colloidal solutions is confirmed by their zero turbidity immediately after synthesis. The measurements showed that the turbidity of these solutions after 2, 5, 8, 11, 16, 24, 32, 40, 60, and 100 days is still equal to 0.

According to the DLS data, the size of Ag_2S quantum dots in colloidal solutions 8–19 is not larger than 20 nm (see Table 4.10). Such small nanoparticles do not deposit almost and practically do not change their sizes. The DLS zeta potential measurements of colloidal solutions 8–19 confirmed that these solutions remain stable for more than 100 days. The particle size distributions for the colloidal solutions from 8 to 19 with different size of Ag_2S quantum dots and the appearance of these colloidal solutions are shown in Fig. 4.45a, b.

It is known that the zeta potential ζ of nanoparticles in a solution is an indicator of the system stability. The absolute values $\pm(35 \pm 15)$ mV of zeta potential are indicative of electrostatic stability of colloidal solutions. The particles with highly negative or positive surface electric charge are considered as stable particles. The DLS measurements revealed that three days after synthesis of solutions 8–19 their zeta potential ζ was -45 to -28 mV, and the nanoparticle size was equal to 2–13 nm. The zeta potential ζ and the size of Ag_2S nanoparticles measured 100 days after synthesis of colloidal solutions remained almost unchanged. Small variation of the zeta potential during long-term storage of colloidal solutions and a large negative value of the zeta potential of solutions 8–19 confirm their stability. The comparison of the zeta potential ζ with the average size D_{DLS} of quantum dots for

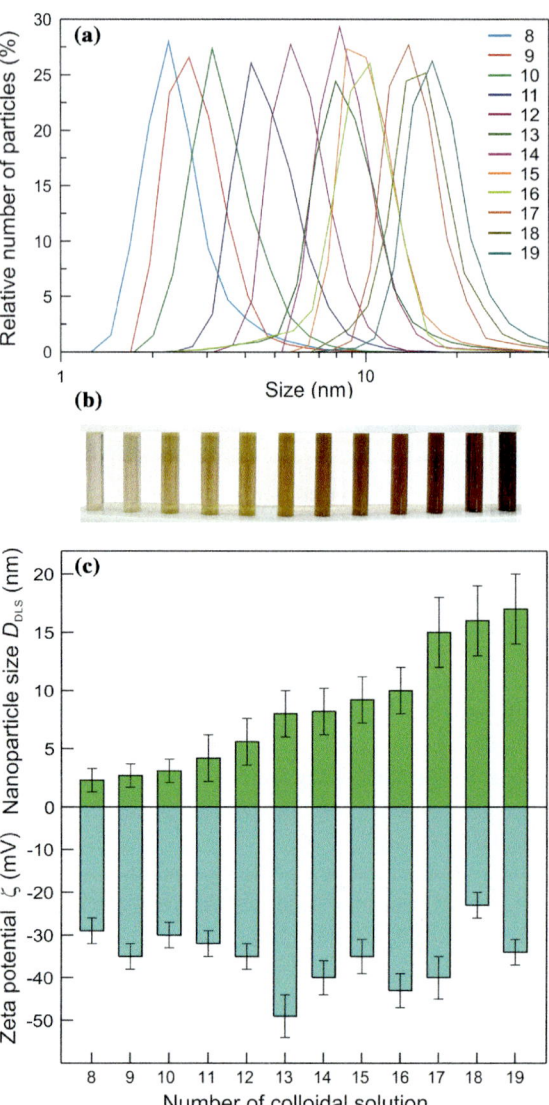

Fig. 4.45 **a** The particle size distributions measured by dynamic light scattering for the colloidal solutions 8–19 with different average size of Ag$_2$S quantum dots, **b** the appearance of these colloidal solutions, and **c** the average size D_{DLS} and zeta potential ζ of Ag$_2$S quantum dots in colloidal solutions 8–19 (see Table 4.10) measured 100 days after synthesis

colloidal solutions 8–19 100 days after synthesis is displayed in Fig. 4.45c. The size D_{DLS} of quantum dots is 2–17 nm, the value of ζ varies from −49 to −29 mV, and the average value of the zeta potential is equal to -35 ± 10 mV. It is seen that the smaller is the absolute value of the zeta potential, the larger is the size of Ag_2S quantum dots.

Figure 4.46 demonstrates the appearance of the colloidal solution 10 as an example. The glass contains opalescent colloidal solution 10 stored for 100 days. Solution 10 sidewise looks translucent and bluish (Fig. 4.46a). However in straight transmitted light, i.e. at direct through illumination it has light-brown color and is completely transparent (Fig. 4.46b). The optical absorption band maximum of the solution corresponds to 385 nm. Opalescence of the solution is indicative of noticeable fluctuations of its density, at which light scattering occurs. The fluctuations are due to the presence of quantum dots of size under 20 nm in the solution. Measurements of the quantum dot size in colloidal solution 10 by the DLS method right after synthesis and later 35, 70 and 100 days showed that the average size is from 3 ± 1 to 4 ± 2 nm (Fig. 4.46c), i.e. with allowance for measurement error the quantum dot size remained practically constant. The value of zeta potential ζ of quantum dots in solution 10 changed little if at all over the whole storage period and was -30 ± 3 mV. The invariability of the quantum dot size and zeta potential is indicative of the stability of colloidal solution 10.

Figure 4.47 presents the size-dependent photoluminescence (PL) emission spectra of Ag_2S colloidal solutions 8–13, 16, and 17 in which the fluorescence of Ag_2S quantum dots is tunable from ~ 1176 to ~ 960 nm by decreasing the nanoparticle size D_{DLS} from 15.0 to 2.3 nm.

According to [59], the PL peak for Ag_2S quantum dot with a size about 1.5 nm is observed at a wavelength of ~ 640 nm (see Fig. 4.47); in study [59] water-soluble Ag_2S quantum dots have been synthesized at heating of mixed solution of mercaptopropionic acid, ethylene glycol, and silver nitrate $AgNO_3$. The PL emission peaks shift from ~ 960 to ~ 1170 nm with the size of Ag_2S quantum dots increasing from ~ 2.3 to 4.2 nm and keep constant at 1166–1176 nm with increase of the quantum dot size from ~ 4.2 to 15 nm and further. The continuous blue shift of the PL emission of Ag_2S quantum dots from ~ 1176 to ~ 640 nm may be attributed to the strengthened quantum confinement effect and increasing the band gap E_g which resulted from the decreasing Ag_2S quantum dots size. This agrees with experimental data [164] on the size-dependent band gap of Ag_2S nanopowders (see Fig. 4.44). Almost constant position of the PL emission peaks at ~ 1166–1176 nm for the Ag_2S quantum dots with a boundary value of size 4.2 nm and larger is evidence for transition from the strong quantum confinement regime to the weak quantum confinement regime. According to this, we experimentally estimated Ag_2S exciton radius R_{exc} as less than half of boundary value of size 4.2 nm, i.e. ≤ 2.1 nm.

According to [145], radius of the exciton in the macroscopic (bulk) semiconducting crystal is determined by the equation

$$R_{exc} = (\varepsilon m_0 / \mu_{ex}) a_B, \tag{4.21}$$

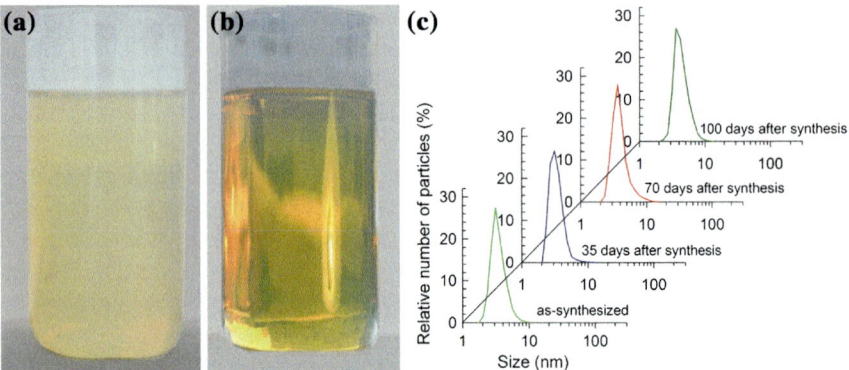

Fig. 4.46 The appearance of colloidal solution 10 later 100 days after synthesis [163]:
a opalescent colloidal solution sidewise looks translucent and bluish; **b** in straight transmitted
light the solution has light-brown color and is completely transparent; **c** size distributions of Ag$_2$S
quantum dots for as-synthesized colloidal solution 10 and for the same colloidal solution later 35,
70, and 100 days after synthesis. Reproduced from [163] with permission from Wiley

where $a_B = \hbar^2/m_0 e^2 = 0.053$ nm is the Bohr radius, $\hbar = 1.055 \times 10^{-27}$ cm^2 g s^{-1}
is the reduced Planck constant, $e = 4.803 \times 10^{-10}$ cm$^{3/2}$ g$^{1/2}$ s^{-1} is the charge on
the electron, $m_0 = 9.1095 \times 10^{-28}$ g is the free electron mass. For Ag$_2$S, dielectric
constant ε is about 6 [176], $m_e = 0.286 m_0$ and $m_h = 1.096 m_0$ [37], respectively. At
such values of ε, m_e and m_h, the Ag$_2$S exciton radius R_{exc} which is calculated by the
formula (4.21) is equal to $(26 \pm 1)a_B$ or $\sim 1.4 \pm 0.1$ nm, and the exciton diameter
is about 3 nm. According to analogous estimation [177], the Ag$_2$S exciton diameter
is equal from 3.0 to 4.4 nm (in study [177] Ag$_2$S quantum dots were obtained by
thermal decomposition of single-source Ag(DDTC) precursor where DDTC is
diethyldithiocarbamate). Strong blue shift for a quantum dot with a size 1.5 nm
agrees with data [177] on Ag$_2$S exciton diameter. Estimated exciton diameter
~ 3 nm of silver sulfide is in satisfactory agreement with experimental result
4.2 nm [163] which follows from the size-dependent PL emission spectra (see
Fig. 4.47). As confined excitons are responsible for optical transitions, Ag$_2$S
quantum dots with a size less than 3–4 nm can be used for NIR applications in the
biomedical field as luminophores or biomarkers.

Until lately there were no experimental works on the studying of stability of
silver sulfide colloidal solutions on the duration of storage. Meanwhile, preparation
of stable colloidal solutions is evidently the first method of producing disperse
nanoparticles. Visitors to the Royal Institution's Faraday Museum in London may

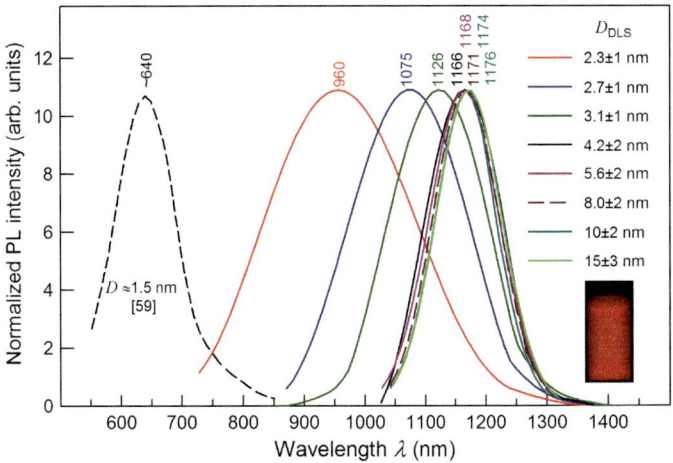

Fig. 4.47 The size-dependent PL emission spectra of Ag$_2$S colloidal solutions 8–13, 16, and 17 with quantum dot size D_{DLS} from 2.3 to 15.0 nm under an excitation of 658 nm. For comparison, dashed line show the position of PL emission peak for Ag$_2$S quantum dots with a size ~1.5 nm [59]. The wavelengths corresponding to the maxima of the PL peaks are indicated. The *inset* presents fluorescence image of Ag$_2$S colloidal solution 13 (see Table 4.10) with quantum dot size about 8 nm

see two glass vessels with a colloidal solution of gold produced by M. Faraday in the first half of the 19th century. These solutions have retained their stability for almost 200 years. The preparation and optical properties of colloidal solutions, particles and films of gold, silver and other metals were described by Faraday in 1857 [178].

Recently, the stability of silver sulfide colloidal solutions and UV-vis absorption spectra of these solutions were studied by Sadovnikov et al. [164]. In study [164], colloidal solutions of Ag$_2$S nanoparticles were prepared by chemical condensation method from aqueous solutions of silver nitrate AgNO$_3$ and sodium sulfide Na$_2$S containing sodium citrate Na$_3$Cit or Trilon B as complexing and stabilizing agents. The concentrations of AgNO$_3$, Na$_2$S and complexing agents (sodium citrate or Trilon B) in the reaction mixtures are given in Table 4.11. Besides, colloidal solutions were obtained by washing of deposited Ag$_2$S powders with distilled water with subsequent removal of the powder by filtration.

When the solutions of reagents are mixed, the reaction mixture first turns black, and then deposition of Ag$_2$S particles begins. The largest particles deposit during 1 h, the average-size (50–80 nm) particles agglomerate and also deposit. The deposition ends in one day, and the solution becomes transparent. The color of the solution changes from pale yellow to dark brown depending on the concentrations of reagents.

The size (hydrodynamic diameter D_{DLS}) and zeta potential ζ of Ag$_2$S nanoparticles in the as-synthesized colloidal solutions and in the solutions obtained by washing of deposited Ag$_2$S powders were determined by non-invasive Dynamic

Light Scattering (DLS) on a Zetasizer Nano ZS facility at 298 K. UV-vis absorption spectra of colloidal solutions of Ag$_2$S nanoparticles were recorded on a Shimadzu UV-2401 PC spectrophotometer at a temperature of 298 K.

The images of some colloidal solutions obtained by the transmission electron microscopy method are displayed in Fig. 4.48. Colloidal solutions I and III (Fig. 4.48a, b) are synthesized from reaction mixtures I and III and contain only silver sulfide Ag$_2$S nanoparticles of the size 15–35 and 12–18 nm, respectively. The HRTEM image of Ag$_2$S nanoparticle from colloidal solution I and its diffraction pattern obtained by Fast Fourier Transform (FFT) are demonstrated in Fig. 4.48e. The observed diffraction reflections and the interplanar distances correspond to monoclinic (space group $P2_1/c$) acanthite α-Ag$_2$S. Colloidal solution VIII (Fig. 4.48c) was synthesized with the use of Trilon B as a complexing agent, and along with Ag$_2$S nanoparticles, this solution contains nanoparticles of metallic impurity silver Ag (the Ag nanoparticles on the TEM image of colloidal solution VIII (see Fig. 4.48c) look darker). Colloidal solution VII (Fig. 4.48d) is synthesized in light from reaction mixture VII with small concentrations of AgNO$_3$ and Na$_2$S and a large concentration of Na$_3$Cit and it also contains nanoparticles of impurity Ag along with silver sulfide nanoparticles. The HRTEM image of Ag nanoparticle from colloidal solution VIII and its diffraction pattern are shown in Fig. 4.48f. The observed diffraction reflections and the interplanar distances correspond to cubic (space group $Fm\overline{3}m$) silver Ag. Microtwinning in the direction of the family planes (111) is well visible in the silver nanoparticle.

Crystal structure, interplanar distances, and indices (hkl) of electron diffraction reflections were determined for single Ag$_2$S nanoparticles by HRTEM method. In study [164], selected area of electron diffraction (SAED) was obtained by standard FFT of selected area of HRTEM image (see Sect. 4.3.3). The scheme of sequence of operation for determination of the interplanar distances and indices (hkl) for the diffraction reflections, observed on SAED, is shown earlier in Fig. 4.43. The indices (hkl) were determined by comparison of found interplanar distances with diffraction data [30] for nanocrystalline monoclinic silver sulfide.

The results of transmission electron microscopy combined with EDX analysis and XRD data allow one to divide all the synthesized colloidal solutions into two groups (see EDX and XRD data in Supplementary material for study [164]). The first group contains colloidal solutions 0, I, II, III, and IV containing only Ag$_2$S nanoparticles. Colloidal solutions V, VI, VII, VIII, and IX, in which impurity metallic silver Ag nanoparticles are present together with silver sulfide nanoparticles, belong to the second group.

The size of the nanoparticles in all the colloidal solutions was measured with the use of the DLS method (see Table 4.11). The average size D of unagglomerated particles in the deposited powders was estimated from the value of specific surface area S_{sp} measured by the Brunauer-Emmett-Teller (BET) method which gives the particle size averaged with respect to the volume (see Table 4.11). In the approximation that all particles have similar size and spherical shape, the average particle size D is equal to $6/\rho \, S_{sp}$ ($\rho = 7.25$ g cm^{-3} is the density of silver sulfide).

Table 4.11 Composition of the reaction mixtures, average particle size D of silver sulfide Ag_2S in the synthesized colloidal solutions and powders [164]

No.	Concentration of reagents in the reaction mixture (mmol l⁻¹)				D_{dls} (nm) in colloidal solutions			Powders[d]	
	AgNO₃	Na₂S	Na₃Cit	Trilon B	Filtered solutions		Unfiltered solutions after 60 days of storage	$S_{sp} \pm 0.2$ (m² g⁻¹)	D (nm)
					b	c			
0	50	25 + δ[a]	50	–	64 ± 10	65 ± 10	15 ± 5	19.2	43 ± 5
I	50	25 + δ[a]	25	–	60 ± 10	60 ± 10	13 ± 3	19.0	44 ± 5
II	50	25 + δ[a]	12.5	–	55 ± 10	55 ± 10	16 ± 5	14.9	56 ± 5
III	50	25 + δ[a]	10	–	53 ± 10	53 ± 10	18 ± 5	14.0	59 ± 5
IV	50	25 + δ[a]	5	–	50 ± 10	50 ± 10	17 ± 5	9.3	89 ± 6
V	2.5	1.25	7.5	–	–	–	26 ± 8	–	–
VI	3.75	1.875	2.5	–	–	–	19 ± 5	–	–
VII	0.625	0.3125	5.0	–	–	–	8 ± 2	–	–
VIII	50	25	–	50	65 ± 10	–	18 ± 5	13.1	63 ± 5
IX	50	25	–	35	60 ± 10	–	17 ± 5	13.8	60 ± 5

[a] δ = 0.1 mmol l⁻¹ (small excess of sodium sulfide Na₂S is necessary for synthesis of silver sulfide colloidal solutions without an impurity of metallic Ag)
[b] 1 day after filtration
[c] 60 days after filtration
[d] BET measurements

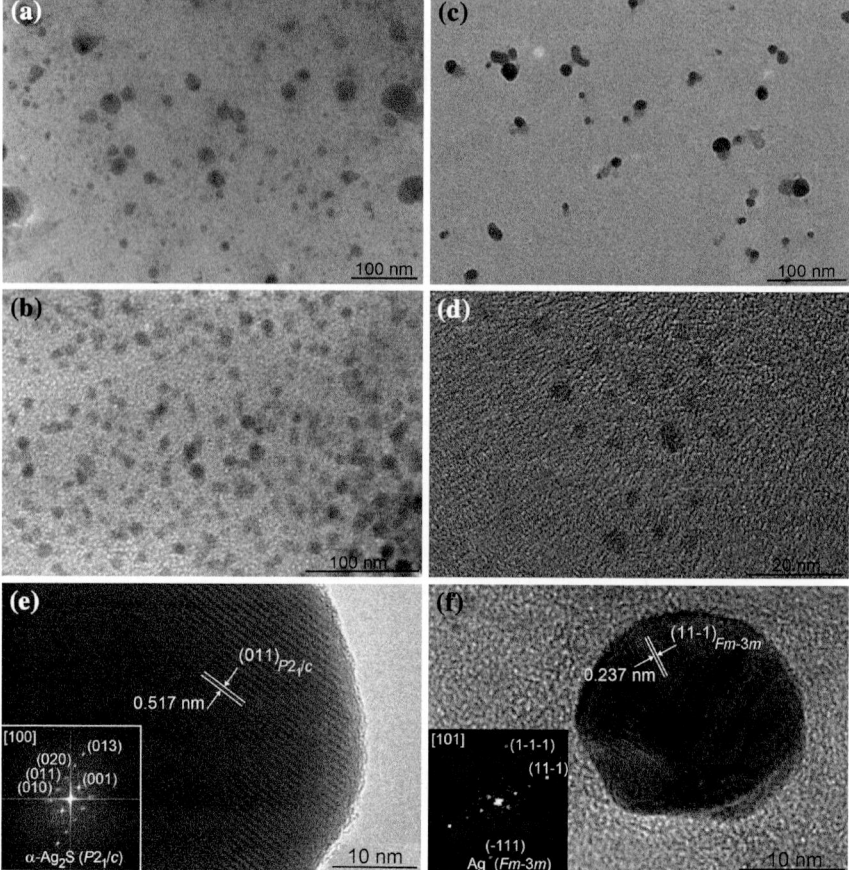

Fig. 4.48 The TEM images of colloidal solutions **a** I and **b** III containing only Ag$_2$S silver sulfide nanoparticles, and colloidal solutions **c** VIII and **d** VII containing impurity silver Ag nanoparticles along with Ag$_2$S nanoparticles. **e** The HRTEM image of a nanoparticle of monoclinic (space group $P2_1/c$) silver sulfide with acanthite structure α-Ag$_2$S from colloidal solution I and its diffraction pattern obtained by FFT method (the zone axis [100]). **f** The HRTEM image of a nanoparticle of cubic Ag from colloidal solution VIII and diffraction pattern obtained by FFT of the HRTEM image of this Ag nanoparticle (the zone axis [101]). Reprinted from [164] with permission from Elsevier

Figure 4.49 shows the isotherms of nitrogen adsorption by some Ag$_2$S silver sulfide nanopowders obtained from different reaction mixtures.

DLS measurements in colloidal solutions allow one to determine how the particle size varies as a function of the solution storage duration. The colloidal solutions containing only Ag$_2$S nanoparticles are most interesting. Let us consider the size of silver sulfide nanoparticles in such solutions as a function of storage time using colloidal solutions I and II (Fig. 4.50) as an example.

Right after synthesis by the dynamic light scattering method, authors of study [164] measured the particle size D_{DLS} in the solutions I and II prepared from the reaction mixtures I and II. The Ag_2S nanoparticles formed in the solution immediately agglomerate; in time some agglomerates deposit and therefore the size of the particles remaining in the solution changes.

Initially, the both synthesized solutions had bimodal distributions of particles (see Fig. 4.50a, b). In solution I, the size D_{DLS} of agglomerated particles corresponding to the size distribution maxima was ~ 170 and ~ 660 nm; the relative amount of agglomerates corresponding to these peaks was equal to ~ 55 and $\sim 45\%$ (Fig. 4.50a). In solution II, the size of particles corresponding to the size distribution maxima was ~ 120 and ~ 570 nm at their quantitative ratio of ~ 70 and $\sim 30\%$ (Fig. 4.50b). After three days, most of large particles deposited leading to monomomodal size distributions of particles in both solutions. In solution I, the size of agglomerates remaining in the solution in suspended state, which corresponded to the maximum of monomadal size distribution, was equal to ~ 150 nm, and the average size of agglomerates was ~ 210 nm (Fig. 4.50a). The size of the majority of agglomerates remaining in suspended state in solution II was ~ 230 nm, and the average size of agglomerates was ~ 280 nm (Fig. 4.50b).

Then the synthesized solutions above the deposit were decanted.

The deposits were washed three times with distilled water; the resulting aqueous suspension was treated with ultrasound to break the agglomerates and after that it was filtrated. The produced filtered solutions are aqueous colloid solutions of silver sulfide nanoparticles. Therefore size of nanoparticles was afterwards determined in the filtered solutions.

The measurements showed that one day after filtration (or four days after synthesis) both filtered solutions I and II contained nanoparticles with average size of 60 ± 10 and 55 ± 10 nm, respectively (Fig. 4.50c, d).

The measurements performed eight days after filtration revealed repeated agglomeration of nanoparticles taking place in the solutions. The size of agglomerates corresponding to the size distribution maxima in solutions I and II was equal to ~ 300 and ~ 320 nm, respectively, and the average size of agglomerates in solutions I and II was ~ 380 and ~ 520 nm (Fig. 4.50e, f). However no visible deposition was observed. Apparently, agglomerated Ag_2S particles were in suspended state, the largest agglomerates deposited, but the amount of such large agglomerates was small and insufficient for visual observation of deposition. Deposition was visually observed only 30 days after filtration.

60 days after filtration, the DLS measurements showed that in both solutions most particles were ~ 50 nm in size, and the average particle size in solutions I and II was equal to 60 ± 10 and 55 ± 10 nm (Fig. 4.50g, h), respectively. The average size of nanoparticles coincided with that determined one day after filtration. After synthesis, the solutions contained a large amount of silver sulfide nanoparticles (~ 1 wt.%); about 98 wt.% of particles deposited in the form of agglomerates during the first day. Thus, about 2 wt.% Ag_2S remained in filtered solutions. The reduction of the amount of particles in filtered solutions resulted in slowing-down of agglomeration and in deposition 30 days later filtration. The size of nanoparticles

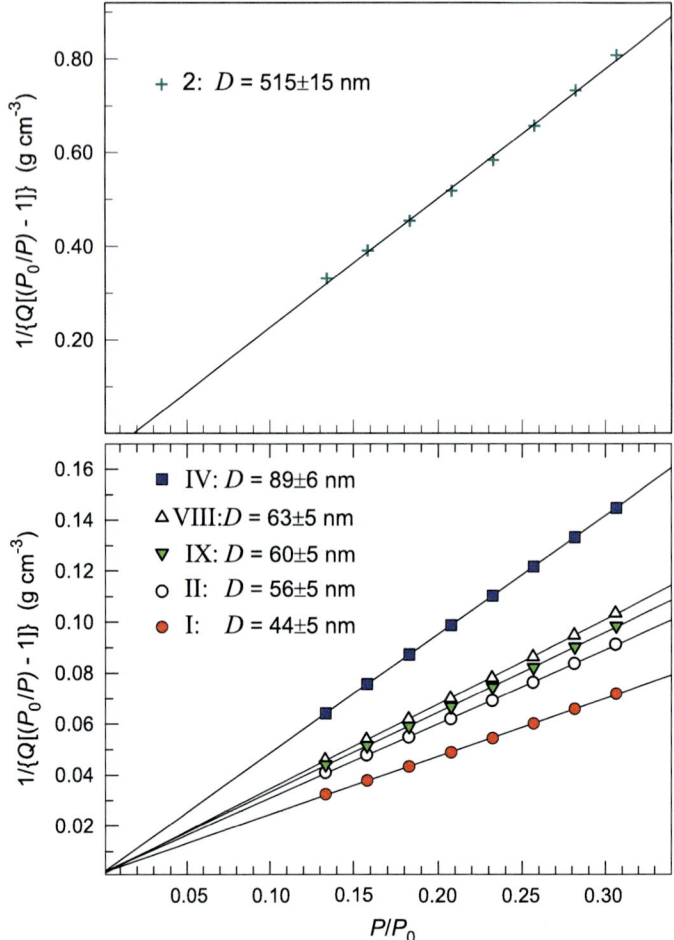

Fig. 4.49 The adsorption isotherms ($T = 77$ K) of molecular nitrogen N$_2$ by the surface of Ag$_2$S powders obtained by hydrochemical deposition from the reaction mixtures I (*filled red circle*), II (*open circle*), IV (*filled blue square*), VIII (*open triangle*), and IX (*inverted green filled triangle*) (see Table 4.11). The adsorption isotherm for coarse-crystalline Ag$_2$S powder 2 (see Table 4.10) is shown in the top for comparison

remaining in the aqueous colloidal solution does not change, and the solution remains stable for more than two years.

Figure 4.51 demonstrates the variation in the appearance of solution II at different stages of storage. In the left cell there is as-synthesized solution. This solution contains suspended agglomerates and therefore it has dark-brown or almost black color. The cell in the middle contains filtered solution II 60 days after filtration: it has brown color and is ideally transparent. The third cell contains unfiltered opalescent colloidal solution II after 60 days of storage. In straight transmitted light,

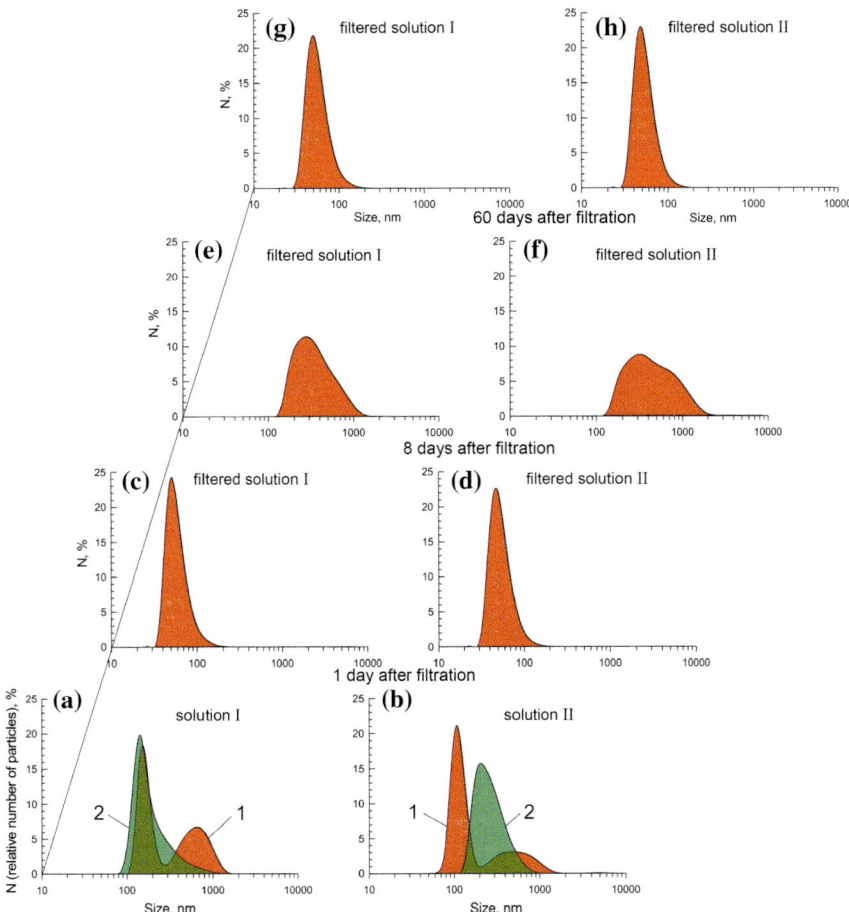

Fig. 4.50 The variation of size distributions of Ag$_2$S particles in the synthesized solutions I and II, and their filtered colloidal solutions versus the duration of storage: (*1*) bimodal distributions in as-synthesized solutions; (*2*) monomodal distributions in solutions after 3 day storage. All size distributions are determined at a temperature of 298 K. Reprinted from [164] with permission from Elsevier

i.e. at direct through illumination it has light-brown color and is completely transparent. Opalescence of the solution is indicative of noticeable fluctuations of its density. These fluctuations are due to the presence of nanoparticles of size under 20 nm in the solution.

Indeed, the DLS measurements showed that the particle size in unfiltered solution I varies from 8 to 30 nm, and the average size of nanoparticles is equal to ~13 nm, which is close to the particle size corresponding to the size distribution maximum (Fig. 4.52a). Similar size distributions of Ag$_2$S nanoparticles in unfiltered colloidal solutions II and III 60 days after synthesis is displayed in Fig. 4.52b,

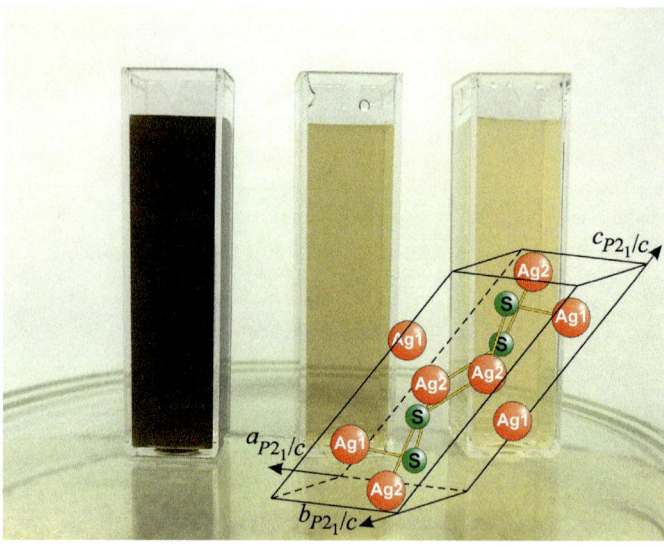

Fig. 4.51 The variation of the appearance of solution II (from *left* to *right*): as-synthesized solution, filtered colloid solution after 60 days of storage, unfiltered colloid matrix solution later 60 days after synthesis. The *inset* shows the unit cell of monoclinic (space group $P2_1/c$) silver sulfide which is contained in colloidal solutions

c. The particle size in solution II varies from 9 to 35 nm, and the average size of nanoparticles is ~ 16 nm, the particle size in solution III varies within from 12 to ~ 60 nm, the average size of nanoparticles is ~ 18 nm. The size distributions of nanoparticles in unfiltered colloidal solutions V, VI, and VII have two peaks with different intensities (Fig. 4.52d–f). Considering the EDX, XRD, and TEM data on the presence of impurity silver in these colloidal solutions, it can be assumed that the low-intensity peaks on the size distributions (Fig. 4.52d–f) correspond to silver nanoparticles, while the peaks of large intensity correspond to Ag$_2$S nanoparticles. The average size of Ag nanoparticles in unfiltered colloidal solutions V, VI, and VII is equal to ~ 18, ~ 13, and ~ 6 nm, respectively.

The absolute values $\pm(35 \pm 15)$ mV of zeta potential ζ are indicative of electrostatic stability of colloidal solutions. The particles with highly negative or positive surface electric charge are stable particles. The DLS measurements revealed that three days after synthesis of solutions V, VI, and VII their zeta potential ζ was -35 to -28 mV, and the nanoparticle size was equal to 6–22 nm. The zeta potential ζ and the size of Ag$_2$S nanoparticles measured 60 days after synthesis of colloidal solutions remained almost unchanged: the value of ζ varies from -37 to -29 mV, the size of Ag$_2$S nanoparticles is 8 to 26 nm. Small variation of the zeta potential during long-term storage of colloidal solutions and a large negative value of the zeta potential confirm the stability of solutions V–VII.

The addition of 3-mercaptopropyltrimethoxysilane (MPTMS) C$_6$H$_{16}$O$_3$SSi in colloidal solutions in order to create MPTMS-capped Ag$_2$S particles not led to a

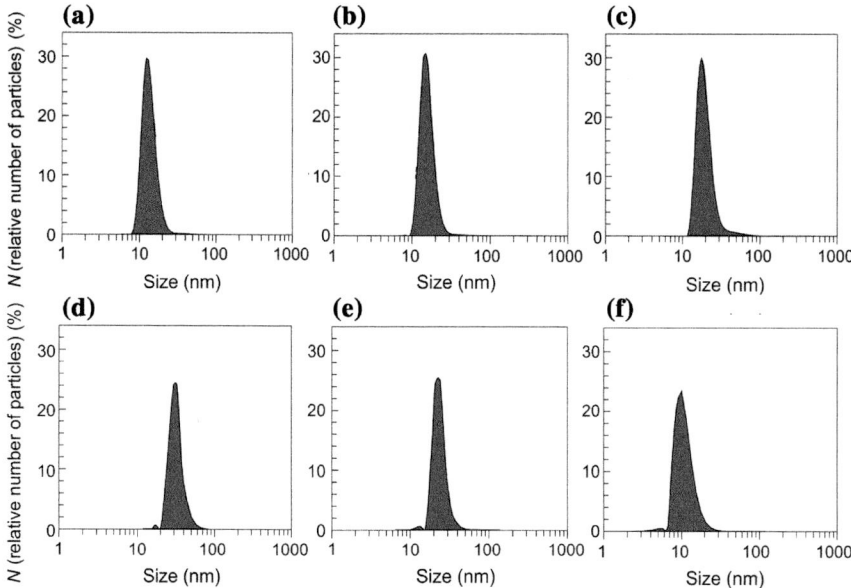

Fig. 4.52 The size distribution of Ag₂S particles and impurity Ag particles in unfiltered colloidal solutions 60 days after synthesis: **a**, **b**, and **c** solutions I, II, and III containing only Ag₂S nanoparticles; **d**, **e**, and **f** solutions V, VI, and VII containing Ag₂S nanoparticles and impurity Ag nanoparticles. The average size of impurity Ag nanoparticles in colloidal solutions V, VI, and VII is equal to ∼18, ∼13, and ∼6 nm, respectively. Reprinted from [164] with permission from Elsevier

noticeable stabilization of a size of small nanoparticles. However, preliminary study has shown that it is possible to synthesize Ag₂S nanoparticles with stable average size of ∼15 nm in colloidal solutions with MPTMS.

The technology for the preparing stable aqueous Ag₂S colloidal solutions has been patented [179].

As was noted, the synthesized solutions I–IX (see Table 4.11) can be divided into two groups: colloidal solutions containing only Ag₂S nanoparticles and colloidal solutions, in which nanoparticles of metallic silver Ag are present together with silver sulfide nanoparticles.

The UV-vis absorption spectra have been obtained directly from the unfiltered colloidal solutions of both groups (Fig. 4.53).

The UV-vis absorption spectra of unfiltered colloidal solutions I, II, and III are displayed in Fig. 4.53a. According to the DLS data, the average size of Ag₂S nanoparticles in these solutions is equal to ∼13, ∼16, and ∼18 nm, respectively. The absorption peak observed at ∼316–320 nm corresponds to Ag₂S nanoparticles [68, 70]. The relative intensity of the absorption peak in the first approximation is proportional to the volume content of silver sulfide in the solution, which decreases with the reduction of the concentration of Na₃Cit in the reaction mixtures. The reduction of the concentration of Na₃Cit in the reaction mixtures leads also to weak

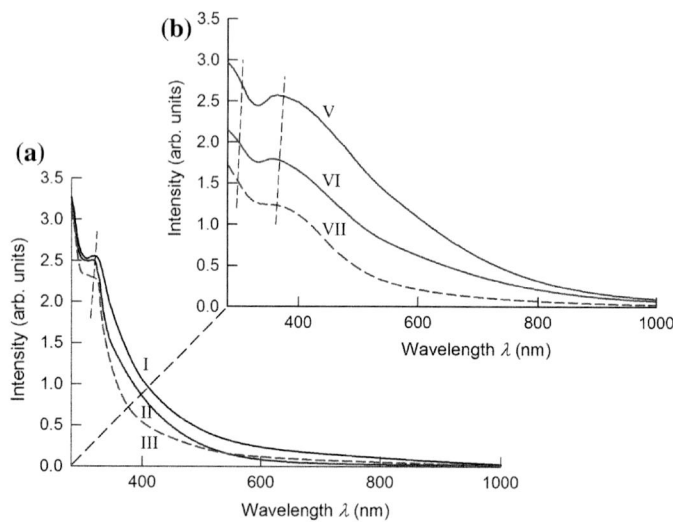

Fig. 4.53 The UV-vis absorption spectra of unfiltered colloidal solutions 60 days after synthesis: **a** colloidal solutions I, II, and III containing only Ag$_2$S nanoparticles; **b** colloidal solutions V, VI, and VII containing Ag$_2$S nanoparticles and impurity Ag nanoparticles. The absorption peak observed in the spectra of solutions I, II, and III at $\sim$316–320 nm corresponds to Ag$_2$S nanoparticles. The broad absorption peak in the spectra of solutions V, VI, and VII at $\sim$360–460 nm corresponds to impurity Ag nanoparticles. Reprinted from [164] with permission from Elsevier

smearing of the absorption peak of colloidal solutions and to a small displacement of this peak into the short-wave region. Studying Ag$_2$S nanofibers with a diameter of less than 20 nm, the authors of study [180] observed an optical absorption band in the region of $\sim$530 nm and supposed that this absorption band is due to Ag$_2$S nanoparticles. However the absorption band observed in work [180] at $\sim$530 nm is more likely to be due to impurity silver nanoparticles.

The UV-vis absorption spectra of unfiltered colloidal solutions V, VI, and VII containing Ag$_2$S nanoparticles and impurity Ag nanoparticles are presented in Fig. 4.53b. A typical feature of these spectra is the presence of a broad intensive absorption band in the region of $\sim$360–460 nm, which corresponds to metallic silver nanoparticles and is due to surface plasmon resonance (SPR) [58, 181, 182]. According to [68, 70], if Ag nanoparticles are present in the colloidal solutions of silver sulfide, the absorption band of impurity Ag nanoparticles is retained in the absorption spectra of these colloidal solutions in ethanol and other solvents.

It is known that silver nanoparticles in the spectral SPR region exhibit the greatest absorption as compared with any other nanoparticles of the same size, and the increase in the size of Ag nanoparticles manifests itself in the growth of the SPR band intensity [182]. Indeed, the absorption band of Ag particles (Fig. 4.53b) is most intensive for colloidal solution V with the largest ($\sim$ 18 nm) impurity Ag nanoparticles and it decreases in the transition to colloidal solution VII having the smallest Ag nanoparticles with size about 6 nm (see Fig. 4.52d–f). The absorption of Ag_2S nanoparticles in the optical absorption spectra of colloidal solutions V–VII weakly shows up in the variation of the inclination of the short-wave part of the spectrum in the region of ~ 298–306 nm.

Sadovnikov et al. [164] have also examined preliminarily the antibacterial activity of colloidal solutions of silver sulfide with respect to the Gram-negative bacteria "Escherichia coli". Study [164] revealed that the Ag_2S colloidal solutions possess the potential antibacterial activity. Experimental results [164] on the antibacterial activity of Ag_2S colloidal solutions and nanopowders are in good agreement with data [57, 58].

The performed studies [163, 164] of the colloidal solutions showed that in the presence of sodium citrate Na_3Cit stable Ag_2S quantum dots with size ~ 3–17 nm can be prepared from the reaction mixtures with silver nitrate concentration C_{AgNO_3} ranging from 0.3125 to 2.5 mmol l^{-1} and sodium sulfide concentration $C_{Na_2S} = (C_{AgNO_3}/2) + \delta_1$, where $\delta_1 = 0.01$ mmol l^{-1} (see Table 4.10). The concentration of sodium citrate necessary to maintain the small size of Ag_2S quantum dots lies within the interval between 1 and 15 mmol l^{-1}.

4.3.5   Core-Shell $Ag_2S@C$ Nanoparticles

Lately, much attention is devoted to the production of hybrid heteronanostructures of the core-shell type (for example, $Ag_2O@Ag_2S$, $PbS@S$, etc.) [75, 183, 184]. The $Ag_2S@C$ core-shell nanoparticles have been synthesized for the first time in our works [64, 167].

According to [64, 163, 167], the colloidal solutions above the silver sulfide deposits contain the finest nanoparticles. These solutions were used for electron-microscopy examination of nanoparticles contained in them.

The TEM studies of colloidal solutions prepared from the reaction mixtures without Na_3Cit and with different content of Na_3Cit revealed the Ag_2S nanoparticles without a shell and $Ag_2S@C$ nanoparticles with a carbon-containing citrate shell.

The TEM image of the nanoparticles contained in the solution prepared from the reaction mixture 1 without Na_3Cit is displayed in Fig. 4.54a. These nanoparticles have no shell. The size of the most part of the nanoparticles is ~ 7–15 nm, but there are also larger particles to 20 nm is size. This agrees with the DLS measurements of Ag_2S nanoparticle size distribution in this solution (Fig. 4.54d). As an example, HRTEM image of the quantum dot having a size about 4 nm is shown in Fig. 4.54b. The diffraction pattern (or selected area of electron diffraction SAED) (Fig. 4.54c) is calculated using Fast Fourier Transform (FFT) of HRTEM image of this quantum dot. The observed spots (011) and (10–2) corresponds to the [2–11] plane of the reciprocal lattice of the monoclinic (space group $P2_1/c$) silver sulfide with α-Ag_2S acanthite structure.

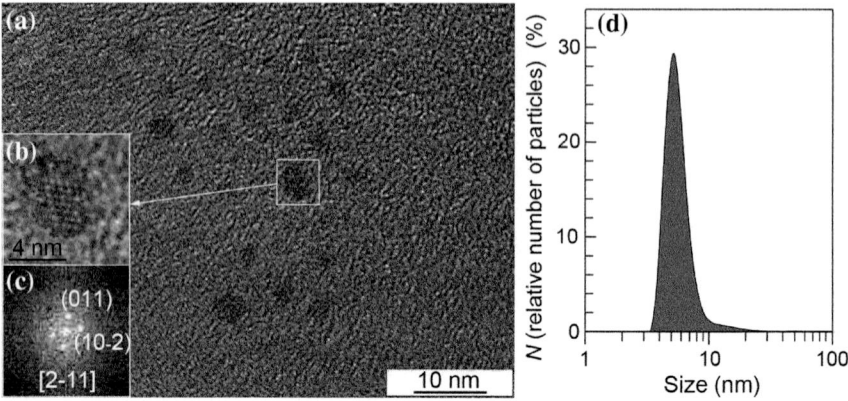

Fig. 4.54 **a** TEM image of Ag$_2$S colloidal solution prepared from the reaction mixture 1 without Na$_3$Cit (see Table 4.10), **b** HRTEM image of Ag$_2$S quantum dot with a size about 4 nm, **c** diffraction pattern obtained by FFT of this HRTEM image, and **d** size distribution of Ag$_2$S quantum dots in colloidal solution prepared from the reaction mixture 1. Reproduced from [163] with permission from Wiley

The Ag$_2$S@C nanoparticles with a carbon-containing citrate shell were found in the colloidal solutions prepared from the reaction mixtures 20–25 synthesized with Na$_3$Cit (see Table 4.10).

XRD patterns of Ag$_2$S nanopowders deposited from the reaction mixture 20 (see Table 4.10) with different dwell time of nanopowders in the solution are shown in Fig. 4.55a. These nanopowders are distinguished by their dwell time in the solution (from 20 to 1200 min). The quantitative analysis of the XRD patterns and comparison with data [30] have shown that the observed set of diffraction reflections corresponds to nonstoichiometric monoclinic (space group $P2_1/c$) acanthite $\sim$Ag$_{1.93}$S (see also Sect. 4.2.2). The amorphous carbon-containing shell is not visible on the XRD patterns.

It is seen on HRTEM images that the silver sulfide nanoparticles extracted from the colloidal solutions which were prepared with sodium citrate have an amorphous shell (Fig. 4.55b). Other things being equal, the thickness of the shell grows when the nanoparticle dwell time in the colloidal solution containing citrate ions C$_6$H$_5$O$_7^{3-}$ (Fig. 4.55b(a–d)) increases, and when the concentration of Na$_3$Cit in the solution increases. Filtered image of the area 1 isolated by white square is shown in Fig. 4.55b(a) also. The determination of the interplanar distances of cores confirmed a monoclinic structure of the colloidal silver sulfide nanoparticles. Figure 4.55(e) shows as an example the diffraction pattern of the core of the nanoparticle presented in Fig. 4.55b(a). The diffraction pattern (or selected area of electron diffraction SAED) (Fig. 4.55c(e)) is obtained by FFT from area 1 of HRTEM image of nanoparticle core. The observed set of spots ($-1-11$), (-111), (020), and (-212) corresponds to the [101] plane of the reciprocal lattice of the monoclinic (space group $P2_1/c$) α-Ag$_2$S phase with acanthite structure. The interplanar distances corresponding to these reflections are equal to ~0.356, ~0.356, ~0.347, and ~0.200 nm. These data are in good agreement with the results on the determination of the structure of nanocrystalline nonstoichiometric acanthite α-$Ag_{1.93}S$ [30] (see also Table 4.15 of Appendix).

The information about the composition of the core-shell nanoparticles is provided by the EDX analysis (Fig. 4.55f–h). According to the EDX results, the content of silver Ag and sulfur S in the colloid core-shell nanoparticles corresponds to sulfide $Ag_{1.95-1.98}S$. Besides silver and sulfur, the EDX spectra contain a $K\alpha$ line of carbon C, a weak $K\alpha$ line of oxygen O, and a $K\alpha$ line of copper Cu from a copper grid onto which colloid solutions with examined nanoparticles were applied. A copper grid with an aperture diameter of 50 μm was used for TEM observing of nanoparticles. One or two layers of hydrocarbon-glue were applied to Cu grid. After

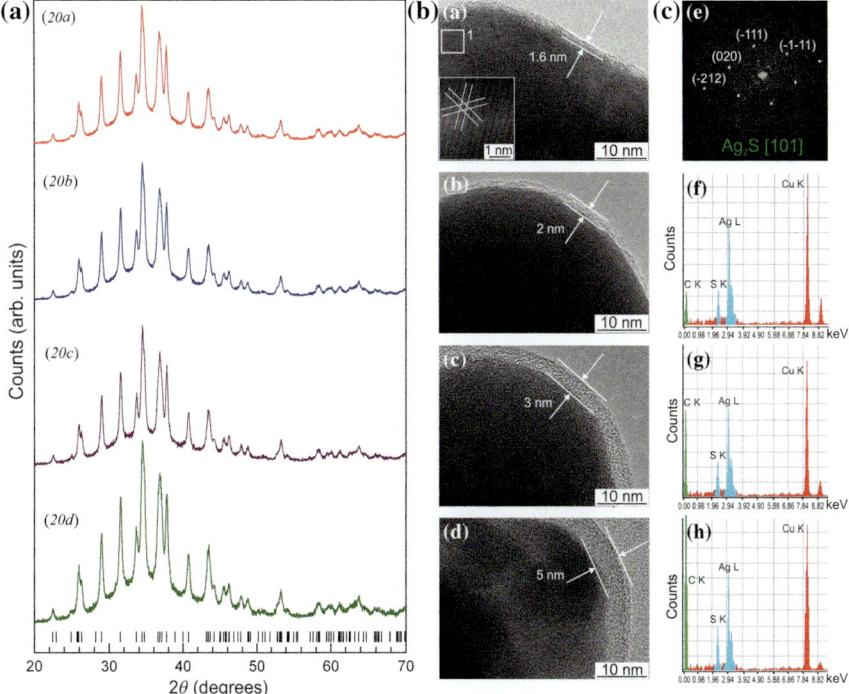

Fig. 4.55 **a** The XRD patterns of monoclinic (space group $P2_1/c$) Ag_2S nanoparticles deposited from the initial reaction mixture 20 and **b** HRTEM images of silver sulfide nanoparticles and the growth of the carbon-containing citrate shell thickness as a function of the nanoparticle dwell time in the solution: (*a*) 20 min, (*b*) 40 min, (*c*) 420 min, (*d*) 1200 min. Vertical marks on XRD patterns indicate the positions of diffraction reflections of nonstoichiometric monoclinic α-$Ag_{1.93}S$ phase. **c** (*e*) selected area of electron diffraction (SAED obtained from the area 1 of HRTEM nanoparticle (*a*); (*f*), (*g*), and (*h*) cumulative elemental EDX patterns of nanoparticles (*b*), (*c*), and (*d*), respectively. Weak $K\alpha$ lines of oxygen O are observed on EDX patterns at 0.525 keV. Reproduced from [163] with permission from Wiley

the drying of the glue coating a carbon grid with voids is formed. For studying nanoparticles with shells, we used such regions where the nanoparticles are located in the carbon network voids. The TEM images (Fig. 4.56) show the arrangement of agglomerated Ag$_2$S@C nanoparticles in the carbon network voids.

It is exactly such regions that we used for producing TEM images and for EDX analysis of nanoparticles. This decreased considerably the possible effect of the carbon network on the desired carbon signal in the EDX spectra. Note that the EDX spectra in Fig. 4.55 are displayed only for qualitative estimation of the carbon concentration growth occurring with an increase in the shell thickness. The EDX results show that oxygen is distributed on the surface of particles. In the first approximation it can be supposed that it is oxygen from the citrate shell or oxygen belonging to adsorbed water. No other elements have been found in the core-shell nanoparticles.

The content of carbon is proportional to the intensity of the C $K\alpha$ line and increases with the growth of the shell thickness (Fig. 4.55f–h). This means that the shell of sulfide nanoparticles contains carbon and is citrate shell.

Indeed, the three carboxylate groups of Na$_3$Cit have strong affinity to silver ions, which favors the attachment of citrate groups on the surface of the silver sulfide nanoparticles and prevents them from aggregating into large particles. In other words, C$_6$H$_5$O$_7^{3-}$ ions are adsorbed on a surface of nanoparticles and form a citrate carbon-containing shell that prevents the growth and agglomeration of the nanoparticles.

The TEM image of the silver sulfide nanoparticles with a shelland its schematic structure are demonstrated in Fig. 4.57. The nanoparticle of Ag$_2$S with a diameter of 37 nm is a core covered by a carbon-containing shell of 3.2 nm thickness. The total diameter of the Ag$_2$S nanoparticle with carbon-containing shell is equal to about 40 nm.

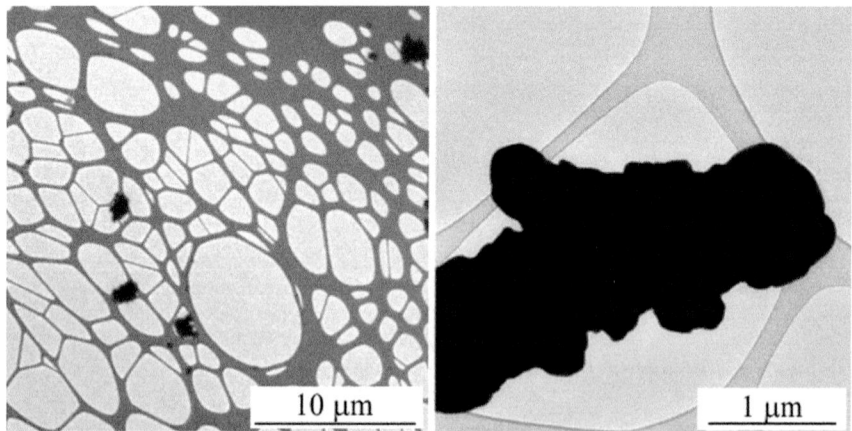

Fig. 4.56 The arrangement of agglomerated Ag$_2$S@C nanoparticles in the carbon grid voids

The variation in the thickness of the carbon-containing citrate shell of silver sulfide nanoparticles versus the nanoparticle dwell time in the colloid solution 20 containing citrate $C_6H_5O_7^{3-}$ ions is demonstrated in Fig. 4.58. When the citrate shell thickness increases, the content of Na_3Cit in the colloid solution lowers and the growth of the shell slows down.

The schematic representation of the mechanism of formation of a carbon-containing citrate shell on the Ag_2S core is shown in Fig. 4.59. In the solutions with sodium citrate, the $C_6H_5O_7^{3-}$ ions are adsorbed on the surface of Ag_2S nanoparticles and first form an uneven discontinuous shell. As the Ag_2S nanoparticle dwell time in the solution increases, the discontinuities are gradually filled with citrate complexes, and a continuous carbon-containing shell is formed. Gradual adsorption of the citrate complexes by the formed coating promotes smoothing of the shell surface and the growth of the shell thickness [64].

The formation of a shell depends on the ratio between the concentrations of silver and sulfur precursors, on the one hand, and the sodium citrate concentration $C_{Na_3Cit} \equiv C_{Cit^{3-}}$, on the other hand (Fig. 4.60). If in the solution $C_{S^{2-}} = C_{Ag^+}/2$ and the citrate concentration is $C_{Cit^{3-}} < C_{Ag^+}/4$, then the amount of citrate ions is not sufficient to form a continuous citrate shell, and an uneven discontinuous coating appears (Fig. 4.60a). A continuous shell is formed when $C_{Ag^+}/4 \leq C_{Cit^{3-}} \leq C_{Ag^+}$ (Fig. 4.60b). When the concentration of citrate is large and $C_{Cit^{3-}} > > C_{Ag^+}$, an amorphous citrate matrix with incorporated nanoparticles of size from ~ 12 to ~ 40 nm is formed is formed in the solution (Fig. 4.60c).

The presence of a protective citrate shell is important for photoluminescence of Ag_2S nanoparticles. For core-shell nanocomposites, the core material plays an important role for tuning the optical emission. Decrease of nanoparticle size should be accompanied by a blue shift of photoluminescence peak. However, observed shift may be less than expected blue shift. It is due to the formation of surface trap states in the band gap and to electron-phonon coupling. To induce a blue shift, one should suppress the formation of surface trap states in the band gap of Ag_2S nanoparticles. The most widely used and efficient method for elimination of surface

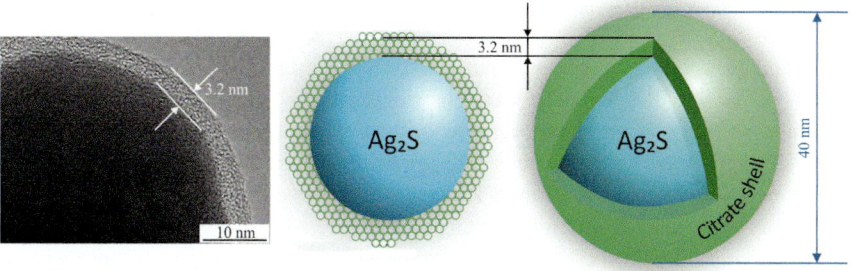

Fig. 4.57 HRTEM image of $Ag_2S@C$ core-shell nanoparticle and its schematic section and representation [64]: the solid core Ag_2S with about 37 nm inner diameter is covered by a citrate C-containing shell of 3.2 nm thickness. The total diameter of the nanoparticle is about 40 nm

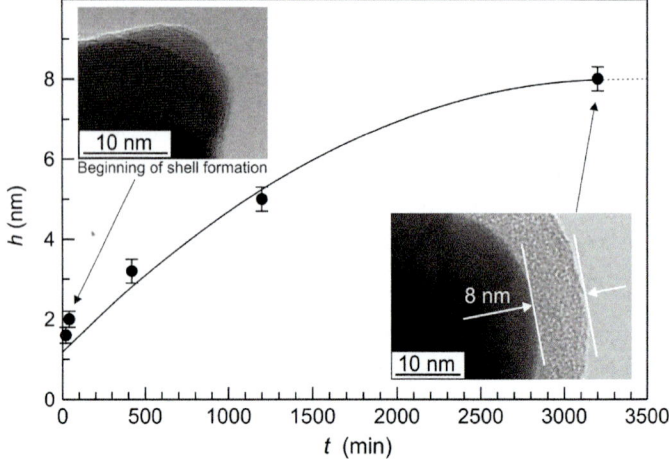

Fig. 4.58 The thickness of the carbon-containing citrate shell versus the dwell time of silver sulfide nanoparticles in the colloid solution 20 (see Table 4.10). Reprinted from [64] with permission from Elsevier

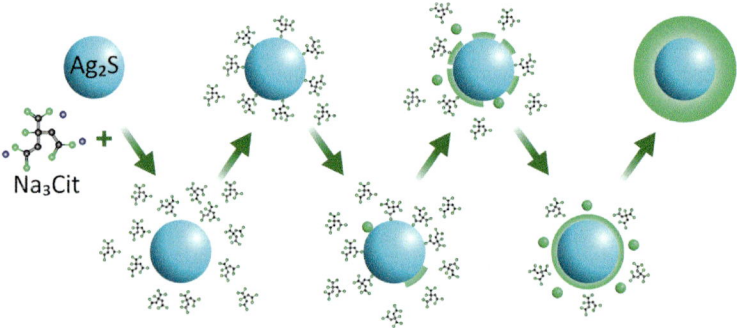

Fig. 4.59 The schematic image of the citrate shell growth mechanism on silver sulfide core of Ag$_2$S@C nanoparticles. Reproduced from [163] with permission from Wiley

states is a passivation of the nanoparticle surface by a protective shell surrounding the Ag$_2$S core. If the core size is constant, a growth of the protective carbon-containing citrate shell thickness leads to weak intensity enhancement of the PL peaks and small shift of peaks into region of lesser wavelength (Fig. 4.61). It should be noted that the position of PL peaks does not depend almost on the size of Ag$_2$S core equal to 12 nm and more.

By varying the initial concentrations of reagents in the reaction mixture and the deposition conditions it is possible to prepare Ag$_2$S@C nanoparticles with different Ag$_2$S core size and different controllable thickness of the carbon-containing non-toxic citrate nanosized shell. Figure 4.62 shows the dependences of the core

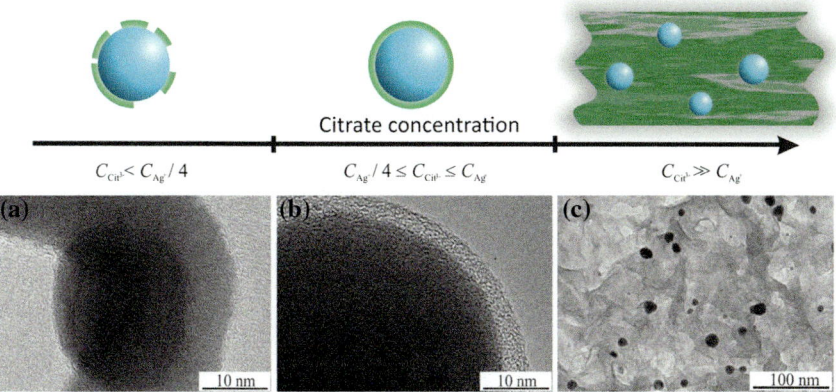

Fig. 4.60 The effect of sodium citrate concentration $C_{Cit^{3-}}$ in colloidal solutions with $C_{S^{2-}} = C_{Ag^+}/2$ on the formation of shell of Ag$_2$S@C nanoparticle: **a** uneven discontinuous citrate shell at insufficient citrate concentration $C_{Cit^{3-}} < C_{Ag^+}/4$; **b** continuous citrate shell; **c** amorphous citrate matrix with Ag$_2$S silver sulfide nanoparticle inclusions of size from ~ 12 to ~ 40 nm at a large excessive citrate concentration $C_{Cit^{3-}} \gg C_{Ag^+}$. Reproduced from [163] with permission from Wiley

size d_{core} and citrate shell thickness h_{shell} of Ag$_2$S@C nanoparticles on the Na$_3$Cit concentration C_{Na_3Cit} and dwell time t of silver sulfide particles in the colloid solutions with concentrations of AgNO$_3$ and Na$_2$S equal to 50 and 25 mmol l^{-1}. It is seen that the core size and the shell thickness grow when the concentration of Na$_3$Cit and the duration of a presence of nanoparticles in colloid solutions increase.

When the concentrations of silver and sulfur ions decreases from 50 and 25 mmol l^{-1} to 5 and 2.5 mmol l^{-1}, the particle size in the colloidal solutions changes from 10 to 40 nm. The reduction on the concentration of reagents leads to the formation of a smaller quantity of nuclei. In that case, the rate of nanoparticle growth is higher than the rate of their formation, and larger particles appear.

Thus, hydrochemical bath deposition method allows preparing Ag$_2$S@C core-shell nanoparticles with pre-assigned size of Ag$_2$S core from 10 and 50 nm and pre-assigned carbon-containing citrate shell thickness from 1.5 to 10 nm.

The art of manufacture of Ag$_2$S@C core-shell nanoparticles with a protective citrate carbon-containing shell by hydrochemical deposition has been patented [185].

4.3.6   Ag$_2$S/Ag Heteronanostructures

The synthesis of multifunctional nanocomposite systems is of great interest. In particular, such heteronanostructures as Ag$_2$S/Ag consisting of a semiconductor and a noble metal are very useful for a variety of applications.

Fig. 4.61 Effect of citrate shell thickness on the PL emission spectra of Ag₂S@C core-shell nanoparticles prepared from the reaction mixtures 20 and 25 (see Table 4.10). The wavelengths corresponding to the maxima of the PL peaks are indicated. The wavelength of excitation is 658 nm

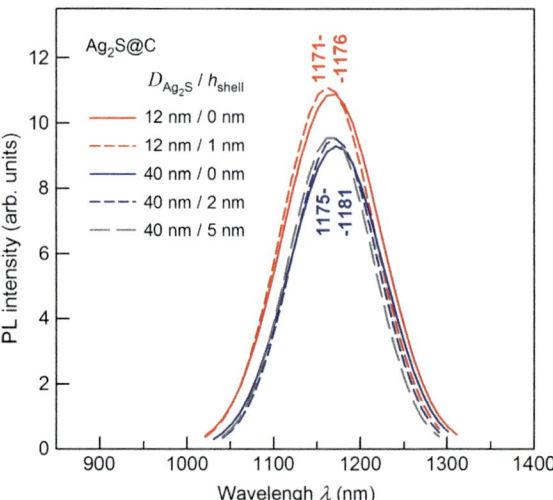

As early as in 1833, Faraday found that heated silver sulfide possesses high ion conductivity comparable to the conductivity of metals in a wide temperature range [186]. He wrote: "I formerly described a substance, sulfuret of silver, whose conducting power was increased by heat... When a piece of that substance, which had been fused and cooled, was introduced into the circuit of a voltaic battery, it stopped the current. Being heated, it acquired conducting powers..." [187]. In contrast to metals, silver sulfide Ag₂S lost ion conductivity upon cooling.

Among composite heterostructures of silver sulfide, the semiconductor/metal Ag₂S/Ag heteronanostructure attracts special attention. This type of heterostructures that contain Ag and Ag₂S nanowires or a Ag film with Ag₂S/Ag nanoclusters are considered as a potential basis for the production of biosensors [188], resistive switches, and nonvolatile memory devices [48–50, 189–191]. The action of the resistive switch is based on the phase transformation between nonconducting α-Ag₂S acanthite and superionic β-Ag₂S argentite.

It is supposed that combined Ag₂S/Ag heterostructure will possess the improved bactericidal properties in comparison with either component separately [54, 57, 192].

In the examination of heteronanostructures, it is necessary to determine the crystallographic indices of reflections obtained experimentally by electron diffraction method or FFT of the HRTEM images. This is especially important for exact identification of phase components which form a heteronanostructure. For compounds with oblique-angled (triclinic and monoclinic) unit cells, errors are sometimes committed during reflection indexing. In particular, Xu et al. [49] incorrectly determined the diffraction reflection indices of monoclinic acanthite α-Ag₂S for seven selected area of electron diffraction (SAED) (see Figures 2*d*1, 2*d*2, 3*c*3, and 4*d*1–4*d*4 in [49]) of Ag/Ag₂S/W heteronanostructure. Besides, authors of study [49] compared their data on acanthite with stale data [8, 133]. Horvath et al. [193]

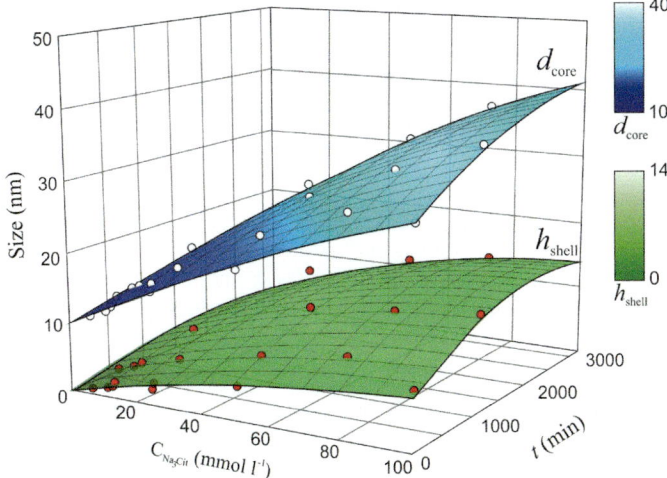

Fig. 4.62 Effect of the Na_3Cit concentration C_{Na_3Cit} and dwell time t of silver sulfide particles in the colloidal solutions on the core size d_{core} and citrate shell thickness h_{shell} of $Ag_2S@C$ nanoparticles (the concentrations of $AgNO_3$ and Na_2S are equal to 50 and 25 mmol l^{-1}, respectively) [64, 163]. Reproduced from [163] with permission from Wiley

incorrectly determined the indices of diffraction reflections and interplanar distances of metallic silver for Ag/polypyrrole composite on Si substrate. Such errors can be found in many studies devoted to the Ag_2S/Ag heteronanostructures in which monoclinic (space group $P2_1/c$) α-Ag_2S acanthite, cubic (space group $Im\overline{3}m$) β-Ag_2S argentite with body centered cubic (bcc) crystal lattice, and cubic (space group $Fm\overline{3}m$) silver Ag can coexist.

Usually, the electron diffraction indices (hkl) are determined by comparing the derived interplanar distances d_{hkl} with the d_{hkl} values corresponding to a unit cell with known parameters. However in low-symmetry structures, the neighbor values of d_{hkl} differ very slightly forming an almost continuous spectrum. That is why the error in determining the indices (hkl) of reflections from the value of d_{hkl} in low-symmetry structures is large. Much more accurately we can determine the angle φ_{refl} between the reflections with assumed indices $(h_ik_il_i)$ on the exprerimental electron diffraction pattern or selected area of electron diffraction, i.e. the angle between the straight lines passing through each reflection and the central spot (000). The coincidence of the estimated and experimental φ_{refl} angles unequivocally proves that the indices $(h_ik_il_i)$ are determined correctly.

In order to calculate the angles φ_{refl}, i.e. the angles between the atomic surface normals, it is necessary to transform the non-orthogonal (triclinic or monoclinic) coordinates into rectangular coordinates and then, using the transformed coordinates, to determine the basis vector of the reciprocal cell. For example, the basis vectors $(100)_{mon}$, $(010)_{mon}$ and $(001)_{mon}$ of the monoclinic unit cell written in rectangular coordinates have the form $\boldsymbol{a} = (a00)$, $\boldsymbol{b} = (0b0)$ and $\boldsymbol{c} = (c\cos\beta\ 0$

$c\sin\beta$), respectively. The basis vectors of the reciprocal lattice found by the known formula have the form $a^* = (1/a \; 0 - \cos\beta/(a\sin\beta))$, $b^* = (0 - 1/b \; 0)$ and $c^* = (0 \; 0 \; 1/(c\sin\beta))$. Accordingly, the arbitrary vector $(hkl)^*_{mon}$ $(hkl)^*_{mon}$ of the reciprocal lattice in the rectangular coordinate system has the explicit form

$$(hkl)^*_{mon} = h\mathbf{a}^* + k\mathbf{b}^* + l\mathbf{c}^* \equiv \left(\frac{h}{a} \quad -\frac{k}{b} \quad \frac{al - hc\cos\beta}{ac\sin\beta} \right), \tag{4.22}$$

where $h_{cub} = h/a$, $k_{cub} = -k/b$ and $l_{cub} = (al - hc\cos\beta)/(ac\sin\beta)$.

The angle φ_{refl} between reflections $(h_1k_1l_1)$ and $(h_2k_2l_2)$ in the rectangular coordinate system is determined by the standard formula

$$\cos\varphi = \frac{h_{1cub}h_{2cub} + k_{1cub}k_{2cub} + l_{1cub}l_{2cub}}{\sqrt{h_{1cub}^2 + k_{1cub}^2 + l_{1cub}^2} \times \sqrt{h_{2cub}^2 + k_{2cub}^2 + l_{2cub}^2}}. \tag{4.23}$$

Replacing in (4.23) the cubic indices h_{cub}, k_{cub} and l_{cub} by their values expressed through monoclinic indices h, k and l, we obtain the formula for determining the angles φ_{refl} between the reflections $(h_1k_1l_1)_{mon}$ and $(h_2k_2l_2)_{mon}$ in the reciprocal lattice of monoclinic structure

$$\cos\varphi = \frac{h_1h_2/a^2 + k_1k_2/b^2}{d_1 \times d_2}$$
$$+ \frac{[l_1l_2a^2 - (h_1l_2 + h_2l_1)ac\cos\beta + h_1h_2c^2\cos^2\beta]/(ac\sin\beta)^2}{d_1 \times d_2} \tag{4.24}$$

where $d_i = \sqrt{(h_i/a)^2 + (k_i/b)^2 + [(l_ia - h_ic\cos\beta)/(ac\sin\beta)]^2}$ with $i = 1$ or 2.

It is easily seen that expression (4.24) at $\beta = 90°$ is transformed into a standard expression suitable for the description of structures with orthogonal (orthorhombic, tetragonal, cubic) unit cells.

Let us consider the incorrect indexing of reflections in the study [49] and the correct determination of the indices of the same reflections.

As an example, at the top of Fig. 4.63 there is figure 2 borrowed from article [49] with a HRTEM image of monoclinic acanthite and an incorrect indexing of reflections on the selected area of electron diffraction (SAED) obtained from regions (1) and (2) of this image by Fast Fourier Transform (FFT). At the bottom of Fig. 4.63, correct indexing of the same reflections with allowance for the experimental values of angles φ_{refl} is shown.

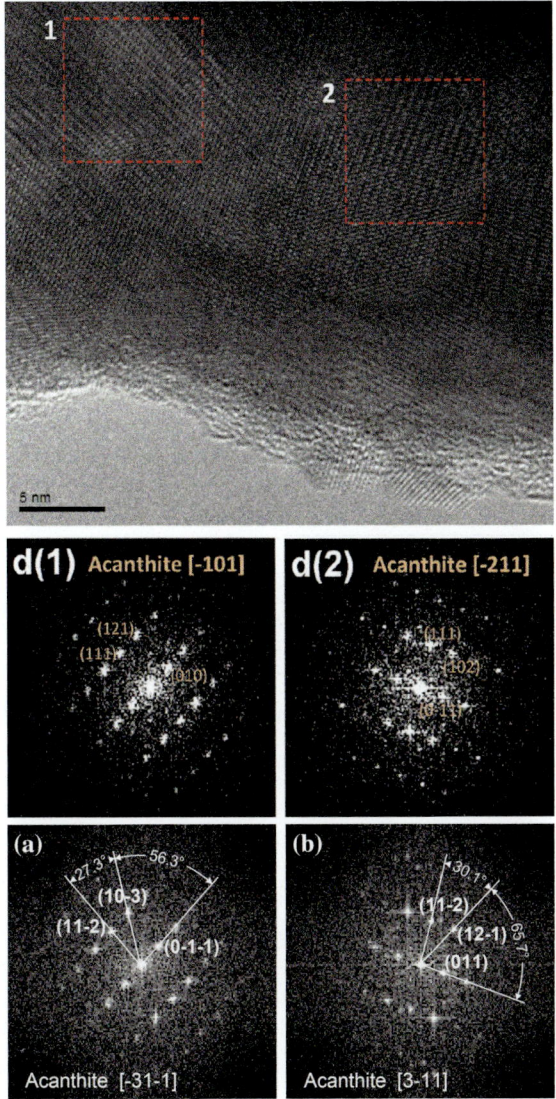

Fig. 4.63 The HRTEM image of monoclinic acanthite from study [49] and indexing of reflections on the selected areas of electron diffraction (SAED) obtained from different regions (1) and (2) of this image by Fast Fourier Transform (FFT). (*d*1) and (*d*2) are incorrect reflection indices reported in study [49]. **a, b** are correct reflection indices on the selected areas of electron diffraction obtained from regions (1) and (2); correct reflection indices are calculated in study [194]

For instance, in Fig. 4.63d1 the experimental angles between the hypothetical reflections (111) and (121) or between (121) and (010) are equal to $\sim 27.3°$ and $\sim 56.3°$ [49]. These indices (111), (121), and (010) have been identified in srudy [49]. Calculated angles between reflections with such indices should be 17.8° and 45.1° and they do not coincide with the experimental values. Consequently, in study [49] the reflection indices are determined incorrectly. We have determined the indices of these reflections using region (1) on the HRTEM image (Fig. 4.63) in article discussed [49]. These reflections have indices (11−2), (10−3) and (0−1−1), the calculated angles between them are equal to 27.0° and 56.6°, which coincides with the experimental values of φ_{refl} (Fig. 4.63a). The analysis showed that these reflections are observed along the [−31−1] zone axis rather than along [−101] as stated by Xu et al. [49].

In Fig. 4.63d2 the angles between the assumed reflections (111) and (102) or between (102) and (011) are $\sim 31°$ and $\sim 67°$ [49]. According to the calculation, the angles between reflections with such indices should be 30.2° and 58.6°. The second angle differs considerably from the experimental value. This means that the reflection indices are determined incorrectly. We have determined the indices of these reflections using region (2) on the HRTEM image (Fig. 4.63). These reflections have indices (11−2), (12−1) and (011), the calculated angles between them are equal to 29.8° and 65.9°, which is close to the experimental values of φ_{refl} (Fig. 4.63b). These reflections are observed along the [3−11] zone axis, rather than along [−211], as stated in paper [49].

It is easily seen that the angles φ_{refl} between the experimental reflections of monoclinic acanthite α-Ag$_2$S on other FFT patterns (Figures 2d1, 2d2, 3c3, 4d1-4d4 in study [49]) do not coincide with the theoretical values of angles between the reflections with those indices that are specified in [49]. This means that the indices (hkl) of observed reflections of monoclinic phase and the zone axes in paper [49] are also determined incorrectly.

Note also that for analysis of the transformation of monoclinic acanthite into cubic argentite, the structure of acanthite would be more properly described in the space group $P2_1/c$ proposed in work [22] and refined for artificial coarse-crystalline acanthite in study [29] and nanocrystalline acanthite in study [30].

Especially we pay attention to the following. In order to exclude the errors at experimental determination of angles φ_{refl}, an area of HRTEM image selected for FFT should have the square form.

As an example, square area of HRTEM image of a nanoparticle of monoclinic acanthite α-Ag$_2$S and electron diffraction pattern obtained by FFT of this image with indexing of observed diffraction spots and the experimental values of angles φ_{refl} are shown in Fig. 4.64. The interplanar distances are determined by inverse FFT of the selected diffraction spots using the Gatan Misroscopy Suite software [172].

The comparison of found interplanar distances 0.229, 0.204, 0.244, and 0.244 nm (see Fig. 4.63b) with data [30] (see also Appendix, Table 4.15) shown that the observed diffraction spots can have following crystallographic indices (030), (023), (013), and (0−13) of monoclinic (space group $P2_1/c$) acanthite.

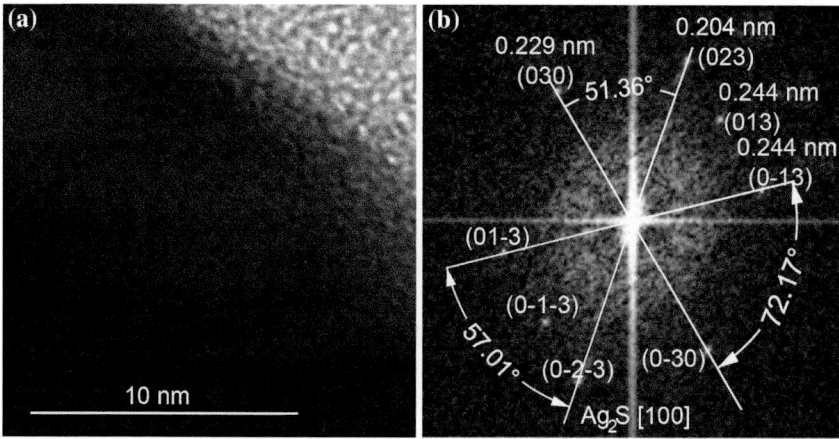

Fig. 4.64 The HRTEM image **a** of nanoparticle of monoclinic acanthite α-Ag$_2$S and **b** electron diffraction pattern obtained by FFT of this image with indexing of observed diffraction spots and the experimental values of angles φ_{refl} [194]

As is seen, the experimental angles φ_{refl} between the hypothetical reflections (030) and (023) or between (023) and (0−13) are equal to ∼51.4° and ∼57.0° (Fig. 4.64b). The calculated angles φ_{refl} between these reflections should be 53.3° and 57.2° and they coincide with the experimental angles within the limits of measurement errors. Analogously, the experimental angles φ_{refl} between the hypothetical reflections (013) and (0−13) or between (0−13) and (0−30) are equal to ∼39.5° and ∼72.2°. The calculated angles φ_{refl} between these reflections are equal 40.9° and 69.5° and coincide with the experimental angles. The analysis showed that these reflections are observed along the [100] zone axis of monoclinic (space group $P2_1/c$) acanthite α-Ag$_2$S.

Let us turn to a discussion of Ag$_2$S/Ag heteronanostructures prepared by hydrochemical deposition method.

Hydrochemical deposition has been applied for the first time to synthesis of Ag$_2$S/Ag heteronanostructures in studies [163, 165, 166, 168, 194].

Ag$_2$S/Ag heteronanostructures have been synthesized by chemical deposition from aqueous solutions of AgNO$_3$, Na$_2$S, and Na$_3$Cit with reduced concentration of sodium sulfide $C_{Na_2S} < C_{AgNO_3}/2$ (see Table 4.10). Synthesis was carried out under illumination of solutions by light emitting diode with a 450 nm wavelength and the irradiation intensity of 15 mW cm^{-2}. Two processes take place simultaneously in such mixtures: formation of Ag$_2$S sulfide and appearance of Ag nanoparticles as a result of photochemical reduction reaction (4.17) of Ag^{2+} ions by C$_6$H$_5$O$_7^{3-}$ citrate ions. At certain synthesis conditions, Ag and Ag$_2$S nanoparticles are united in Ag$_2$S/Ag nanocomposites.

The scheme of deposition of Ag$_2$S/Ag heteronanostructures from aqueous solutions of AgNO$_3$, Na$_2$S and Na$_3$Cit, TEM and HRTEM images of Ag$_2$S, Ag and

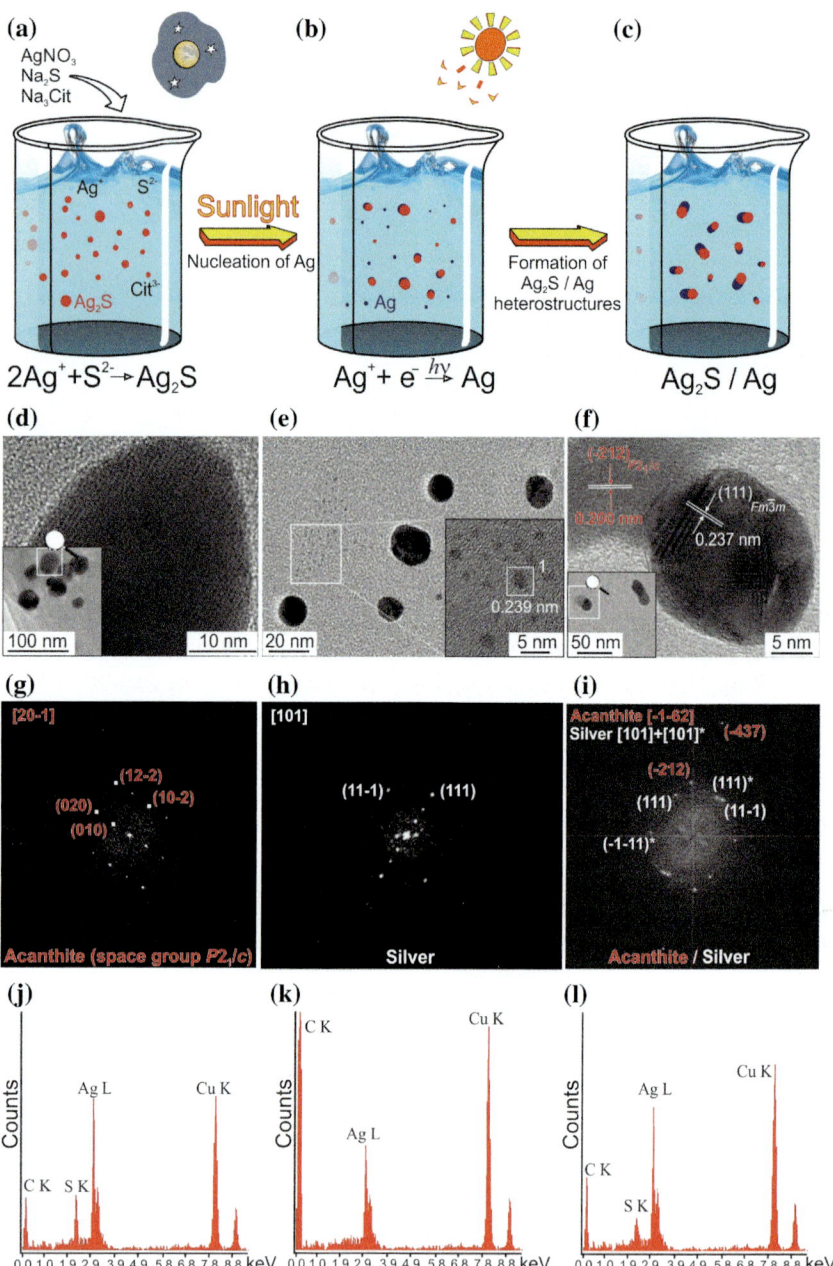

Fig. 4.65 Hydrochemical bath deposition of Ag$_2$S/Ag heteronanostructures. *1st row* scheme of deposition. *2nd row* TEM and HRTEM images of **d** Ag$_2$S nanoparticles, **e** Ag nanoparticles, and **f** Ag$_2$S/Ag heteronanostructures. *3rd row* FFT patterns **g–i** obtained from the **d, e, f** HRTEM images, respectively. *4th row* EDX analysis of **j** Ag$_2$S nanoparticle, **k** Ag nanoparticle, and **l** Ag$_2$S/Ag heteronanostructure. Reproduced from [165] with permission of Springer

Ag_2S/Ag particles and their diffraction patterns, as well as EDX analysis of Ag_2S, Ag particles and Ag_2S/Ag heteronanostructure are shown in Fig. 4.65.

Synthesis was carried out in the following sequence: a complexing agent was added to silver nitrate in the dark; then a solution of Na_2S was poured into the prepared solution (Fig. 4.65a). As a result, deposition of silver sulfide powder takes place. Further the solution was irradiated with monochromatic light with wavelength 450 nm. In accordance with photochemical reaction (4.17), the citrate ions $C_6H_5O_7^{3-}$ reduce the Ag^+ ions to Ag nanoparticles in aqueous solutions (Fig. 4.65b). The reduction of silver at the surface of Ag_2S nanoparticles leads to the formation of the Ag_2S/Ag heteronanostructures (Fig. 4.65c).

As is seen, sodium citrate plays a triple role in synthesis of Ag_2S/Ag heteronanostructures. Firstly, it is a complexing and stabilizing agent during deposition of Ag_2S sulfide nanoparticles. Secondly, during deposition in the light sodium citrate reduce the Ag^+ ions to metallic silver nanoparticles. Thirdly, citrate is absorbed on nanoparticles and prevents their agglomeration.

The HRTEM images of deposited Ag_2S, Ag and Ag_2S/Ag particles are shown in Fig. 4.65d–f, respectively. The diffraction patterns (selected area of electron diffraction SAED) (Fig. 4.65g–i) of these particles are obtained by Fast Fourier Transformation (FFT) of their HRTEM images. In studies [163, 165, 166, 194], the indices (hkl) of the electron diffraction reflections have been determined with taking into account interplanar distances d_{hkl} and angles φ_{refl} between observed reflections.

The observed set (Fig. 4.65g) of diffraction reflections and interplanar distances of silver sulfide nanoparticle corresponds to monoclinic (space group $P2_1/c$) nanocrystalline acanthite α-$Ag_{1.93}S$ [30]. The Ag nanoparticle with cubic (space group $Fm\bar{3}m$) structure clearly exhibits microtwinning in the direction of the [111] planes. The selected area of electron diffraction of the HRTEM image of silver nanoparticle confirms the observed twinning (Fig. 4.65h). Diffraction (Fig. 4.65i) obtained by FFT of HRTEM image (Fig. 4.65f) of Ag_2S/Ag heteronanostructure revealed reflections of monoclinic silver sulfide and twinned reflections of cubic silver.

According to the EDX results (Fig. 4.65j–l), the content of silver Ag and sulfur S in Ag_2S nanoparticle is equal $\sim 86.3 \pm 0.4$ and $\sim 12.9 \pm 0.1$ wt.% and corresponds to $\sim Ag_{1.95-1.98}S$ sulfide. Ag nanoparticle contains silver only, and Ag_2S/Ag heteronanostructure contains about 87.8 and 11.5 wt.% of Ag and S, respectively. Besides silver and sulfur, the EDX spectra contain a $K\alpha$ lines of carbon C and copper Cu. These $K\alpha$ lines appear from a carbon-containing collodium-glue film which covered copper grid onto which colloidal solutions with examined nanoparticles were applied.

All the Ag_2S/Ag heteronanostructures were examined by XRD method on a Shimadzu XRD-7000 and STADI-P (STOE) diffractometers in $CuK\alpha_1$ radiation. The XRD measurements were performed in the angle interval $2\theta = 20$–$95°$ with a

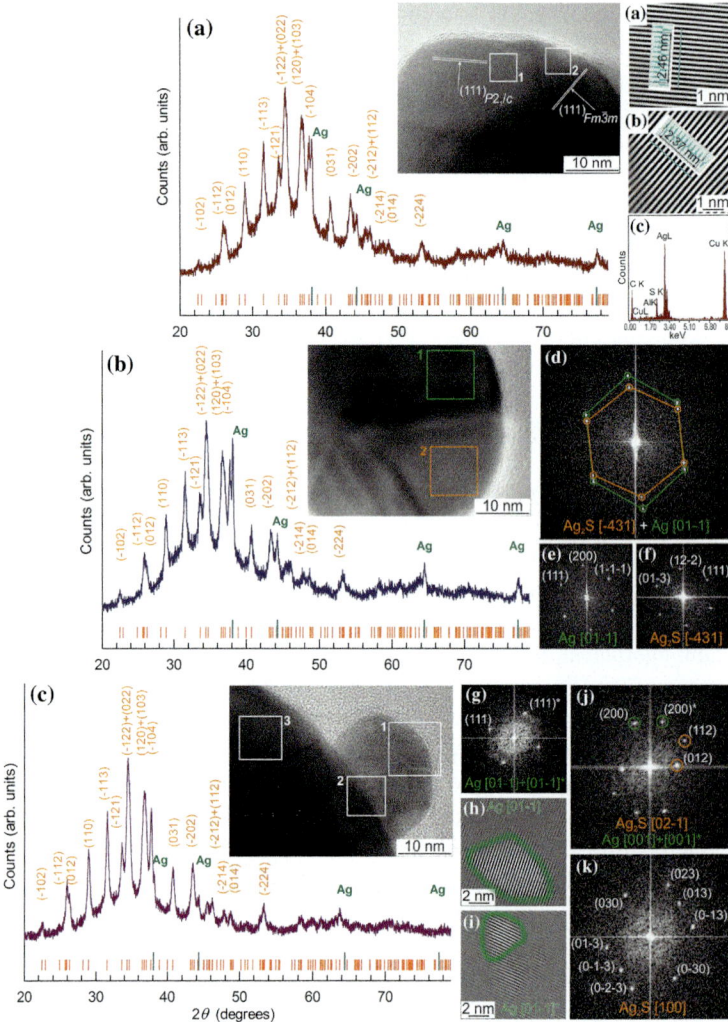

Fig. 4.66 The XRD patterns and HRTEM images of Ag₂S/Ag heteronanostructures. The long and short ticks on XRD patterns correspond to reflections of cubic metallic Ag and monoclinic Ag₂S silver sulfide, respectively. *Row A* Ag₂S/Ag heteronanostructures produced from the reaction mixture 26; (*a*), (*b*), and (*c*) filtered images of areas (*1*) and (*2*) of heteronanostructure (**a**) and its EDX analysis. *Row B* Ag₂S/Ag heteronanostructures produced from the reaction mixture 27; (*d*), (*e*), and (*f*) selected areas of electron diffraction obtained by FFT of HRTEM image of the whole composite heteronanostructure (**b**) and its areas (*1*) and (*2*). *Row C* Ag₂S/Ag heteronanostructures produced from the reaction mixture 28; (*g*), (*j*), and (*k*) selected areas of electron diffraction obtained by FFT of areas (*1*), (*2*), and (*3*) of heteronanostructure (**c**). (*h*) and (*i*) filtered images of area (*1*) containing twinning Ag₂S particle. Twinned reflections of silver are marked by a sign *. Reproduced from [165] with permission of Springer

step of $\Delta(2\theta) = 0.02°$ and scanning time of 10 s in each point. The determination of the crystal lattice parameters and final refinement of the structure of synthesized heteronanostructures were carried out with the use of the X'Pert Plus [143] and X'Pert HighScore Plus [157] software packages.

The XRD patterns of Ag_2S/Ag heteronanostructures produced from reaction mixtures 26, 27, and 28 are shown in rows A, B, and C of Fig. 4.66, respectively. The quantitative analysis of the XRD patterns and comparison with data [30] have shown that the observed set of diffraction reflections corresponds to nanocrystalline nonstoichiometric monoclinic (space group $P2_1/c$) acanthite α-$Ag_{1.93}S$ and cubic (space group $Fm\overline{3}m$) silver Ag. The diffraction reflection broadening is indicative of the nanosized state of the both phases. The content of Ag and Ag_2S in the nanopowders deposited from reaction mixtures 26, 27, and 28 is equal to ~ 5.0 and ~ 95.0, ~ 7.5 and ~ 92.5, ~ 2.0 and ~ 98.0 wt.%, respectively.

For the electron microscopy study of the two-phase nanoparticles, authors of studies [163, 165, 166, 194] used the colloidal solutions above the deposited powders. According to the HRTEM data, the Ag_2S and Ag nanoparticles are in direct contact and, thus, form the heteronanostructures. The size of Ag_2S and Ag nanoparticles is 45–50 and 15–20 nm, respectively.

HRTEM image of Ag_2S/Ag heteronanostructure produced from the reaction mixture 26 (see Table 4.10), filtered inversed FTT images (*a*) and (*b*) of areas (1) and (2) isolated by white quadrates in this HRTEM image, and cumulative EDX analysis of this heteronanostructure are shown in row A of Fig. 4.66. The observed interplanar distances for area (1) correspond to monoclinic (space group $P2_1/c$) acanthite α-Ag_2S, whereas interplanar distances for area (2) correspond to cubic (space group $F m\overline{3}m$) metallic silver. According to the EDX results (Fig. 4.66c), the content of silver Ag and sulfur S in this whole heteronanostructure is equal about 12.3 and 87.7 wt.%, respectively, and corresponds to $\sim Ag_{2.12}S$. XRD and HRTEM image of Ag_2S/Ag heteronanostructure produced from the reaction mixture 27, and also selected areas of electron diffraction calculated by FFT is shown in row B of Fig. 4.66. The diffraction pattern (*d*) obtained by FFT of HRTEM image of the whole this composite heteronanostructure contains two set of diffraction reflections corresponding to monoclinic silver sulfide and cubic silver. The selected areas of electron diffraction (*e*) and (*f*) are obtained by FFT from areas (1) and (2) isolated by green and orange quadrates. The observed set (*e*) of spots (111), (200), and (1-1-1) corresponds to the [01-1] plane of the reciprocal lattice of cubic Ag. The interplanar distances for area (2) and the set (*f*) of spots (01-3), (12-2), and (111) correspond to monoclinic α-Ag_2S acanthite. Nanoparticle Ag_2S has zone axis $[-431]_{P2_1/c}$.

HRTEM image of Ag_2S/Ag heteronanostructure produced from the reaction mixture 28 (see Table 4.10) is shown in row C of Fig. 4.66. Areas (1) and (3) correspond to Ag and Ag_2S, respectively. The Ag nanoparticle clearly exhibits microtwinning in the direction of the [01-1] planes. The performed FFT (Fig. 4.66g) of area (1) of silver nanoparticle and filtered inversed FFT images (Fig. 4.66h, i) confirms the observed twinning. Area (2) corresponds to that part of heteronanostructure where the Ag₂S and Ag nanoparticles are in direct contact. Indeed, set of spots (Fig. 4.66j) contains diffraction reflections of both phases Ag₂S and Ag. The interplanar distances for area (3) and the set of eight spots (Fig. 4.66k) corresponds to monoclinic (space group $P2_1/c$) α-Ag₂S acanthite. Nanoparticle Ag₂S has zone axis $[100]_{P2_1/c}$.

The DLS measurements of the particle size in colloidal solutions 26, 27, and 28 showed that the size and volume distributions are bimodal (Fig. 4.67). This means that these colloidal solutions contain small particles and larger particles. Taking into consideration the TEM data for Ag₂S/Ag heteronanostructures, it can be supposed that small particles are Ag particles and larger particles are Ag₂S particles. The Ag particles are 2–3 times smaller in size than the Ag₂S particles. Therefore, the volume of individual Ag particle is $\sim$ 10–20 times smaller than the volume of Ag₂S particle. According to the DLS data, volume content of the small particles is much smaller than that for the larger particles. Because of the small volume of Ag particles their amount is comparable with the amount of Ag₂S particles.

A typical Ag₂S/Ag heteronanostructure contained in the colloidal solution prepared from reaction mixture 28 (see Table 4.10) is shown in Fig. 4.68 as an example. As is seen, the Ag₂S and Ag nanoparticles are in immediate contact. The Ag nanoparticle clearly exhibits microtwinning in the direction of the [111] planes. The selected area of electron diffraction obtained by FFT of the HRTEM image of silver nanoparticle confirms the observed twinning. According to the EDX analysis data, the nanoparticle contains only silver; there are also lines from the copper mesh, on which the nanoparticle was during investigation. The interplanar distances observed for silver sulfide nanoparticle correspond to monoclinic acanthite. The zone axis of the Ag₂S nanoparticle matrix is $[101]_{P2_1/c}$.

The produced Ag₂S/Ag heteronanostructures combine ionic and electronic conductors. Such heterostructures can be used to create biosensors, resistive switches and nonvolatile memory devices [48–50, 188–191, 195]. The resistive

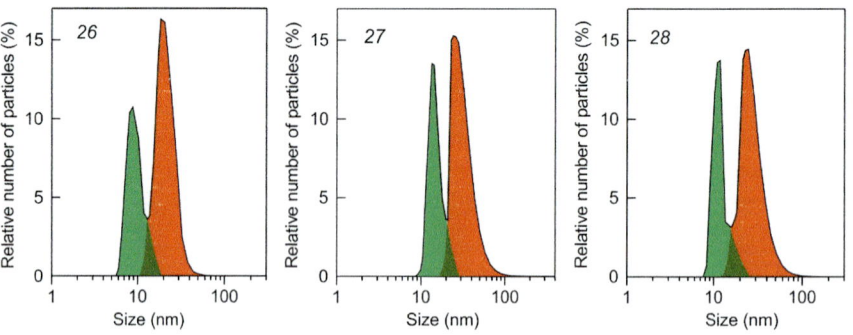

Fig. 4.67 Bimodal size distributions of nanoparticles for colloidal solutions 26, 27, and 28 (see Table 4.10). Maxima of distributions at $\sim$7–15 nm correspond to Ag nanoparticles mainly, maxima of distributions in the region of 20–50 nm correspond to Ag₂S nanoparticles predominantly

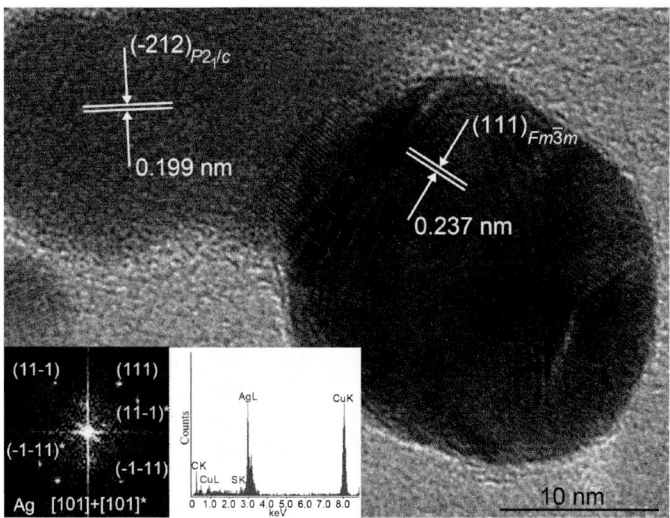

Fig. 4.68 The HRTEM image of Ag₂S/Ag heteronanostructure. Monoclinic (space group $P2_1/c$) Ag₂S nanoparticle with α-Ag₂S acanthite structure (*upper left corner*) is in direct contact with cubic (space group $Fm\bar{3}m$) Ag particle. The electron diffraction pattern obtained by FFT of HRTEM image of Ag particle, and cumulative elemental EDX pattern of this Ag particle are shown (*lower left corner*). Observed interplanar distance 0.199 nm coincides with the distance between (−212) atomic planes of monoclinic acanthite, and interplanar distance 0.237 nm coincides with the distance between (111) atomic planes of cubic silver, respectively. Microtwinning is observed distinctly in Ag nanoparticle in direction of [111] planes, and this is confirmed by FFT. Twinned reflections of silver are marked by a sign *, zone axes are [101] + [101]*. Reproduced from [163] with permission from Wiley

switches of this type consist of a superionic conductor located between two metal electrodes. In the case of Ag₂S/Ag heterostructures, one of the electrodes is silver, and the second electrode can be such metals as Pt, Au, Cu, and W. The conductivity of silver is $6.3 \times 10^5 \ \Omega^{-1} \ cm^{-1}$, that of the β-Ag₂S argentite and α-Ag₂S acanthite phases are about 1.6×10^3 and only $2.5 \times 10^{-3} \ \Omega^{-1} \ cm^{-1}$ at room temperature [11], respectively. Thus, the conductivity of acanthite is 6 orders of magnitude lower than that for the argentite phase. It is thus reasonable to conclude that the conducting channel in Ag₂S/Ag heterostructures can be made of a mixture of Ag and β-Ag₂S argentite.

In studies [163, 165, 166, 194], Ag₂S/Ag heteronanostructures formed by Ag₂S and Ag nanoparticles have been produced by a simple method of hydrochemical bath deposition. Deposition of Ag₂S/Ag heterostructures on a substrate coated with a thin conducting metallic layer will make it possible to form a structure, which can work as a resistive switch. The action of the switch is based on the phase transformation of nonconducting α-Ag$_2$S acanthite into β-Ag$_2$S argentite exhibiting superionic conduction. The transition into a high-conduction state is due to abrupt disordering of the cationic sublattice. Authors of studies [196, 197] have shown that a high-conduction state of a crystal can be achieved by external electric field induced "melting" of the cationic sublattice taking place without heating of the crystal. Such transformation occurring as a result of applied external electric field was confirmed with respect to nanocrystalline silver sulfide in studies [49, 50, 189, 190]. The effect of external electric field induced abrupt disordering allows the realization of the superionic state of silver sulfide at room temperature. This opens up the possibilities for practical use of materials based on silver sulfide.

Authors of works [165, 166, 194] have studied preliminarily the switching processes in Ag$_2$S/Ag heteronanostructure. For this purpose, a metallic Pt micro-contact was supplied to Ag$_2$S/Ag heteronanostructure and bias voltage was impressed so that Ag electrode was charged positively. When positive bias voltage increases to 500 mV, the conduction of the heteronanostructure grows and the nanodevice transforms into the conducting state, i.e., the on-state. The bias back to negative values decreases the conduction and the nanodevice transforms into the off-state.

Figure 4.69 displays a region of Ag$_2$S/Ag heteronanostructure where change of crystal structure at the transition from the off-state (Fig. 4.69a) to the on-state (Fig. 4.69b) can be observed. Using FFT of HRTEM images, authors of studies [164, 165, 193] obtained the selected areas of electron diffraction (SAEDs) (Fig. 4.69c, d).

The electron diffraction pattern of Ag$_2$S/Ag heteronanostructure in the off-state is shown in Fig. 4.69c. This SAED contains (111), (11$-$1) spots and twinning reflection (111)* corresponding to cubic (space group $Fm\bar{3}m$) silver, as well as (2 $-$12) and (030) spots corresponding to monoclinic (space group $P2_1/c$) α-Ag$_2$S acanthite. The observed angle of 100.3° between (2$-$12) and (030) spots of monoclinic acanthite coincides within measurement error with the theoretical value 100.7°. Experimental angles between diffraction spots of cubic silver coincide with the theoretical values.

Then, a positive bias was applied to the Ag$_2$S/Ag heteronanostructure in order to turn it on. HRTEM image of Ag$_2$S/Ag heteronanostructure in the on-state and its selected electron diffraction pattern are presented in Fig. 4.69b, d, respectively. The electron diffraction pattern (Fig. 4.69d) contains two sets of spots corresponding to two cubic phases. The (111), (200), (1$-$1$-$1) spots and the twinning spot (00$-$2)* correspond to cubic (space group $Fm\bar{3}m$) silver, and the (011) and (112) spots correspond to cubic (space group $Im\bar{3}m$) β-Ag$_2$S argentite. The observed angle of 30.1° between the (011) and (112) spots of cubic β-Ag$_2$S argentite coincides with the theoretical value 30°. Experimental angles between diffraction spots of cubic silver (Fig. 4.69c, d) coincide with theoretical values.

Thus, the applied bias really leads to the appearance of conducting β-Ag$_2$S argentite instead of nonconducting α-Ag$_2$S acanthite and the formation of conductive channel from argentite β-Ag$_2$S and silver Ag.

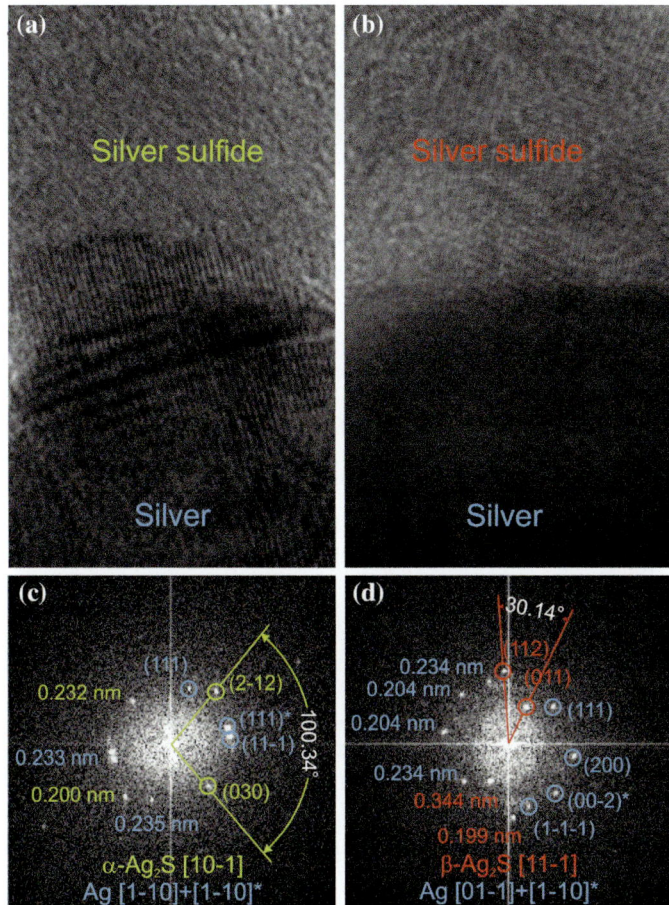

Fig. 4.69 HRTEM images of region of transition between Ag and Ag$_2$S for off-state (**a**) and on-state (**b**) of Ag$_2$S/Ag heteronanostructure. The on-state arises as a result of applied external positive bias voltage to this Ag$_2$S/Ag heteronanostructure. The Pt electrode is located on the top part of the image, and Ag electrode is in the bottom part. Electron diffraction patterns **c**, **d** are obtained by FFT of HRTEM images **a**, **b**, respectively. When Ag$_2$S/Ag heteronanostructure is transformed from the off-state into the on-state, along with Ag spots, the (011) and (112) spots of β-Ag$_2$S argentite appear on the diffraction pattern **d** instead of acanthite spots. Reproduced from [165] with permission of Springer

Current-voltage $I(V)$ characteristics of the resistive switches based on Ag$_2$S/Ag heteronanostructures which were produced by different methods are presented in studies [49, 50, 189, 190]. As a rule, the bias voltage is equal from ±150 to ±500 mV. According to [49], minimum energy barrier for an Ag cation jumping from one atomic site to the other is about 130 meV.

A current-voltage $I(V)$ characteristic of the resistive switch based on Ag$_2$S/Ag heteronanostructure produced by hydrochemical deposition in studies [165, 166,

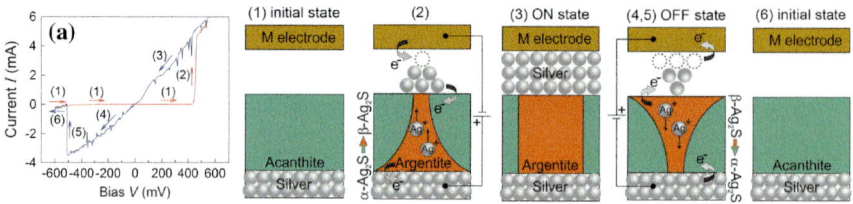

Fig. 4.70 Generalized scheme of the operation of an Ag$_2$S/Ag heteronanostructure-based switch [168, 194]: **a** typical current-voltage characteristic of the switching; (*1*) initial nonconducting state; (*2*) the appearance of a conductive channel upon the application of an external electric field that induces the transformation of acanthite α-Ag$_2$S into argentite β-Ag$_2$S; (*3*) work on-state with a continuous conductive channel formed from argentite β-Ag$_2$S and silver Ag; (*4, 5*) off-state and break down of the conductive channel upon the application of negative bias and the transformation of argentite into initial acanthite; (*6*) initial nonconducting state after disappearance of the conductive channel and turning-off of the switch. M is metal electrode. The numbers from 1 to 6 for different states of switching correspond to the order of the events on **a** current voltage characteristic. Reprinted from [168] with permission from Elsevier

194] is shown in Fig. 4.70a. The schematic operation of this switch is shown in the same Fig. 4.70.

The initial Ag$_2$S phase is a nonconducting acanthite α-Ag$_2$S (Fig. 4.70(1)). When a positive bias is applied, Ag$^+$ cations start to move toward the negatively charged cathode M and are reduced to Ag atoms during their transport. At the same time, the α-Ag$_2$S phase transforms into superionic β-Ag$_2$S argentite (Fig. 4.70(2)), and a continuous conductive channel is formed (Fig. 4.70(3)). The continuous conductive channel which is formed from argentite β-Ag$_2$S and silver Ag is retained, when the external field is turned off. This phenomenon can be considered as a memory effect (Fig. 4.70(3)). If a negative (reverse) bias is applied to the switch, the Ag nanocrystals start dissolving in argentite, the Ag$^+$ cations move to the anode, argentite transforms into the initial acanthite again, the conductive channel breaks down, and off-state is realized (Fig. 4.70(4, 5)). Because of the formation of nonconducting acanthite, the conductive channel disappears, the switch transforms into the initial state and is turned off (Fig. 4.70(6)). If positive bias is applied once again, the destroyed conductive channel is restored due to the appearance of argentite and the formation of silver.

According to [49, 50, 194], the bias voltage which is sufficient to turn on and off the switch is in range from ± 0.2 to ± 10.0 V depending on the metal M used as the second electrode.

The UV-vis optical absorption spectra of colloidal solutions 26, 27, and 28 containing Ag$_2$S/Ag composite nanoparticles are shown in Fig. 4.71. The UV-vis optical absorption spectrum of colloidal solution X containing only Ag$_2$S nanoparticles without an impurity of metallic silver nanoparticles is also shown for comparison. This colloidal solution has been prepared from the reaction mixture with concentrations of AgNO$_3$, Na$_2$S, and Na$_3$Cit equal to 50, (25 + δ), and 10 mmol l^{-1}, respectively. Small excess $\delta = 0.1$ mmol l^{-1} of sodium sulfide Na$_2$S

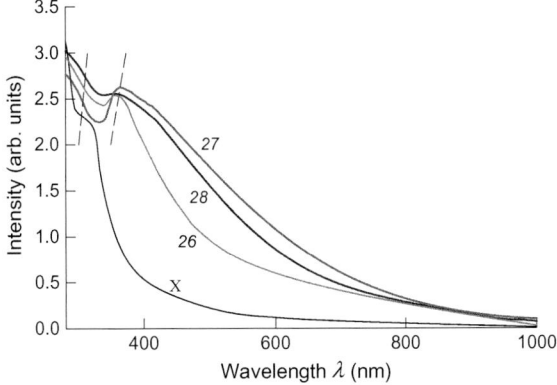

Fig. 4.71 The UV-vis optical absorption spectra of colloidal solutions 26, 27, and 28 (see Table 4.10). The spectrum of colloidal solution X containing only Ag_2S nanoparticles without an impurity of metallic silver nanoparticles is shown for comparison. Colloidal solutions 26, 27, and 28 contain Ag_2S/Ag composite nanoparticles. The broad absorption peak observed in the spectra of solutions 26–28 at ~ 360–450 nm corresponds to Ag nanoparticles. The weak absorption peaks observed in the spectrum of solutions 26, 27, 28, and X at ~ 312–320 nm correspond to Ag_2S nanoparticles. The position of the absorption peaks are marked by *dotted lines*. Reproduced from [165] with permission of Springer

is necessary for synthesis of colloidal solution without an impurity of metallic Ag. According to [68, 70], the weak diffused absorption peak observed in UV-vis optical spectra at ~ 312–320 nm corresponds to Ag_2S nanoparticles. The relative intensity of this absorption peak is proportional to the volume content of silver sulfide in the solution.

The spectra 26, 27, and 28 contain a broad intensive absorption band in the region of ~ 360–450 nm, which corresponds to metallic silver nanoparticles and is due to surface plasmon resonance (SPR) [58]. It is known that the absorption band of Ag nanoparticles is retained in the absorption spectra of silver sulfide colloidal solutions containing impurity Ag nanoparticles [68, 70]. Silver nanoparticles in the spectral SPR region exhibit the greatest absorption as compared with any other nanoparticles of the same size, and the increase in the size of Ag nanoparticles manifests itself in the growth of the SPR band intensity [182]. According to DLS data (see Fig. 4.67), the average Ag nanoparticle size in the colloidal solutions 26, 27, and 28 is 9 ± 2, 14 ± 3, and 11 ± 3 nm, respectively. Indeed, the absorption band of Ag particles (Fig. 4.71) is most intensive for colloidal solution 27 with the largest (~ 14 nm) silver nanoparticles and it decreases in the transition to colloidal solution 26 having the smallest Ag nanoparticles with size about 9 nm.

Ag_2S/Ag nanocomposites can be applied as promising biosensing probes. It is known that silver nanoparticles can be used as biosensors owing to their unique surface plasmon resonance, which depends on the size and shape of particles [188, 198, 199]. However, Ag nanoparticles are easily oxidized; that is why they should be protected. The combination of Ag nanoparticles and chemically stable silver sulfide Ag$_2$S allows one to increase the stability of Ag nanoparticles and to use Ag$_2$S/Ag nanocomposites and Ag@Ag$_2$S core-shell structures for biosensing applications in the future.

According to [192, 200], Ag$_2$S/Ag heteronanostructures possess considerable antibacterial activity and can be used in biology and medicine. This agrees with data [57, 58, 164] on the antibacterial activity of Ag$_2$S colloidal solutions and nanopowders.

4.4 Hydrochemical Bath Deposition: Advantages and Challenges

Investigations of last years have shown that the chemical bath deposition from aqueous solutions of silver nitrate, sodium sulfide, and sodium citrate allows synthesizing different types of nanostructured silver sulfide Ag$_2$S. By varying the ratio between the concentrations of reagents in the reaction mixtures it is possible to prepare Ag$_2$S particles with a pre-assigned average size ranging in the interval from $\sim$500 to $\sim$10–20 nm, colloidal solutions of Ag$_2$S quantum dots, stable Ag$_2$S@C core-shell nanoparticles with Ag$_2$S core coated with a continuous protective carbon-containing citrate shell, and Ag$_2$S/Ag heteronanostructures. For the preparation of colloidal solutions and deposition of silver sulfide powders without Ag impurity, it is necessary to use the reaction mixtures with certain excess of sodium sulfide Na$_2$S. The size of Ag$_2$S particles in deposited powders and in colloidal solutions depends on the initial concentrations of reagents.

The synthesized powders of silver sulfide Ag$_2$S have monoclinic (space group $P2_1/c$) structure of α-Ag$_2$S acanthite type. The silver sulfide nanopowders with the particle size smaller than 50 nm are nonstoichiometric and have the composition $\sim$Ag$_{1.93}$S.

Silver sulfide quantum dots with a size from 2–3 to $\sim$20 nm were prepared from weak aqueous solution of silver nitrate AgNO$_3$ and sodium sulfide Na$_2$S.

Ag$_2$S@C nanoparticles with a continuous shell are formed when the concentration of citrate ions in the reaction mixture is within the interval $C_{Ag^+}/4 \leq C_{Cit^{3-}} \leq C_{Ag^+}$. Ag$_2$S/Ag heteronanostructures are formed in aqueous solutions of AgNO$_3$, Na$_2$S, and Na$_3$Cit with decreased concentration of sodium sulfide, $C_{Na_2S} < C_{AgNO_3}/2$, during synthesis in the light.

Decreasing average nanoparticle size from $\sim$500 to $\sim$17 nm is accompanied by increase in the band gap of the Ag$_2$S nanopowders from $\sim$0.88 $\pm$ 0.1 to $\sim$1.67 $\pm$ 0.1 eV.

Experimental studying of the quantum confinement effect of silver sulfide quantum dots has shown that the exciton diameter of Ag$_2$S is about 4.2 nm.

The important advantage of a discussed hydrochemical bath deposition of nanostructured silver sulfide is the use of harmless substances only. No substances which are hazardous to human health and the environment are used or generated during in this synthesis. One more important advantage of hydrochemical deposition in comparison with other methods is a reproducibility of obtained results.

Synthesized forms of nanostructured Ag_2S have a lot of possible applications. First of all, stable Ag_2S colloidal solutions and Ag_2S quantum dots are suitable for biological and medical application as biomarkers because these solutions are non-toxic and do not contain any hazardous substances. The Ag_2S quantum dots exhibited bright photoluminescence and excellent photostability. Therefore, Ag_2S quantum dots with a size less than 3–4 nm can be used for near-infrared applications in the biomedical field as luminophores or biomarkers. A creation of a protective carbon-containing citrate shell on Ag_2S quantum dots allows to prevent agglomeration and growth of quantum dots and to keep their optical properties.

Nanocrystalline Ag_2S powders can be used for production of such electronic devices as photovoltaic cells, photoconductors, and infra-red detectors. Besides, Ag_2S nanoparticles possess antibacterial action, and can be used as effective agent for asepsis and disinfection.

Ag_2S/Ag heteronanostructure unites ionic and electronic conductors. A high-conducting state of such heteronanostructure can be induced by external electric field without heating of this composite owing to phase transformation of nonconducting acanthite into argentite exhibiting superionic conduction. Ag_2S/Ag heteronanostructures are intended for application in fast-acting resistive switches and nonvolatile memory devices of modern electronics. Also Ag_2S/Ag nanocomposites can be applied as promising biosensing probes.

All the forms of nanostructure silver sulfide possess the significant potential antibacterial activity.

It should be noted that the discussed universal method of hydrochemical bath deposition can be used to synthesize all forms of nanostructured sulfides of other metals (Cd, Zn, Pb, and Cu) by replacing precursor of silver by salt of other metal. For example, detailed description and analysis of application of hydrochemical bath deposition for preparing nanostructured cadmium and lead sulfides is given in review articles [3, 5] (see also Chaps. 2 and 3).

The discussion presented illustrates some of the challenges that scientists and technologists are facing in synthesis of nanostructured silver sulfide.

The concentration of Ag_2S in as-synthesized solutions is equal to ~ 1 wt.% $\cong$ 10 mg mL^{-1} and is rather high. When the solutions of reagents are mixed, deposition of silver sulfide Ag_2S occurs almost instantly, in fractions of a second. About 98% of particles are deposited in the form of agglomerates, and only ~ 0.02 wt.% Ag_2S remain in colloidal solution. The stable colloidal solutions above the silver sulfide deposits contain the finest nanoparticles with size about 8–16 nm but the concentration of such Ag_2S nanoparticles is equal to 0.01–0.02 mg mL^{-1} and rather small. That is why an important problem of synthesis of different forms of nanostructured silver sulfide remains to be preparation of more concentrated colloidal solutions retaining their long-term stability and small size of nanoparticles.

The obtaining of narrow size distribution of synthesized silver sulfide nanoparticles, i.e. the synthesis of uniform-sized nanoparticles is another important challenge. Indeed, the synthesis of monodisperse Ag_2S nanoparticles (with size variation <5%) is of key importance, because the properties of the nanoparticles depend strongly on their sizes and uniformity. The inhibition of additional nucleation during growth, in other words, the complete separation of nucleation and growth of nuclei, is critical for the successful synthesis of monodisperse nanoparticles without a further size-sorting process.

The recognition of these challenges and of their importance for preparing high-quality materials and devices on a base of nanostructured silver sulfide is a first and necessary step in the emergence of safe environmental industrial production. However, the journey in this direction has only just started. Much research is still needed to understand the deeper fundamentals of these challenges. The coming decade promises many exciting and outstanding news and reports in this area of nanoscience.

The study of nanostructured silver sulfide was supported by the Russian Science Foundation (project No, 14-23-00025) via the Institute of Solid State Chemistry, Ural Branch of the Russian Academy of Sciences.).

Appendix

See Tables 4.12, 4.13, 4.14, 4.15, 4.16 and 4.17.

Table 4.12 Crystal data, experimental conditions and characteristics of the refinement for the artificial coarse-crystalline α-Ag$_2$S phase with acanthite-type crystal structure [29]

Formula sum	Ag$_8$S$_4$
Formula weight	991.18
Crystal system	Monoclinic
Space group	$P12_1/c1$ (No. 14)
a (nm)	0.42264(2)
b (nm)	0.69282(3)
c (nm)	0.95317(3)
α (°)	90
β (°)	125.554(2)
γ (°)	90
Z	4
V (nm^3)	0.227065
D_{calc} (g cm^{-3})	7.25
Coefficients for peak FWHM	
U	0.000
V	0.017(3)
W	0.013(1)
Diffractometer	Shimadzu XRD-7000
Radiation type	X-ray Cu$K\alpha_1$

(continued)

Table 4.12 (continued)

Formula sum	Ag_8S_4
Wavelength (nm)	0.154056
Temperature (K)	300
2θ Range (°)	20–95
Step-scan (°)	0.02
Counting time (s)	10
Preferred orientation	(011)
Number of reflections measured	176
R_I (R_{Bragg}) (%)	2.465
R_p (%)	7.194
ωR_p (%)	9.735
R_{expect} (%)	5.247

Table 4.13 Diffraction reflections of synthesized artificial coarse-crystalline monoclinic (space group $P2_1/c$) silver sulfide α-Ag_2S with acanthite-type crystal structure [29]

d (Å)	2θ	I_{calc}	I_{obs}	hkl
3.95686	22.451	77	76	-1 0 2
3.87734	22.918	<1	<1	0 0 2
3.57074	24.916	30	29	-1 1 1
3.46410	25.696	4	4	0 2 0
3.43843	25.891	113	115	1 0 0
3.43597	25.910	181	183	-1 1 2
3.38351	26.319	191	196	0 1 2
3.16287	28.192	7	6	0 2 1
3.07998	28.967	593	605	1 1 0
2.83712	31.508	795	825	-1 1 3
2.66378	33.617	476	480	-1 2 1
2.60640	34.380	1000	990	-1 2 2
2.58328	34.697	700	698	0 2 2
2.45693	36.543	476	483	1 1 1
2.44036	36.800	756	762	1 2 0
2.42182	37.092	486	484	0 1 3
2.38235	37.730	825	865	-1 0 4
2.31412	38.886	25	30	-1 2 3
2.25288	39.987	34	37	-1 1 4
2.21334	40.733	494	502	0 3 1
2.09354	43.177	98	95	1 2 1
2.08339	43.398	466	454	-2 0 2
2.07169	43.656	146	144	0 2 3
2.04839	44.179	161	159	1 0 2
2.01991	44.835	1	1	-1 3 1
2.01155	45.032	4	4	-2 1 3

(continued)

Table 4.13 (continued)

d (Å)	2θ	I_{calc}	I_{obs}	hkl
1.99454	45.437	6	6	−1 3 2
1.99514	45.423	144	145	−2 1 2
1.98412	45.689	48	44	0 3 2
1.97843	45.828	2	2	2 0–4
1.96433	46.176	209	201	1 1 2
1.96295	46.210	51	49	−1 2 4
1.93867	46.823	8	8	0 0 4
1.91712	47.381	20	21	1 3 0
1.90239	47.771	147	149	−2 1 4
1.86695	48.736	159	152	0 1 4
1.86157	48.886	1	1	−2 1 1
1.85407	49.097	17	15	−1 3 3
1.81646	50.183	24	22	−1 1 5
1.79710	50.762	29	26	−2 2 3
1.78537	51.119	22	20	−2 2 2
1.76320	51.808	<1	<1	1 2 2
1.73467	52.726	2	2	1 3 1
1.73205	52.812	62	59	0 4 0
1.72218	53.139	38	36	0 3 3
1.71815	53.273	70	67	−2 1 5
1.71799	53.279	186	179	−2 2 4
1.69175	54.172	21	19	0 2 4
d (Å)	2θ	I_{calc}	I_{obs}	hkl
1.69040	54.219	28	27	0 4 1
1.68775	54.311	5	5	−2 2 1
1.66861	54.986	8	8	2 1 0
1.65818	55.361	2	2	−1 3 4
1.65391	55.517	3	3	−1 2 5
1.60972	57.179	30	27	1 1 3
1.59938	57.583	7	6	−1 4 1
1.58670	58.087	65	66	−1 4 2
1.58143	58.299	59	58	0 4 2
1.57868	58.410	75	73	−2 2 5
1.55491	59.392	12	12	−2 0 6
1.55454	59.407	62	60	−2 3 3
1.54693	59.729	13	12	−2 3 2
1.54688	59.732	22	21	1 4 0
1.54019	60.017	51	51	−1 0 6
1.53999	60.026	40	40	2 2 0
1.53244	60.353	4	4	1 3 2
1.51717	61.024	24	24	−2 1 6

(continued)

Table 4.13 (continued)

d (Å)	2θ	I_{calc}	I_{obs}	hkl
1.51347	61.189	92	94	0 1 5
1.51318	61.202	23	23	−1 4 3
1.50349	61.640	11	11	−1 1 6
1.48484	62.501	17	17	0 3 4
1.48213	62.628	86	91	−2 3 1
1.46960	63.223	80	86	2 1 1
1.45905	63.733	213	216	−1 3 5
1.44615	64.370	48	49	1 4 1
1.43889	64.734	18	19	0 4 3
1.41856	65.778	27	28	−2 2 6
1.41554	65.937	52	53	0 2 5
1.40868	66.298	3	3	−3 0 4
1.40736	66.369	4	4	−1 2 6
1.40662	66.408	26	29	−2 3 5
1.40093	66.713	14	16	−1 4 4
1.38044	67.837	3	3	−3 1 4
1.37999	67.862	19	21	1 0 4
1.37945	67.892	10	11	2 2 1
1.37904	67.915	7	7	2 3 0
1.36404	68.766	4	6	0 5 1
1.36183	68.893	7	7	−3 1 3
1.35633	69.212	34	36	−3 1 5
1.35340	69.383	2	2	1 1 4
d (Å)	2θ	I_{calc}	I_{obs}	hkl
1.34524	69.865	108	114	1 3 3
1.33674	70.375	41	45	−2 4 3
1.33228	70.646	73	78	−2 1 7
1.33189	70.669	16	17	−2 4 2
1.32912	70.838	16	17	−3 0 2
1.31895	71.468	65	65	−3 0 6
1.31385	71.788	2	2	2 0 2
1.30777	72.174	27	27	−1 5 2
1.30482	72.363	6	6	0 5 2
1.30532	72.331	37	37	−3 1 2
1.30492	72.357	34	33	−3 2 4
1.30320	72.468	6	6	−2 4 4
1.29568	72.956	17	18	−3 1 6
1.29164	73.221	1	1	0 4 4
1.29085	73.273	33	35	2 1 2
1.28916	73.385	18	18	−3 2 3
1.28753	73.493	4	4	0 3 5

(continued)

Table 4.13 (continued)

d (Å)	2θ	I_{calc}	I_{obs}	hkl
1.28521	73.648	16	17	1 5 0
1.28201	73.862	1	1	1 2 4
1.28137	73.905	8	8	−1 3 6
1.27521	74.322	31	30	−1 1 7
1.27053	74.643	45	44	0 1 6
1.26569	74.977	34	34	−1 5 3
1.26401	75.094	16	15	−2 2 7
1.26019	75.360	4	3	2 3 1
1.24092	76.742	9	9	−3 2 2
1.23916	76.871	8	7	−2 4 5
1.23263	77.353	2	2	−3 2 6
1.22846	77.665	3	3	2 2 2
1.22567	77.875	19	20	1 5 1
1.22365	78.028	14	15	−1 3 1
1.22124	78.211	12	13	0 5 3
1.22018	78.292	32	33	2 4 0
1.21496	78.694	12	12	−1 2 7
1.21179	78.940	2	2	−3 1 7
1.21091	79.008	17	17	0 2 6
1.20261	79.662	1	1	−3 3 4
1.19778	80.048	3	3	−1 5 4
1.19658	80.144	5	5	1 4 3
1.19118	80.582	26	25	−2 0 8
1.19025	80.658	1	1	−3 3 3
1.18657	80.960	5	5	−3 3 5
1.17395	82.015	9	9	−2 1 8
1.17036	82.321	2	2	−2 3 7
1.17012	82.341	9	9	−3 2 1
1.16309	82.949	18	17	1 1 5
1.15974	83.242	7	7	−3 2 7
1.15691	83.491	1	1	−2 5 3
1.15542	83.623	9	9	0 4 5
d (Å)	2θ	I_{calc}	I_{obs}	hkl
1.15470	83.687	6	6	0 6 0
1.15376	83.771	4	4	−2 5 2
1.15196	83.931	10	11	−3 3 2
1.14771	84.313	7	7	1 5 2
1.14532	84.530	10	9	−3 3 6
1.14211	84.823	4	4	0 6 1
1.14198	84.835	10	10	2 3 2
1.13923	85.089	10	10	2 1 3

(continued)

Table 4.13 (continued)

d (Å)	2θ	I_{calc}	I_{obs}	hkl
1.13554	85.430	6	6	2 4 1
1.13496	85.484	3	3	−2 5 4
1.13318	85.651	11	11	−3 0 8
1.13078	85.877	10	9	3 1 0
1.12784	86.155	4	4	0 3 6
1.12730	86.206	10	10	0 5 4
1.12644	86.288	44	42	−2 2 8
1.11832	87.071	18	18	−3 1 8
1.11808	87.094	6	6	−1 0 8
1.11683	87.215	2	2	1 2 5
1.11276	87.615	2	2	−1 6 1
1.10847	88.042	2	2	−1 6 2
1.10379	88.512	32	30	−1 1 8
1.09566	89.344	11	11	2 2 3
1.09466	89.447	6	6	−3 3 1
1.09463	89.451	12	13	1 6 0
1.09287	89.633	22	24	−3 4 4
1.09190	89.734	4	4	−2 5 5
1.08813	90.130	5	5	3 2 0
1.08614	90.341	38	37	−3 3 7
1.08357	90.615	1	1	−3 4 3
1.08249	90.731	2	2	−1 6 3
1.07931	91.073	3	3	1 4 4
1.07702	91.322	6	6	−3 2 8
1.06850	92.260	3	3	−2 4 7
1.06403	92.763	1	1	−1 2 8
1.06243	92.943	3	3	1 5 3
1.05444	93.861	4	4	−3 4 2
1.05105	94.258	2	3	−4 0 6
1.04934	94.459	1	1	−3 4 6

Table 4.14 Crystal data, experimental conditions and characteristics of the refinement for the nanocrystalline α-Ag$_{1.93}$S phase with acanthite-type crystal structure [30]

Formula sum	Ag$_{7.72}$S$_{4.00}$
Formula weight	960.98
Crystal system	Monoclinic
Space group	$P12_1/c1$ (No. 14)
a (nm)	0.4234(3)
b (nm)	0.6949(3)
c (nm)	0.9549(5)
α (°)	90
β (°)	125.43(6)
γ (°)	90
Z	4
V (nm^3)	0.22891
D_{calc} (g cm^{-3})	6.97
Coefficients for peak FWHM	
U	0.000
V	−0.3(2)
W	0.47(6)
Diffractometer	Shimadzu XRD-7000
Radiation type	X-ray CuKα_1
Wavelength (nm)	0.154056
Temperature (K)	300
2θ Range (°)	20–95
Step-scan (°)	0.02
Counting time (s)	10
Preferred orientation	(101)
Number of reflections measured	122
R_I (R_{Bragg}) (%)	5.551
R_p (%)	11.649
ωR_p (%)	14.305
R_{expect} (%)	4.209

Table 4.15 Diffraction reflections of nanocrystalline monoclinic (space group $P2_1/c$) silver sulfide α-Ag$_{1.93}$S with acanthite-type crystal structure [30]

d (Å)	2θ	I_{calc}	I_{obs}	hkl
3.96216	22.451	134	112	−1 0 2
3.57951	24.854	48	47	−1 1 1
3.47490	25.615	16	18	0 2 0
3.45004	25.803	76	89	1 0 0
3.44207	25.863	109	128	−1 1 2
3.39496	26.229	59	68	0 1 2
3.17291	28.101	9	9	0 2 1
3.09022	28.869	727	693	1 1 0
2.84222	31.450	1000	997	−1 1 3
2.67110	33.522	497	514	−1 2 1
2.61251	34.297	982	989	−1 2 2
2.59173	34.581	318	324	0 2 2
2.46618	36.401	340	351	1 1 1
2.44829	36.677	758	787	1 2 0
2.43011	36.961	140	143	0 1 3
2.38704	37.653	853	859	−1 0 4
2.31930	38.796	28	34	−1 2 3
2.25759	39.901	47	56	−1 1 4
2.22030	40.600	213	243	0 3 1
2.10105	43.015	55	68	1 2 1
2.08801	43.297	238	304	−2 0 2
2.07861	43.503	34	43	0 2 3
2.05653	43.995	56	69	1 0 2
2.01508	44.948	5	5	−2 1 3
1.99985	45.310	20	19	−1 3 2
1.99971	45.313	142	129	−2 1 2
1.99049	45.535	46	40	0 3 2
1.98108	45.763	10	9	2 0–4
1.97200	45.986	170	147	1 1 2
1.96754	46.096	111	97	−1 2 4
1.94539	46.652	25	22	0 0 4
1.92325	47.221	4	4	1 3 0
1.90519	47.697	60	64	−2 1 4
1.87338	48.558	32	32	0 1 4
1.85880	48.964	41	39	−1 3 3
1.82074	50.057	2	2	−1 1 5
1.80075	50.652	21	19	−2 2 3
1.74064	52.532	2	2	1 3 1
1.73745	52.636	31	28	0 4 0
1.72782	52.952	4	4	0 3 3
1.72502	53.044	10	9	2 0 0

(continued)

Table 4.15 (continued)

d (Å)	2θ	I_{calc}	I_{obs}	hkl
1.72103	53.177	118	115	$-2\ 2\ 4$
1.72066	53.189	11	10	$-2\ 1\ 5$
1.69748	53.974	15	13	$0\ 2\ 4$
1.69270	54.139	3	3	$-2\ 2\ 1$
1.65802	55.367	1	1	$-1\ 2\ 5$
1.61604	56.935	3	3	$1\ 1\ 3$
1.60415	57.396	2	2	$-1\ 4\ 1$
1.59118	57.908	6	7	$-1\ 4\ 2$
1.58645	58.097	4	4	$0\ 4\ 2$
1.58139	58.301	9	10	$-2\ 2\ 5$
1.55812	59.257	17	17	$-2\ 3\ 3$
1.55729	59.292	7	7	$-2\ 0\ 6$
1.55178	59.524	5	5	$1\ 4\ 0$
1.55098	59.557	30	30	$-2\ 3\ 2$
1.54511	59.807	3	3	$2\ 2\ 0$
1.54410	59.850	3	3	$-1\ 0\ 6$
1.53795	60.114	23	24	$1\ 3\ 2$
1.51961	60.916	1	1	$-2\ 1\ 6$
1.51870	60.956	4	5	$0\ 1\ 5$
1.51733	61.017	1	2	$-1\ 4\ 3$
1.50734	61.465	6	6	$-1\ 1\ 6$
1.50562	61.543	2	2	$-2\ 3\ 4$
1.48977	62.270	2	2	$0\ 3\ 4$
1.48653	62.421	6	7	$-2\ 3\ 1$
1.47499	62.965	3	4	$2\ 1\ 1$
1.46288	63.547	28	36	$-1\ 3\ 5$
1.45098	64.130	13	15	$1\ 4\ 1$
1.42111	65.645	6	7	$-2\ 2\ 6$
1.42036	65.684	3	3	$0\ 2\ 5$
1.41122	66.164	5	6	$-3\ 0\ 4$
1.40474	66.509	13	13	$-1\ 4\ 4$
1.38543	67.559	2	2	$1\ 0\ 4$
1.38441	67.616	2	2	$2\ 2\ 1$
1.36830	68.521	3	3	$0\ 5\ 1$
1.35844	69.089	2	2	$-3\ 1\ 5$
1.35019	69.572	2	2	$1\ 3\ 3$
1.34010	70.172	5	6	$-2\ 4\ 3$
1.33468	70.499	4	5	$-2\ 1\ 7$
1.32072	71.358	7	8	$-3\ 0\ 6$
1.31159	71.931	17	17	$-1\ 5\ 2$
d (Å)	2θ	I_{calc}	I_{obs}	hkl

(continued)

Table 4.15 (continued)

d (Å)	2θ	I_{calc}	I_{obs}	hkl
1.30894	72.100	7	7	0 5 2
1.30882	72.108	17	16	−3 1 2
1.30751	72.191	3	3	−3 2 4
1.30625	72.272	6	6	−2 4 4
1.29750	72.837	13	12	−3 1 6
1.29586	72.944	3	3	0 4 4
1.29579	72.948	14	13	2 1 2
1.29242	73.170	6	5	−2 3 6
1.29218	73.185	1	1	−3 2 3
1.28926	73.378	1	1	1 5 0
1.28692	73.534	8	7	1 2 4
1.28484	73.672	11	10	−1 3 6
1.27805	74.129	1	1	−1 4 5
1.27492	74.342	7	7	0 1 6
1.26930	74.727	5	5	−1 5 3
1.23456	77.210	3	3	−3 2 6
1.22969	77.573	2	2	1 5 1
1.22414	77.990	9	9	2 4 0
1.19352	80.392	6	6	−2 0 8
1.17630	81.817	2	2	−2 1 8
1.17283	82.111	2	1	−2 3 7
1.15999	83.220	2	2	−2 5 3
1.15704	83.480	4	4	−2 5 2
1.15516	83.646	6	5	−3 3 2
1.15160	83.964	5	5	1 5 2
1.15001	84.106	3	2	3 0 0
1.14736	84.346	6	5	−3 3 6
1.14618	84.452	6	5	2 3 2
1.13476	85.503	3	3	−3 0 8
1.13165	85.794	3	3	0 3 6
1.12879	86.064	7	7	−2 2 8
1.12122	86.789	3	3	−1 0 8
1.10691	88.198	2	2	−1 1 8
1.09807	89.096	4	4	1 6 0
1.09541	89.370	5	5	−3 4 4
1.09178	89.747	5	5	3 2 0
1.08801	90.143	1	1	−3 3 7
1.08321	90.653	4	4	1 4 4
1.07870	91.139	4	5	−3 2 8
1.06705	92.422	3	3	−1 2 8
1.06099	93.108	1	1	−2 3 8

Table 4.16 Crystal data, experimental conditions and characteristics of the refinement for the coarse-crystalline β-Ag$_2$S (β-Ag$_{2.01}$S) phase with argentite-type crystal structure [31, 34, 35]

Formula sum	Ag $_{4.02}$S$_{2.00}$
Formula weight	497.83
Crystal system	Cubic
Space group	$I m\bar{3}m$ (No. 229)
a (nm)	0.4874(1)
b (nm)	0.4874(1)
c (nm)	0.4874(1)
α (°)	90
β (°)	90
γ (°)	90
Z	2
V (nm^3)	0.115821
D_{calc} (g cm^{-3})	7.14
Coefficients for peak FWHM	
U	0.000
V	$-0.01(3)$
W	0.02(1)
Diffractometer	X'Pert PRO MPD (Panalytical)
Radiation type	X-ray Cu$K\alpha$
Wavelength (nm)	0.154056
Temperature (K)	503
2θ Range (°)	20.008–67.484
Step-scan (°)	0.026
Scan time per step (s)	197.37
Preferred orientation	(010)
Number of reflections measured	6
R_I (R_{Bragg}) (%)	2.503
R_p (%)	6.386
ωR_p (%)	7.785
R_{expect} (%)	1.932

Table 4.17 Diffraction reflections of the coarse-crystalline cubic (space group $I m\bar{3}m$) silver sulfide β-Ag$_2$S (β-Ag$_{2.01}$S) with argentite-type crystal structure at 503 K [31, 34, 35]

d (Å)	2θ	I_{calc}	I_{obs}	hkl
3.44678	25.827	9	12	0 1 1
2.43724	36.849	1000	990	0 0 2
1.99000	45.546	303	316	1 1 2
1.72339	53.099	92	87	0 2 2
1.54145	59.964	83	75	0 1 3
1.40714	66.380	34	35	2 2 2

References

1. Tang, A., Wang, Yu., Ye, H., Zhou, C., Yang, C., Li, X., Peng, H., Zhang, F., Hou, Y., Teng, F.: Controllable synthesis of silver and silver sulfide nanocrystals via selective cleavage of chemical bonds. Nanotechnology **24**(35), 355602–355612 (2013)
2. Cui, C., Li, X., Liu, J., Hou, Y., Zhao, Y., Zhong, G.: Synthesis and functions of Ag₂S nanostructures. Nanoscale Res. Lett. **10**, 431–21 (2015)
3. Sadovnikov, S.I., Gusev, A.I., Rempel, A.A.: Nanostructured lead sulfide: synthesis, structure, and properties. Russ. Chem. Rev. **85**(7), 731–758 (2016)
4. Shi, X., Zheng, S., Gao, W., Wei, W., Chem, M., Deng, F., Liu, X., Xiao, Q.: Excitation wavelength and intensity dependence of photo-spectral blue shift in single CdSe/ZnS quantum dots. J. Nanopart. Res. **16**(12), 2741–2749 (2014)
5. Kozhevnikova, N.S., Vorokh, A.S., Uritskaya, A.A.: Cadmium sulfide nanoparticles prepared by chemical bath deposition. Russ. Chem. Rev. **84**(3), 225–250 (2015)
6. Sharma, R.C., Chang, Y.A.: The Ag-S (silver-sulfur) system. Bull. Alloy Phase Diagrams. **7**(3), 263–269 (1986)
7. Sharma, R.C., Chang, Y.A.: Ag-S (Silver-Sulphur). In: Massalski, T.B., Okamoto, H., Kacprzak, L. (eds.) Binary Alloy Phase Diagrams, pp. 86–87. ASM International Publisher, Materials Park (1990)
8. Frueh, A.J.: The crystallography of silver sulfide, Ag₂S. Ztschr. Kristallographie **110**(1), 136–144 (1958)
9. Reye, H., Schmalzried, H.: On the nonstoichiometry of α-Ag2S. Ztschr. Physik. Chemie. Neue Folge **128**(1), 93–100 (1981)
10. Junod, P.: Relations entre la structure crystalline et les propriétiés électroniques des combinaisond Ag₂S, Ag₂Se, Cu₂Se. Phys. Acta **32**(6–7), 567–600 (1959)
11. Junod, P., Hediger, H., Kilchör, B., Wullschleger, J.: Metal-non-metal transition in silver chalcogenides. Philos. Mag **36**(4), 941–958 (1977)
12. Akamatsu, K., Takei, Sh, Mizuhata, M., Kajinami, A., Deki, Sh, Takeoka, Sh, Fujii, M., Hayashi, Sh, Yamamoto, K.: Preparation and characterization of polymer thin films containing silver and silver sulfide nanoparticles. Thin Sol. Films **359**(1), 55–60 (2000)
13. Kashida, S., Watanabe, N., Hasegawa, T., Iida, H., Mori, M., Savrasov, S.: Electronic structure of Ag₂S, band calculation and photoelectron spectroscopy. Sol. State Ionics **158**, 167–175 (2003)
14. Wagner, C.: Investigations on silver sulfide. J. Chem. Phys. **21**(10), 1819–1827 (1953)
15. Rau, H.: Defect equilibria in silver sulphide. J. Phys. Chem. Solids **35**(11), 1553–1559 (1974)
16. Bonnecaze, G., Lichanot, A., Gromb, S.: Proprietes electrogalvaniques et electroniques du sulfure d'argent β: Domaine d'existence. J. Phys. Chem. Solids **39**(3), 299–310 (1978)
17. Bonnecaze, G., Lichanot, A., Gromb, S.: Proprietes electroniques et electrogalvaniques du sulfure d'argent α: Domaine d'existence. J. Phys. Chem. Solids **39**(8), 813–821 (1978)
18. Ditman, A.V., Kulikova, I.N.: Investigation of dissociation of solid and melted silver sulfide by a dew–point method. Zh. Fizich. Khimii. **53**(1), 260–261 (1979). (in Russian)
19. Mitteilung, K.: Zustandsdiagramm Ag-S im Bereich der Verbindung Ag2 ± δ. S. Ztschr. Physik. Chemie. Neue Folge. **119**(2), 251–255 (1980)
20. Van Doorselaer, M.K.: Solid state properties and photographic activiti of crystalline Ag₂S and (Ag, Au)₂S-specks at the surface of silver halide crystalls. J. Photographic Sci **35**(2), 42–52 (1987)
21. Ramsdell, L.S.: The crystallography of acanthite, Ag₂S. Amer. Mineralogist **28**(7–8), 401–425 (1943)
22. Sadanaga, R., Sueno, S.: X-ray study on the & α-β transition of Ag2S. Mineralog. J. Japan **5**(2), 124–148 (1967)

23. Rahlfs, P.: Über die kubischen Hochtemperaturmodifikationen der Sulfide, Selenide und Telluride des Silbers und des einwertigen Kupfers. Ztschr. Physik. Chemie. **31**(3), 157–194 (1936)
24. Strock, L.W.: Kristallstructur des Hochtemperatur-Jodsilbers & α-AgJ. Ztschr. Physik. Chemie. **25**(5/6), 411–459 (1934)
25. Strock, L.W.: Erganzung und Berichtigung zu: "Kristallstruktur des Hochtemperatur-Jodsilbers α-AgJ". Ztschr. Physik. Chemie. **31**(2), 132–136 (1936)
26. Cava, R.J., Reidinger, F., Wuensch, B.J.: Single-crystal neutron diffraction study of the fast-ion conductor β-Ag$_2$S between 186 and 325°C. J. Solid State Chem. **31**(1), 69–80 (1980)
27. Blanton, T., Misture, S., Dontula, N., Zdzieszynski, S.: In situ high-temperature X-ray diffraction characterization of silver sulfide, Ag$_2$S. Powder Diffr. **26**(2), 110–118 (2011)
28. Frueh, A.J.: The use of zone theory in Problems of sulfide mineralogy. Part III: Polymorphism of Ag$_2$Te and Ag$_2$S. Am. Mineral. **46**(5–6), 654–660 (1961)
29. Sadovnikov, S.I., Gusev, A.I., Rempel, A.A.: Artificial silver sulfide Ag2S: crystal structure and particle size in deposited powders. Superlat. Microstr **83**, 35–47 (2015)
30. Sadovnikov, S.I., Gusev, A.I., Rempel, A.A.: Nonstoichiometry of nanocrystalline monoclinic silver sulfide. Phys. Chem. Chem. Phys. **17**(19), 12466–12471 (2015)
31. Sadovnikov, S.I., Gusev, A.I., Rempel, A.A.: An *in situ* high-temperature scanning electron microscopy study of acanthite–argentite phase transformation in nanocrystalline silver sulfide powder. Phys. Chem. Chem. Phys. **17**(32), 20495–20501 (2015)
32. Sadovnikov, S.I., Gusev, A.I., Rempel, A.A.: Nanocrystalline silver sulfide Ag$_2$S. Rev. Adv. Mater. Sci. **41**(1), 7–19 (2015)
33. Sadovnikov, S.I., Gusev, A.I., Rempel, A.A.: Structure and stoichiometry of nanocrystalline silver sulfide. Dokl. Akad. Nauk. **464**(5), 568–573 (2015). (in Russian). (Engl. Transl.: Dokl. Phys. Chem. **464**(2), 238–243 (2015))
34. Sadovnikov, S.I., Chukin, A.V., Rempel, A.A., Gusev, A.I.: Polymorphic transformation in nanocrystalline silver sulfide. Fiz. Tverd. Tela. **58**(1), 32–38 (2016). (in Russian) (Engl. Transl.: Phys. Solid State. **58**(1), 30–36 (2016))
35. Sadovnikov, S.I., Gusev, A.I., Chukin, A.V.: Rempel, A.A: High-temperature X-ray diffraction and thermal expansion of nanocrystalline and coarse-crystalline acanthite α-Ag2S and argentite β-Ag2S. Phys. Chem. Chem. Phys. **18**(6), 4617–4626 (2016)
36. Aliev, F.F., Jafarov, M.B., Tairov, B.A., Pashaev, G.P., Saddinova, A.A., Kuliev, A.A.: Effect of a phase transition on the electron energy spectrum in Ag$_2$S. Fiz. Tekhn. Poluprovodnikov. **42**(10), 1165–1167 (2008). (in Russian) (Engl. Transl.: Semiconductors. **42**(10), 1146–1148 (2008))
37. Ehrlich, S.H.: Spectroscopic studies of AgBr with quantum-size clusters of iodide, silver, and silver sulfides. J. Imaging Sci. Technol. **37**(1), 73–91 (1993)
38. Rickert, H.: Elektrische Eigenschaften von festen Stoffen mit gemischter Elektronen- und Ionenleitung, z.B. Ag2S. In: Madelung, O. (ed.) Festkörperprobleme, vol. VI, pp. 85–105. Braunschweig: F. Vieweg (1967)
39. Lim, W.P., Zhang, Z., Low, H.Y., Chin, W.S.: Preparation of Ag$_2$S nanocrystals of predictable shape and size. Angew. Chem. Int. Ed. **43**(42), 5685–5689 (2004)
40. Yang, J., Ying, J.Y.: Nanocomposites of Ag$_2$S and noble metals. Angew. Chem. Int. Ed. **50**(20), 4637–4643 (2011)
41. Zhu, G.X., Xu, Z.: Controllable growth of semiconductor heterostructures mediated by bifunctional Ag$_2$S nanocrystals as catalyst or source-host. J. Am. Chem. Soc. **133**(1), 148–157 (2011)
42. Kryukov, A.I., Stroyuk, A.L., Zin'chuk, N.N., Korzhak, A.V., Kuchmii, S.Y.: Optical and catalytic properties of Ag$_2$S nanoparticles. J. Mol. Catal. A: Chem. **221**(1–2), 209–221 (2004)
43. Shen, S., Zhang, Y., Liu, Y., Peng, L., Chen, X., Wang, Q.: Manganese-doped Ag$_2$S-ZnS heteronanostructures. Chem. Mater. **24**(12), 2407–2413 (2012)

44. Nasrallah, T.B., Dlala, H., Amlouk, M., Belgacem, S., Bernede, J.C.: Some physical investigations on Ag₂S thin films prepared by sequential thermal evaporation. Synth. Met. **151**(3), 225–230 (2005)
45. Hsu, T-Y., Buhay, H., Murarka, N.P.: Characteristics and applications of Ag₂S films in the milli-meter wavelength region. In: Tanton, G.A. (ed.) Millimeter Optic, pp. 38–45. SPIE Proc. 259 (1980)
46. Karashanova, D., Nihtianova, D., Starbova, K., Starbov, N.: Crystalline structure and phase composition of epitaxially grown Ag₂S thin films. Sol. State Ionics **171**(3–4), 269–275 (2004)
47. Liu, L., Hu, S., Dou, Y.-P., Liu, T., Lin, J., Wang, Y.: Nonlinear optical properties of near-infrared region Ag₂S quantum dots pumped by nanosecond laser pulses. Beilst. J. Nanotechnol **6**, 1781–1787 (2015)
48. Liang, C.H., Terabe, K., Hasegawa, T.: Aono, M: Resistance switching of an individual Ag₂S/Ag nanowire heterostructure. Nanotechnology. **18**(48), 5 (2007). Paper 485202
49. Xu, Z., Bando, Y., Wang, W., Bai, X., Golberg, D.: Real-time *in situ* HRTEM-resolved resistance switching of Ag₂S Nanoscale ionic conductor. ACS Nano **4**(5), 2515–2522 (2010)
50. Belov, A.N., Pyatilova, O.V., Vorobiev, M.I.: Synthesis of Ag/Ag₂S nanoclusters resistive switches for memory cells. Advanc. Nanoparticles **3**, 1–4 (2014)
51. El-Nahass, M.M., Farag, A.A.M., Ibrahim, E.M., Abd-El-Rahman, S.: Structural, optical and electrical properties of thermally evaporated Ag₂S thin films. Vacuum **72**(4), 453–460 (2004)
52. Jadhav, U.M., Patel, S.N., Patil, R.S.: Synthesis of silver sulphide nanoparticles by modified chemical route for solar cell applications. Res. J. Chem. Sci. **3**(7), 69–74 (2013)
53. Leidinger, P., Popescu, R., Gerthsen, D., Feldmann, C.: Nanoscale Ag₂S hollow spheres and Ag₂S nanodiscs assembled to three-dimensional nanoparticle superlattices. Chem. Mater. **25**(21), 4173–4180 (2013)
54. Xie, Y., Heo, S.H., Kim, Y.N., Yoo, S.H., Cho, S.O.: Synthesis and visible-light-induced catalytic activity of Ag₂S-coupled TiO₂ nanoparticles and nanowires. Nanotechnology. **21**(1), 7 (2010). Paper 015703
55. Zhu, L., Meng, Z., Trisha, G., Oh, W.-C.: Hydrothermal synthesis of porous Ag₂S sensitized TiO₂ catalysts and their photocatalytic activities in the visible light range. Chin. J. Catal. **33**(2), 254–2604 (2012)
56. Pourahmad, A.: Ag₂S nanoparticle encapsulated in mesoporous material nanoparticles and its application for photocatalytic degradation of dye in aqueous solution. Superlatt. Mictostr. **52**(2), 276–287 (2012)
57. Pang, M.L., Hu, J.Y., Zeng, H.C.: Synthesis, morphological control, and antibacterial properties of hollow/solid Ag₂S/Ag heterodimers. J. Am. Chem. Soc. **132**(31), 10771–10785 (2010)
58. Xiong, S., Xi, B., Zhang, K., Chen, Y., Jiang, J., Hu, J., Zeng, H.C.: Ag nanoprisms with Ag₂S attachment. Sci. Rep **3**, 2177–2179 (2013)
59. Jiang, P., Zhu, C.-N., Zhang, Z.-L., Tian, Z.-Q., Pang, D.-W.: Water-soluble Ag₂S quantum dots for near-infrared fluorescence imaging in vivo. Biomaterials **33**(20), 5130–5135 (2012)
60. Li, C., Zhang, Y., Wang, M., Zhang, Y., Chen, G., Li, L., Wu, D., Wang, Q.: *In vivo* real-time visualization of tissue blood flow and angiogenesis using Ag₂S quantum dots in the NIR-II window. Biomaterials **35**(1), 393–400 (2014)
61. Jang, J., Cho, K., Lee, S.H., Kim, S.: Synthesis and electrical characteristics of Ag₂S nanocrystals. Mater. Lett. **62**(8–9), 1438–1440 (2008)
62. Han, L., Lv, Y., Asiri, A.M., Al-Youbi, A.O., Tu, B., Zhao, D.Y.: Novel preparation and near-infrared photoluminescence of uniform core-shell silver sulfide nanoparticle@mesoporous silica nanospheres. J. Mater. Chem. **22**(15), 7274–7279 (2012)
63. Sadovnikov, S.I., Rempel, A.A.: Synthesis of nanocrystalline silver sulfide. Neorgan. Materialy. **51**(8), 829–837 (2015). (in Russian) (Engl. Transl.: Inorg. Mater. **51**(8), 759–766 (2015))

64. Sadovnikov, S.I., Gusev, A.I., Gerasimov, EYu., Rempel, A.A.: Facile synthesis of Ag₂S nanoparticles functionalized by carbon-containing citrate shell. Chem. Phys. Lett. **642**, 17–21 (2015)
65. Chen, R., Nuhfer, N.T., Moussa, L., Morris, H.R., Whitmore, P.M.: Silver sulfide nanoparticle assembly obtained by reacting an assembled silver nanoparticle template with hydrogen sulfide gas. Nanotechnology **19**(45), 11 (2008). Paper 455604
66. Zhang, W., Zhang, L., Hui, Z., Zhang, X., Qian, Y.: Synthesis of nanocrystalline Ag₂S in aqueous solution. Sol. State Ionics. **130**(1–2), 111–114 (2000)
67. Qian, X.F., Yin, J., Huang, J.C., Yang, Y.F., Guo, X.X., Zhu, Z.K.: Preparation and characterization of PVA/Ag₂S nanocomposite. Mater. Chem. Phys. **68**(1–3), 95–97 (2001)
68. Qian, X.F., Yin, J., Feng, S., Liu, S.H., Zhu, Z.K.: Preparation and characterization of polyvinylpyrrolidone films containing silver sulfide nanoparticles. J. Mater. Chem. **11**(10), 2504–2506 (2001)
69. Xu, C., Zhang, Z., Ye, Q.: A novel facile method to metal sulfide (metal = Cd, Ag, Hg) nano-crystallite. Mater. Lett. **58**(11), 1671–1676 (2004)
70. Lu, X., Li, L., Zhang, W., Wang, C.: Preparation and characterization of Ag₂S nanoparticles embedded in polymer fibre matrices by electrospinning. Nanotechnology. **16**(10), 2233–2237 (2005)
71. Prabhune, V.B., Shinde, N.S., Fulari, V.J.: Studies on electrodeposited silver sulphide thin films by double exposure holographic interferometry. Appl. Surf. Sci. **255**(5), 1819–1823 (2008)
72. Meherzi-Maghraoui, H., Dachraoui, M., Belgacem, S., Buhre, K.D., Kunst, R., Cowache, P., Lincot, D.: Structural, optical and transport properties of Ag₂S films deposited chemically from aqueous solution. Thin Solid Films **288**(1–2), 217–223 (1996)
73. Li, X.H., Li, J.X., Li, G.D., Liu, D.P., Chen, J.S.: Controlled synthesis, growth mechanism, and properties of monodisperse CdS colloidal spheres. Chem. – Europ. J. **13**(31), 8754–8761 (2007)
74. Dhumure, S.S., Lokhande, C.D.: Chemical deposition of Ag₂S films from acidic bath. Mater. Chem. Phys. **28**(1), 141–144 (1991)
75. Lyu, L.-M., Huang, M.H.: Formation of Ag₂S cages from polyhedral Ag₂O nanocrystals and their electrochemical properties. Chem. Asian J. **8**(8), 1847–1853 (2013)
76. Lismont, M., Paez, C.A., Dreesen, L.: A one-step short-time synthesis of Ag@SiO₂ core-shell nanoparticles. J. Colloid Interface Sci. **447**, 40–49 (2015)
77. Li, Z., Jia, L., Li, Y., He, T., Li, X.-M.: Ammonia-free preparation of Ag@SiO₂ core/shell nanoparticles. Appl. Surf. Sci. **345**, 122–126 (2015)
78. Peng, X., Schlamp, M.C., Kadavanich, A.V., Alivisatos, A.P.: Epitaxial growth of highly luminescent CdSe/CdS core/shell nanocrystals with photostability and electronic accessibility. J. Am. Chem. Soc. **119**(30), 7019–7029 (1997)
79. Pinaud, F., King, D., Moore, H.P., Weiss, S.: Bioactivation and cell targeting of semiconductor CdSe/ZnS nanocrystals with phytochelatin-related peptides. J. Am. Chem. Soc. **126**(19), 6115–6123 (2004)
80. Hota, G., Jain, S., Khilara, K.C.: Synthesis of CdS-Ag₂S core-shell/composite nanoparticles using AOT/*n*-heptane/water microemulsions. Colloids Surf. A: Physicoch. Eng. Asp. **232**(2–3), 119–127 (2004)
81. Demchenko, D.O., Robinson, R.D., Sadtler, B., Erdonmez, C.K., Alivisatos, A.P., Wang, L.-W.: Formation mechanism and properties of CdS-Ag₂S nanorod superlattices. ACS Nano **2**(4), 627–636 (2008)
82. Emamdoust, A., Shayesteh, S.F., Marandi, M.: Synthesis and characterization of aqueous MPA-capped CdS-ZnS core-shell quantum dots. Pramana J. Phys. **80**(4), 713–721 (2013)
83. Gerion, D., Pinaud, F., Williams, S.C., Parak, W.J., Zanchet, D., Weiss, S., Alivisatos, A.P.: Synthesis and properties of biocompatible water-soluble silica-coated CdSe/ZnS semiconductor quantum dots. J. Phys. Chem. **105**(37), 8861–8871 (2001)
84. Gao, X., Cui, Y., Levenson, R.M., Chung, L.W.K., Nie, S.: *In vivo* cancer targeting and imaging with semiconductor quantum dots. Nature Biotechn. **22**(8), 969–976 (2004)

85. Tipcompor, N., Thongtem, S., Thongtem, T.: Characterization of cubic $AgSbS_2$ nanostructured flowers synthesized by microwave-assisted refluxing method. J. Nanomaterials (Hindawi) **2013**, 6 (2013). Paper 970489

86. Xiang, J., Cao, H., Wu, Q., Zhang, S., Zhang, X., Watt, A.A.R.: *L*-cysteine-assisted synthesis and optical properties of Ag_2S nanospheres. J. Phys. Chem. C **112**(10), 3580–3584 (2008)

87. Yu, Y., Zhang, K., Sun, S.: One-pot aqueous synthesis of near infrared emitting PbS quantum dots. Appl. Surf. Sci. **258**, 7181–7187 (2012)

88. Deng, D., Xia, J., Cao, J., Qu, L., Tian, J., Qian, Z., Gu, Y., Gu, Z.: Forming highly fluorescent near-infrared emitting PbS quantum dots in water using glutathione as surface-modifying molecule. J. Coll. Interf. Sci. **367**(1), 234–240 (2012)

89. Sadjadi, M.S., Khalilzadegan, A.: The effect of capping agents, EDTA and EG on the structure and morphology of CdS nanoparticles. J. Non-Oxide Glass. **7**(4), 55–63 (2015)

90. Zeng, J., Zheng, Y., Rycenga, M., Tao, J., Li, Z.Y., Zhang, Q., Zhu, Y., Xia, Y.: Controlling the shapes of silver nanocrystals with different capping agents. J. Am. Chem. Soc. **132**(25), 8552–8853 (2010)

91. Gutierrez, L., Aubry, C., Cornejo, M., Croue, J.-P.: Citrate-coated silver nanoparticles interactions with effluent organic matter: Influence of capping agent and solution conditions. Langmuir **31**(32), 8865–8872 (2015)

92. D'Souza, S., Mashazi, P., Britton, J., Nyokong, T.: Effects of differently shaped silver nanoparticles on the photophysics of pyridylsulfanyl-substituted phthalocyanines. Polyhedron **99**, 112–121 (2015)

93. Philip, D.: Honey mediated green synthesis of gold nanoparticles. Spectrochim. Acta, Part A **73**(4), 650–653 (2009)

94. Philip, D.: Honey mediated green synthesis of silver nanoparticles. Spectrochim. Acta, Part A **75**(3), 1078–1081 (2010)

95. Shenya, D.S., Mathew, J., Philip, D.: Phytosynthesis of Au, Ag and Au–Ag bimetallic nanoparticles using aqueous extract and dried leaf of Anacardium occidentale. Spectrochim. Acta, Part A **79**(1), 254–262 (2011)

96. Ganeshkumar, M., Sathishkumar, M., Ponrasu, T., Girija, Dinesh M., Suguna, L.: Spontaneous ultra fast synthesis of gold nanoparticles using Punica granatum for cancer targeted drug delivery. Colloids Surf. B: Biointerf. **2013**(106), 208–216 (2013)

97. Annamalai, A., Christina, V.L.P., Sudha, D., Kalpana, M., Lakshmi, P.T.V.: Green synthesis, characterization and antimicrobial activity of Au NPs using Euphorbia hirta L. leaf extract. Colloids Surf. B: Biointerf. **108**, 60–65 (2013)

98. Mendoza-Reséndez, R., Gómez-Treviño, A., Barriga-Castro, E.D., Núñez, N.O., Luna, C.: Synthesis of antibacterial silver-based nanodisks and dendritic structures mediated by royal jelly. RSC Advances. **4**(4), 1650–1658 (2014)

99. Ayodhya, D., Veerabhadram, G.: Green synthesis, characterization, photocatalytic, fluorescence and antimicrobial activities of *Cochlospermum gossypium* capped Ag_2S nanoparticles. J. Photochem. Photobiol. B: Biology. **157**, 57–69 (2016)

100. Yang, H.-Y., Zhao, Y.-W., Zhang, Z.-Y., Xiong, H.-M., Yu, S.-N.: One-pot synthesis of water-dispersible Ag_2S quantum dots with bright fluorescent emission in the second near-infrared window. Nanotechnology. **24**(5), 055706–055710 (2013)

101. Esteves, A.C.C., Trindade, T.: Synthesis studies on II/VI semiconductor quantum dots. Curr. Opinion Solid State Mater. Sci. **6**(4), 347–353 (2002)

102. Tang, Q., Yoon, S.M., Yang, H.J., Lee, Y., Song, H.J., Byon, H.R., Choi, H.C.: Selective degradation of chemical bonds: From single source molecular precursors to metallic Ag and semiconducting Ag_2S nanocrystals via instant thermal activation. Langmuir **22**(6), 2802–2805 (2006)

103. Wang, T.X., Xiao, H., Zhang, Y.C.: Simple solid state synthesis of Ag_2S crystallites using a single source molecular precursor. Mater. Lett. **62**(21–22), 3736–3738 (2008)

104. Zhang, C.L., Zhang, S.M., Yu, L.G., Zhang, Z.J.: Size-controlled synthesis of monodisperse Ag_2S nanoparticles by a solventless thermolytic method. Mater. Lett. **85**, 77–80 (2012)

105. Burda, C., Chen, X.B., Narayanan, R., Sayed, E.I.: Chemistry and properties of nanocrystals of different shapes. Chem. Rev. **105**(4), 1025–1102 (2005)
106. Zhao, Y., Zhang, D.W., Shi, W.F.: A gamma-ray irradiation reduction route to prepare rod-like Ag₂S nanocrystallines at room temperature. Mater. Lett. **61**(14–15), 3232–3234 (2007)
107. Xu, C.G., Zhang, Z.C., Ye, Q.: A novel facile method to metal sulfide (metal = Cd, Ag, Hg) nanocrystallite. Mater. Lett. **58**(11), 1671–1676 (2004)
108. Chen, M.H., Gao, L.: Synthesis of leaf-like Ag₂S nanosheets by hydrothermal method in water alcohol homogenous medium. Mater. Lett. **60**(8), 1059–1062 (2006)
109. Zhai, H.J., Wang, H.S.: Ag₂S morphology controllable via simple template-free solution route. Mater. Res. Bull. **43**(8–9), 2354–2360 (2008)
110. Wang, X.B., Liu, W.M., Hao, J.C., Fu, X.G., Xu, B.S.: A simple large-scale synthesis of well-defined silver sulfide semiconductor nanoparticles with adjustable size. Chem. Lett. **34** (12), 1664–1665 (2005)
111. Dong, L.H., Chu, Y., Liu, Y.: Synthesis of faceted and cubic Ag₂S nanocrystals in aqueous solution. J. Colloid. Interf. Sci. **317**(2), 485–492 (2008)
112. Fang, Y., Bai, C., Zhang, Y.: Preparation of metal sulfide-polymer composite microspheres with patterned surface structures. Chem. Commun. **7**, 804–805 (2004)
113. Sun, Y.Z., Zhou, B.B.: Single-crystalline Ag2S hollow nanohexagons and their assembly into ordered arrays. Mater. Lett. **64**(12), 1347–1349 (2010)
114. Zhuang, Z., Peng, Q., Wang, X., Li, Y.: Tetrahedral colloidal crystals of Ag₂S nanocrystals. Angew. Chem. Int. Ed. **46**(43), 8174–8177 (2007)
115. Chaudhuri, R.G., Paria, S.: A novel method for the templated synthesis of Ag₂S hollow nanospheres in aqueous surfactant media. J. Colloid. Interf. Sci. **369**(1), 117–122 (2012)
116. Liu, M.Y., Xu, Z.L., Li, B.N., Lin, C.M.: Synthesis of worm-like Ag₂S nanocrystals in W/O reverse microemulsion. Mater. Lett. **65**(3), 555–558 (2011)
117. Lv, L.Y., Wang, H.: Ag₂S nanorice: hydrothermal synthesis and characterization study. Mater. Lett. **121**, 105–108 (2014)
118. Yarema, M., Pichler, S., Sytnyk, M., Seyrkammer, R., Lechner, R.T., Fritz-Popovski, G., Jarzab, D., Szendrei, K., Resel, R., Korovyanko, O., Loi, M.A., Paris, O., Hesser, G., Heiss, W.: Infrared emitting and photoconducting colloidal silver chalcogenide nanocrystal quantum dots from a silylamide-promoted synthesis. ACS Nano **5**(5), 3758–3765 (2011)
119. Du, Y., Xu, B., Fu, T., Cai, M., Li, F., Zhang, Y., Wang, Q.: Near-infrared photoluminescent Ag₂S quantum dots from a single source precursor. J. Am. Chem. Soc. **132**(5), 1470–1471 (2010)
120. Cai, W., Shin, D.W., Chen, K., Gheysens, O., Cao, Q., Wang, S.X., Gambhir, S.S., Chen, X.: Peptide-labeled near-infrared quantum dots for imaging tumor vasculature in living subjects. Nano Lett. **6**(4), 669–676 (2006)
121. Chen, J., Zhang, T., Feng, L.L., Zhang, M., Zhang, X., Su, H., Cui, D.: Synthesis of ribonuclease-A conjugate Ag₂S quantum dots clusters via biomimetic route. Mater. Lett. **96**, 224–227 (2013)
122. Siva, C., Iswarya, C.N., Baraneedharan, P., Sivakumar, M.: *L*-cysteine assisted formation of mesh like Ag₂S and Ag₃AuS₂ nanocrystals through hydrogen bonds. Mater. Lett. **134**, 56–59 (2014)
123. Koneswaran, M., Narayanaswamy, R.: *L*-cysteine-capped ZnS quantum dots based fluorescence sensor for Cu²⁺ ion. Sens. Actuator B: Chem. **139**(1), 104–109 (2009)
124. Hou, X.M., Zhang, X.L., Yang, W., Liu, Y.: Synthesis of SERS active Ag₂S nanocrystals using oleylamine as solvent reducing agent and stabilizer. Mater. Res. Bull. **47**(9), 2579–2583 (2012)
125. Shakouri-Arani, M., Salavati-Niasari, M.: Structural and spectroscopic characterization of prepared Ag₂S nanoparticles with a novel sulfuring agent. Spectrochim. Acta A: Mol Biomol Spectrosc. **133**, 463–471 (2014)

126. Yan, Zhang: Hong G., Zhang Y., Chen G., Li F., Dai H., Wang Q. Ag_2S quantum dot: A bright and biocompatible fluorescent nanoprobe in the second near-infrared window. ACS Nano **6**(5), 3695–3702 (2012)

127. Hocaoglu, I., Çizmeciyan, M.N., Erdem, R., Ozen, C., Kurt, A., Sennaroglu, A., Acar, H.Y.: Development of highly luminescent and cytocompatible near-IR-emitting aqueous Ag_2S quantum dots. J. Mater. Chem. **22**(29), 14674–14681 (2012)

128. Wang, C., Zhang, X., Qian, X., Wang, W., Qian, Y.: Ultrafine powder of silver sulfide semiconductor prepared in alcohol solution. Mater. Res. Bull. **33**(7), 1083–1086 (1998)

129. Krylova, V., Samuolaitiene, L.: Investigation of optical and electrical properties of silver sulfide films deposited on polyamide substrates. Mat. Sci. (Lithuania) **19**(1), 10–14 (2013)

130. Grocholl, L., Wang, J., Gillan, E.G.: Synthesis of sub-micron silver and silver sulfide particles via solvothermal silver azide decomposition. Mater. Res. Bull. **38**(5), 213–220 (2003)

131. Kim, B., Park, C.-S., Murayama, M., Hochella, M.F.: Discovery and characterization of silver sulfide nanoparticles in final sewage sludge products. Environ. Sci. Technol. **44**(19), 7509–7514 (2010)

132. Martínez-Castañón, G.A., Sánchez-Loredo, M.G., Dorantes, H.J., Martínez-Mendoza, J.R., Ortega-Zarzosa, G.: Ruiz Facundo. Characterization of silver sulfide nanoparticles synthesized by a simple precipitation method. Mater. Lett. **59**(4), 529–534 (2005)

133. JCPDS Card No. 14-0072

134. JCPDS Card No. 65–2356

135. Trandafilović, L.V., Djoković, V., Bibić, N., Georges, M.K., Radhakrishnan, T.: Confined growth of Ag_2S semiconductor nanocrystals in the presence of PDMAEMA-co-AA polyampholyte co-polymer. Mater. Lett. **64**(9), 1123–1126 (2010)

136. Buerger, M.J.: Elementary Crystallography, pp. 15–16. John Wiley @ Sons, New York (1956)

137. Vainshtein, B.K.: Modern Crystallography. Vol. 1. Fundamentals of Crystals. Symmetry, and Methods of Structural Crystallography. 2nd edn, pp. 480. Springer-Verlag, Berlin (1994)

138. Delaunay, B.: Sur la généralisation de la théorie des paralléloèdres. Bulletin de l'Académie des Sciences de l'URSS. Classe des sciences mathématiques et naturelles. 5, 641–646 (1933)

139. Delaunay, B.: Neue Darstellung der geometrischen Kristallographie. Erste Abhandlung. Ztschr. Kristallographie. **84**(1), 109–149 (1933)

140. Delaunay, B.: Sur la sphère vide. A la mémoire de Georges Voronoï. Bulletin de l'Académie des Sciences de l'URSS. Classe des sciences mathématiques et naturelles. **6**, 793–800 (1934)

141. Delaunay, B.N.: The geometry of positive quadratic forms. Uspekhi Mat. Nauk. **3**, 16–62 (1937) Uspekhi Mat. Nauk. **4**, 102–164 (1938) (in Russian)

142. Patterson, A.L., Lowe, W.E.: Remarks on the Delaunay reduction. Acta Crystallogr. **10**(2), 111–116 (1957)

143. X'Pert Plus Version 1.0. Program for Crystallography and Rietveld analysis Philips Analytical B. V. Koninklijke Philips Electronics N. V. (1999)

144. Hall, W.H., Williamson, G.K.: The diffraction pattern of cold worked metals: I. The nature of extinction. Proc. Phys. Soc. London. Sect.B. **64**(11). Part 11 **383B**, 937–946 (1951)

145. Gusev, A.I., Rempel, A.A.: Nanocrysnalline Materials, p. 351. Cambridge Intern. Science Publ, Cambridge (2004)

146. Gusev, A.I., Kurlov, A.S.: Certification of nanocrystalline materials on the size of particles (grains). Metallofizika i Noveishie Tekhnologii. **30**(5), 679–694 (2008) (in Russian)

147. Sadovnikov, S.I., Gusev, A.I.: Chemical deposition of nanocrystalline lead sulfide powders with controllable particle size. J. Alloys Comp. **586**, 105–112 (2014)

148. Perrott, C.M., Fletcher, N.H.: Heat capacity of silver sulfide. J. Chem. Phys. **50**(6), 2344–2350 (1969)

149. Thompson, W.T., Flengas, S.N.: Drop calorimetric measurements on some chlorides, sulfides, and binary melts. Can. J. Chem. **49**(9), 1550–1563 (1971)

150. Mills, K.C.: Thermodynamic Data for Inorganic Sulfides, Selenides, and Tellurides, p. 845. Butterworths, London (1974)

151. Okazaki, H., Takano, A.: The specific heat of Ag2S in & α-phase. Ztsch. Naturforsch. A. **40** (10), 986–988 (1985)

152. Grønvold, F., Westrum, E.F.: Silver(I) sulfide: Ag$_2$S Heat capacity from 5 to 1000 K, thermodynamic properties, and transitions. J. Chem. Therm. **18**(4), 381–401 (1986)

153. Match! Version 1.10. Phase Identification from Powder Diffraction © 2003–2010 Crystal Impact

154. https://summary.ccdc.cam.ac.uk/structure-summary?ccdc=1062400

155. Gusev, A.I., Sadovnikov, S.I., Chukin, A.V., Rempel, A.A.: Thermal expansion of nanocrystalline and coarse-crystalline silver sulfide Ag$_2$S. Fiz. Tverd. Tela. **58**(2), 246–251 (2016). (in Russian). (Engl. Transl.: Phys. Solid State. **58**(2), 251–257 (2016))

156. Honma, K., Iida, K.: Specific heat of superionic conductor Ag2S, Ag2Se and Ag2Te in α-phase. J. Phys. Soc. Japan. **56**(5), 1828–1836 (1987)

157. X'Pert HighScore Plus. Version 2.2e (2.2.5). 2009 PANalytical B. V. Almedo, the Netherlands

158. Sadovnikov, S.I., Kozhevnikova, N.S., Rempel, A.A., Magerl, A.: Thermal expansion of a lead sulfide nanofilm. Thin Solid Films **548**, 230–234 (2013)

159. Sadovnikov, S.I., Gusev, A.I.: Effect of particle size on the thermal expansion of nanostructured lead sulfide films. J. Alloys Comp. **610**, 196–202 (2014)

160. Sadovnikov, S.I., Gusev, A.I.: Thermal expansion of nanostructured PbS films and anharmonicity of atomic vibrations. Fiz. Tverd. Tela. **56**(11), 2274–2278 (2014). (in Russian). (Engl. Transl.: Phys. Sol. State. **56**(11), 2353–2358 (2014))

161. Montrol, E.W.: Size effect in low temperature heat capacities. J. Chem. Phys. **18**(2), 183–185 (1950)

162. Gmelin's Handbuch der anorganischen Chemie. 5th edn. In: Silber. Verlag Chemie GmbH, Weinheim, Teil B3 (1973)

163. Sadovnikov, S.I., Gusev, A.I.: Universal approach to the synthesis of silver sulfide in the forms of nanopowders, quantum dots, core-shell nanoparticles, and heteronanostructures. Eur. J. Inorg. Chem. **2016**(31), 4944–4957 (2016)

164. Sadovnikov, S.I., Kuznetsova, YuV, Rempel, A.A.: Ag$_2$S silver sulfide nanoparticles and colloidal solutions: Synthesis and properties. Nanostr. Nano-Obj. **7**, 81–91 (2016)

165. Sadovnikov, S.I., Gusev, A.I.: Facile synthesis, structure, and properties of Ag$_2$S/Ag heteronanostructure. J. Nanopart. Res. **18**(9), 277–12 (2016)

166. Gusev, A.I., Sadovnikov, S.I.: Acanthite-argentite transformation in nanocrystalline silver sulfide and the Ag$_2$S/Ag nanoheterostructure. Fiz. Tekhn. Poluprovodnikov. **50**(5), 694–699 (2016). (in Russian). (Engl. Transl.: Semiconductors. **50**(5), 682–687 (2016))

167. Sadovnikov, S.I., Gusev, A.I., Gerasimov, E.Yu., Rempel, A.A.: Silver sulfide nanoparticles with a carbon-containing shell. Neorgan. Materialy. **52**(5), 487–492 (2016). (in Russian). (Engl. Transl.: Inorg. Mater. **52**(5), 441–446 (2016))

168. Gusev, A.I., Sadovnikov, S.I.: Structure and properties of nanoscale Ag$_2$S/Ag heterostructures. Mater. Lett. **188**, 351–354 (2017)

169. Anastas, P., Eghbali, N.: Green chemistry: Principles and practice. Chem. Soc. Rev. **39**(1), 301–312 (2010)

170. Patnaik P.: Dean's Analytical Chemistry Handbook. 2nd edn., p. 1280. McGraw-Hill, New York (2004) Table 4.2.

171. Lee, P.C., Meisel, D.: Adsorption and surface-enhanced Raman of dyes on silver and gold sols. J. Phys. Chem. **86**(17), 3391–3395 (1982)

172. Gatan Misroscopy Suite. Gatan Inc Version 2.31.734.0

173. http://www.gatan.com

174. Sadovnikov. S.I., Rempel. A.A.: Method of production of nanocrystalline powder of silver sulfide. Patent No. 2572421 of Russian Federation. 1–4 (2016)

175. Kayanuma, Y.: Quantum-size effects of interacting electrons and holes in semiconductor microcrystals with spherical shape. Phys. Rev. B. **38**(14), 9797–9805 (1988)

176. Chang, S., Li, Q., Xiao, X., Wong, K.Y., Chen, T.: Enhancement of low energy sunlight harvesting in dye-sensitized solar cells using plasmonic gold nanorod. Energy Environm. Sci. **5**(11), 9444–9448 (2012)

177. Zhang, Y., Liu, Y., Li, C., Chen, X., Wang, Q.: Controlled synthesis of Ag_2S quantum dots and experimental determination of the exciton Bohr radius. J. Phys. Chem. C **118**(9), 4918–4923 (2014)

178. Faraday, M.: The Bakerian lecture: Experimental relations of gold (and other metals) to light. Philosoph. Trans. Roy. Soc. (London) **147**, 145–181 (1857)

179. Sadovnikov, S.I., Kuznetsova, Yu.V., Gusev A.I., Rempel A.A.: Method of production of aqueous colloidal solutions of silver sulfide nanoparticles. Patent No. 2600761 of Russian Federation. 1–11 (2016)

180. Wang, H., Qi, L.: Controlled synthesis of Ag_2S, Ag_2Se, and Ag nanofibers by using a general sacrificial template and their application in electronic device fabrication. Adv. Func. Mater. **18**(8), 1249–1256 (2008)

181. Henglein, A.: Nanoclusters of semiconductors and metals: Colloidal nano-particles of semiconductors and metals: Electronic structure and processes. Ber. Bunsenges. Phys. Chem. **101**(4), 1562–1572 (1997)

182. Krutyakov, Y.A., Kudrinskiy, A.A., Olenin, A.Y., Lisichkin, G.V.: Synthesis and properties of silver nanoparticles: advances and prospects. Usp. Khim. **77**(3), 242–269 (2008). (in Russian). (Engl. Transl.: Rus. Chem. Rev. **77**(3), 233–257 (2008))

183. Lukashin, A.V., Eliseev, A.A., Zhuravleva, N.G., Vertegel, A.A., Tretyakov, YuD, Lebedev, O.I., van Tendeloo, G.: One-step synthesis of shelled PbS nanoparticles in a layered double hydroxide matrix. Mend. Commun. **14**(4), 174–176 (2004)

184. Hayes, R., Ahmed, A., Edge, T., Zhang, H.: Core–shell particles: Preparation, fundamentals and applications in high performance liquid chromatography. J. Chromatogr. A **1357**, 36–52 (2014)

185. Sadovnikov, S.I., Gusev, A.I., Rempel, A.A.: Silver sulfide nanoparticles in ligand organic shell and method of its production. Patent No. 2603666 of Russian Federation. 1–13 (2016)

186. Faraday, M.: Experimental researches in electricity. Fourth series. Phil. Trans. Royal Soc. London. **123**, 507–522 (1833) Art. 433–438

187. Faraday M.: Experimental researches in electricity. Twelfth series. Phil. Trans. Royal Soc. London. **128**, 83–123 (1938) Art. 1340

188. Liu, B., Ma, Z.: Synthesis of Ag_2S-Ag nanoprisms and their use as DNA hybridization probes. Small **7**(11), 1587–1592 (2011)

189. Wang D., Liu L., Kim Y., Huang Z., Pantel D., Hesse D., Alexe M.: Fabrication and characterization of extended arrays of Ag_2S/Ag nanodot resistive switches. Appl. Phys. Lett. **98**(24), 3 (2011) Paper 243109

190. Morales-Masis, M., Molen, S.J., Fu, W.T., Hesselberth, M.B., Ruitenbeek,J.M.: Conductance switching in Ag_2S devices fabricated by *in situ* sulfurization. Nanotechnology **20**(9), 6 (2009) Paper 095710

191. Tanaka, H., Akai, T., Tanaka, D., Ogawa, T.: Sequential phase transition during fabricating β-Ag_2S film on Ag electrode by wet chemical process. e-J. Surf. Sci. Nanotechn. **12**, 185–188 (2014)

192. Ma, X., Zhao, Y., Jiang, X., Liu, W., Liu, S., Tang, Z.: Facile preparation of Ag_2S/Ag semiconductor/metal heteronanostructures with remarkable antibacterial properties. ChemPhysChem **13**(10), 2531–2535 (2012)

193. Horvath, B., Kawakita, J., Chikyow, T.: Diffusion barrier and adhesion properties of SiO_xN_y and SiO_x layers between Ag/polypyrrole composites and Si substrates. ACS Appl. Mat. Interf. **6**(12), 9201–9206 (2014)

194. Sadovnikov, S.I., Gusev, A.I.: Structure and properties of Ag_2S/Ag semiconductor/metal heteronanostructure. Biointerf. Res. Appl. Chem. **6**(6), 1797–1804 (2016)

195. Terabe, K., Hasegawa, T., Nakayama, T., Aono, M.: Quantized conductance atomic switch. Nature **433**(7021), 47–50 (2005)

196. Kharkats, YuI: Electric-field induced transition to superionic conductive state. Fiz Tverd Tela. **23**(7), 2190–2192 (1981). (in Russian)
197. Gurevich, Y.Y., Kharkats, Y.I.: Features of the thermodynamics of superionic conductors. Usp. Fiz. Nauk. **136**(4), 693–728 (1982). (in Russian). (Engl. Transl.: Sov. Phys. Uspekhi. **25**(4), 257–276 (1982))
198. Hu, M., Chen, J.Y., Li, Z.Y., Au, L., Hartland, G.V., Li, X.D., Marquez, M., Xia, Y.N.: Gold nanostructures: engineering their plasmonic properties for biomedical applications. Chem. Soc. Rev. **35**(11), 1084–1094 (2006)
199. Song, S.P., Yu, Q., He, Y., Huang, Q., Fan, C.H., Chen, H.Y.: Functional nanoprobes for ultrasensitive detection of biomolecules. Chem. Soc. Rev. **39**(11), 4234–4243 (2010)
200. Yang, J., Liu, H.: Metal-Based Composite Nanomaterials, vol. 4, pp. 93–114. Springer, Cham (2015)

Name Index

© Springer International Publishing AG 2018
S.I. Sadovnikov et al., *Nanostructured Lead, Cadmium, and Silver Sulfides*,
Springer Series in Materials Science 256, DOI 10.1007/978-3-319-56387-9

Subject Index

A

Acanthite, 189–193, 218, 274
  coarse-crystalline, 208, 214, 238, 244, 245, 248, 292, 302
  nanostructured nosstoichiometric, 214, 218, 240, 243
  thermal expansion, 228, 229, 233–235, 237
Anharmonicity of atomic vibrations, 102, 107
Annealing
  isothermal, 150, 151
  virtual, 86, 165
Argentite, 189, 190, 192, 209, 218, 275, 288
  thermal expansion, 228

B

Band gap, 6, 31, 66, 89, 90, 94, 96, 152, 173–176, 192, 199, 251, 252, 271, 290
Boltzmann distribution, 102
Broadening of diffraction reflection
  deformation (strain), 18, 20
  inhomogeneity, 12, 14, 22
  size, 11, 12, 17, 19
Brunauer, Emmett and Teller theory, 26

C

Chalcogenide, 2–4, 31, 33, 57, 64, 130, 131
Chemical vapor deposition, 196
Constant
  dielectric, 91, 96, 129, 193, 252
  Grüneisen, 98, 103, 234
  Planck, 152, 256
  Scherrer, 16, 17, 22
Correlation
  many-particle, 5
  parameter, 83–85
Coulomb interaction, 95, 175

D

Degree of space filling, 156–159
Density of electronic states, 174
Density of states, 6, 175
Deposition
  electrochemical, 56, 64, 194, 240
  hydrochemical, 4, 34, 48, 55, 67, 194, 203, 240, 242, 262, 273, 279, 280, 287
  solvothermal, 58, 194
Diffusion, 25, 61, 145
Dynamic Light Scattering (DLS), 25, 132, 139, 242, 258

E

Effective mass, 6
  electron, 88, 95, 129, 189, 243
  hole, 88, 95, 129, 193, 252
Effect of particle size
  heat capacity, 103, 234
  phonon spectrum, 101, 103
  thermal expansion coefficient, 98, 100, 102, 103, 105, 106, 229
Electron microscopy
  scanning, 9, 10, 52, 71, 143, 180, 212, 218, 243
  transmission, 9, 11, 71, 143, 165, 170, 219, 243, 258
Epitaxy
  gas phase, 64
  molecular beam, 64, 67
Exciton, 3, 8
  energy, 92, 164, 172
  radius (Bohr radius), 4, 31, 81, 95, 96, 129, 193, 252, 256
  reduced mass, 91, 129, 130, 152, 173, 252

F

Films, 2, 5, 31, 43, 56, 64, 66, 67, 69, 70, 72, 76–78, 80, 88–90, 92–94, 97, 98, 107, 109, 113, 129, 130, 145, 148, 165
Formula

© Springer International Publishing AG 2018
S.I. Sadovnikov et al., *Nanostructured Lead, Cadmium, and Silver Sulfides*,
Springer Series in Materials Science 256, DOI 10.1007/978-3-319-56387-9